Lecture Notes in Mathematics

Edited by A. Dold and B. Eckmann

685

Knot Theory

Proceedings, Plans-sur-Bex, Switzerland 1977

Edited by J. C. Hausmann

Springer-Verlag
Berlin Heidelberg New York 1978

Editor
Jean-Claude Hausmann
Institut de Mathématiques
Université de Genève
CH-1211 Genève 24

AMS Subject Classifications (1970): 57C45, 55A25, 15A63

ISBN 3-540-08952-7 Springer-Verlag Berlin Heidelberg New York
ISBN 0-387-08952-7 Springer-Verlag New York Heidelberg Berlin

Printing and binding: Beltz Offsetdruck, Hemsbach/Bergstr.
2141/3140-543210

This volume is dedicated to the memory of

Christos Demetriou PAPAKYRIAKOPOULOS

1914 - 1976

PREFACE

This volume contains mostly the texts of the lectures presented at the seminar on Knot Theory held in Plans-sur-Bex in 1977. This seminar was organized by M. KERVAIRE and Cl. WEBER under the auspices of the IIIème Cycle romand de Mathématique.

The first two articles are general surveys presenting recent developments in the theory of knots. The first article treats "classical" knots ($S^1 \subset S^3$) and the second is concerned with multidimensional knots ($S^{n-2} \subset S^n$, $n \geq 4$) . I would like to express my thanks to the authors of these contributions for their important and excellent work.

In the name of all the participants of the seminar I express my thanks to Mr. and Mrs. AMIGUET and their staff for taking care of the excellent living and eating accomodations for the participants during their stay in Plans-sur-Bex. Finally my thanks to Mrs. MAULINI for preparing the typescript of several articles in this volume.

Jean-Claude HAUSMANN

Geneva, June 1978

TABLE OF CONTENTS

SOME ASPECTS OF CLASSICAL KNOT THEORY

by

C. McA. Gordon

0. Introduction

Man's fascination with knots has a long history, but they do not appear to have been considered from the mathematical point of view until the 19th century. Even then, the unavailability of appropriate methods meant that initial progress was, in a sense, slow, and at the beginning of the present century rigorous proofs had still not appeared. The arrival of algebraic-topological methods soon changed this, however, and the subject is now a highly-developed one, drawing on both algebra and geometry, and providing an opportunity for interplay between them.

The aim of the present article is to survey some topics in this theory of knotted circles in the 3-sphere. Completeness has not been attempted, nor is it necessarily the case that the topics chosen for discussion and the results mentioned are those that the author considers the most important: non-mathematical factors also contributed to the form of the article.

For additional information on knot theory we would recommend the survey article of Fox [43], and the books of Neuwirth [112] and Rolfsen [128]. Reidemeister's book [125] is also still of interest. As far as problems are concerned, see [44], [112], [113], [75], as well as the present volume. Again, we have by no means tried to include a complete bibliography, although we hope that credit for ideas has been given where it is due. For a more extensive list of early references, see [26].

In the absence of evidence to the contrary, we shall be working in the smooth category (probably), and homology will be with integer coefficients.

I should like to thank Rick Litherland for helpful discussions and suggestions concerning this article.

Contents

1. Enumeration

It seems that the first mathematician to consider knots was Gauss, whose interest in them began at an early age [31, p. 222]. Unfortunately, he himself wrote little on the subject [49, V, p. 605; VIII, pp. 271-286], despite the fact that he regarded the analysis of knotting and linking as one of the central tasks of the 'geometria situs' foreseen by Leibniz [49, V, p. 605]. His student Listing, however, devoted a considerable part of his monograph [88] to knots, and in particular made some attempt to describe a notation for knot diagrams.

A more successful attack, inspired by Lord Kelvin's theory of vortex atoms, was launched in the 1860's(1) by the Scottish physicist Tait. His first papers on knots were published in 1876-77 (see [145]). Later, with the help of the 'polyhedral diagrams' of the Reverend Kirkman, Tait and Little (the latter had done some earlier work [90]) made considerable progress on the enumeration ('census') problem, so that by 1900 there were in existence tables of prime knots up to 10 crossings and alternating prime knots of 11 crossings [91], [92], [93], [145].

Essentially nothing was done by way of extending these tables until about 1960, when Conway invented a new and more efficient notation which enabled him to list all (prime) knots up to 11 crossings and all links up to 10 crossings [19], (revealing, in particular, some omissions in the 19th century tables).

There are two main aspects of this kind of enumeration: completeness and non-redundancy. One wants to know (i.e. prove) that one has listed all knots up to a given crossing number, and also that the knots listed are distinct. The former belongs to combinatorial mathematics, and although a proof of completeness throughout the range of the existing tables would no doubt be long and tedious, it is not hard to envisage how such a proof would go. Indeed, implicit in the compilation of the tables is the possession of at least the outline of such a proof. Although some omissions in Conway's tables have recently been brought to

---

(1) see Maxwell's letter of 1867 quoted in [77, p. 106]

light by Perko (see [117] and references therein), it seems safe to assume that essentially all knots up to 11 crossings have now been listed. (The author understands that we may soon see a proof of completeness in this range.)

As regards the question of non-redundancy, methods for proving that two knot diagrams represent different knots became available only with the advent of algebraic topology, and as a consequence the compilers of the early tables, as they themselves were aware, had to rely on purely empirical evidence that their listed knots were distinct.

Proofs of the existence of non-trivial knots, based on the fundamental group, were known at least as early as 1906 (see [146]), but not until 1927 was there any systematic attempt to establish the non-redundancy of the tables. Then, Alexander and Briggs [3], using the torsion numbers of the first homology of the 2- and 3-fold branched cyclic covers, distinguished all the tabled knots up to 8 crossings and all except 3 pairs up to 9 crossings. (Alexander had pointed out in 1920 (see [3]) that any topological invariant of the k-fold branched cyclic cover of a knot, in particular the Betti and torsion numbers, will be an invariant of the knot, an observation which was made independently by Reidemeister [122].) The Alexander polynomial, introduced in [2], also suffices to distinguish all knots up to 8 crossings, and all except 6 pairs up to 9 crossings. For each of the 3 remaining 9-crossing pairs not distinguished by Alexander and Briggs, the two knots in question have isomorphic $\mathbb{Z}[t,t^{-1}]$-module structures in their infinite cyclic covers, so new methods are necessary to distinguish them. This was done by Reidemeister, by means of the mutual linking numbers of the branch curves in certain (irregular) p-fold dihedral covers, and, more recently, Perko has used these linking invariants, in branched covers associated with representations on dihedral groups and the symmetric group on 4 letters, to distinguish all tabled knots up to 10 crossings [115].

It would now appear that the number of prime knots with crossing number ≤ 10 is 249, as tabulated below.

crossing number	3	4	5	6	7	8	9	10
number of prime knots	1	1	2	3	7	21	49	165

(See [3] for pictures of knots up to 9 crossings, and [115] for those with 10 crossings.) There are 550 11-crossing knots now known [117], and although there is a good chance that these might be all, the task of proving them distinct is a formidable one that has not yet been completed. Indeed, as intimated in [117] (which contains some partial results), invariants more delicate than those which suffice up to 10 crossings are now required.

2. The Group

The knot problem becomes discretized when looked at from the point of view of combinatorial topology. It is noted in [30], for example, that it can be formulated entirely in terms of arithmetic. However, this kind of 'reduction' seems to be of no practical value, nor does it seem to have any theoretical consequences (for decidability, for example). There are also many natural numerical invariants of a knot which may be defined, such as the minimal number of crossing points in any projection of the knot, the minimal number of crossing-point changes required to unknot the knot (the 'gordian number' [160]), the maximal euler characteristic of a spanning surface (orientable or not), and so on (see [125, pp. 16-17]). But these tend to be hard to compute.

The first successful algebraic invariant to be attached to a knot was the fundamental group of its complement, (the <u>group</u> of the knot), and presentations of certain knot groups appear fairly early in the literature (see [146]). General methods for writing down a presentation of the knot group from a knot projection were given by Wirtinger (unpublished (?) ;see [125, III, §9]) and Dehn [27]. Actually it was soon recognized [28] that a knot contains (at least a priori) more information than just its group, as we now explain. Let $K \subset S^3$ be our given (smooth) knot, and let X be its <u>exterior</u>, that is, the closure of the complement of a tubular neighbourhood N of K. (The exterior and the complement are equivalent invariants: clearly the exterior determines the complement, and the converse follows from [33].) Choosing orientations for S^3 and K determines a longitude-meridian pair λ, $\mu \in \pi_1(X)$ in the usual way (λ and μ are represented by oriented curves ℓ and m on ∂X which intersect (transversely) only at the base-point, where ℓ is homologous to K in N and null-homologous in X, and m is null-homologous in N and inherits its orientation from that of K and S^3). If two (oriented) knots $K_1, K_2 \subset S^3$ are equivalent in the strongest possible sense that there is an orientation-preserving homeomorphism of S^3 (or, equivalently, an isotopy) taking K_1 to K_2, preserving their orientations, then there is an isomorphism $\pi_1(X) \xrightarrow{\cong} \pi_1(X_2)$ taking (λ_1,μ_1) to (λ_2,μ_2). If we ignore the orientations of K_1 and K_2 in our definition of equivalence, then we have an isomorphism $\pi_1(X_1) \xrightarrow{\cong} \pi_1(X_2)$ taking (λ_1,μ_1) to either (λ_2,μ_2) or $(\lambda_2^{-1},\mu_2^{-1})$. If, in addition, we ignore the orientation of S^3, then our isomorphism merely takes λ_1 to $\lambda_2^{\pm 1}$ and μ_1 to $\mu_2^{\pm 1}$. Using this additional peripheral information, Dehn [28] proved for example that the trefoil is not isotopic to its mirror-image, a fact which had long been 'known' empirically. (Incidentally the knot tables list only one representative from each class under the weakest equivalence, leaving the amphicheirality and (much harder) invertibility questions to be decided separately [19], [115], [118].)

The natural question arises as to what extent the peripheral structure is determined by the group alone. Thus Dehn asks [28, p. 413] whether every automorphism of a knot group preserves the peripheral structure, and in [2, p, 275] Alexander suggests that 'many, if not all, of the topological properties of a knot are reflected in its group.' In 1933, however, Seifert showed [135], using linking invariants of their cyclic branched covers, that the granny knot and the reef (or square) knot, although they have isomorphic groups, are inequivalent, even ignoring orientations. (Although there seems to be an implicit assumption to the contrary in [38], where an alternative proof is given, it follows from Seifert's proof that in fact the two knots have non-homeomorphic complements. Fox's proof does show, however, that there is no isomorphism between the groups of the two knots preserving the peripheral structure.)

Despite such examples, the group is still a powerful invariant. It was shown by Dehn [27], for example, (modulo his 'lemma', which was introduced specifically for this purpose) that the only knot with group $\mathbb{Z}$ is the unknot. This finally became a theorem in 1956 when Dehn's lemma was established by Papakyriakopoulos [114]. At the same time, Papakyriakopoulos also proved the first version of the sphere theorem, and as a consequence, the asphericity of knots, that is, the fact that the complement of a knot is a $K(\pi,1)$. It follows that the group of a knot determines the homotopy type of its complement.

The role of the peripheral structure was finally completely clarified by Waldhausen's work [155] on irreducible, sufficiently large, 3-manifolds (this work in turn being based on earlier ideas of Haken). Specializing to the case that concerns us here, Waldhausen showed that if K_1 and K_2 are knots with exteriors X_1, X_2, then any homotopy equivalence of pairs $(X_1,\partial X_1) \to (X_2,\partial X_2)$ is homotopic to a homeomorphism. This implies, for example, that knots (under the strongest form of equivalence, which takes both the ambient orientation and that of the knot into account), are classified by (isomorphism classes of) their associated triples $(\pi K,\lambda,\mu)$. We may remark that it is a purely algebraic exercise to pass from such a classifying triple to a classifying group [20]. Other, more complicated, but more geometric, ways of nailing down the peripheral structure within a single group are given in [140], [163] and [37]. (The classifying groups obtained there are, respectively, the free product of the groups of two cables about $K \# K_0$ (where K_0 is, say, the figure eight knot), the group of the double of K, and the group of the (p,q)-cable of K where $|p| \geq 3$ and $|q| \geq 2$.)

The situation may to some extent be summarized by the following diagram, where, for simplicity, $\sim$ now denotes the weak form of knot equivalence which disregards orientations, (and P_i denotes the peripheral subgroup $\pi_1(\partial X_i)$).

$$\begin{array}{ccccc}
K_1 \sim K_2 \Rightarrow X_1 \cong X_2 & \Leftrightarrow & (X_1,\partial X_1) \simeq (X_2,\partial X_2) & \Rightarrow & X_1 \simeq X_2 \\
\Updownarrow & & \Updownarrow & & \Updownarrow \\
(\pi K_1,\lambda_1,\mu_1) \cong (\pi K_2,\lambda_2^{\pm 1},\mu_2^{\pm 1}) & \Rightarrow & (\pi K_1,P_1) \cong (\pi K_2,P_2) & \Rightarrow & \pi K_1 \cong \pi K_2
\end{array}$$

The two upward implications on the right are consequences of asphericity.

The question of the reversibility of the implications on the left, that is, whether a knot is determined by its complement, was rasied by Tietze in 1908 [146], and is still unsettled. It is related to the following question, asked by Bing and Martin [9]:

Question (P). If a tubular neighbourhood of a non-trivial knot K in S^3 is removed and sewn back differently, is the resulting 3-manifold ever simply-connected? (Here, 'differently' has to be interpreted in the obvious way.)

This may be broken down into the following 2 questions:

(1) Do we ever get a fake 3-sphere?

(2) Do we ever get S^3?

One may further ask

(3) If 'yes' in (2), do we get the same knot?

Knots are determined by their complement if and only if an affirmative answer to (2) is always accompanied by an affirmative answer to (3). There is much evidence that the answer to Question (P) is negative. In particular, it is known that this is the case for torus knots [134], composite knots [9], [53], doubled knots [9], [53], most cable knots [53], [139], knots in knotted solid tori with winding number ≥ 3 [89], and many others; (see [75] for additional references). (One says that these knots 'have Property P'.) Also, Thurston has recently shown (unpublished) that if K has a hyperbolic structure (more precisely, the complement of K has a complete Riemannian metric with constant negative sectional curvature and finite volume), then all except possibly finitely many resewings of the tubular neighbourhood of K yield non-simply-connected manifolds. (The existence of a hyperbolic structure is equivalent to the group-theoretic condition that every free abelian subgroup of πK of rank 2 be conjugate to the peripheral subgroup P, and this in turn is satisfied if and only if K has no companions and is not a torus knot.) A proof that all knots have Property P, however, (or even a proof of a negative answer to either (1) or (2)), seems beyond the scope of existing techniques. Question (3) may be easier. (Indeed it follows from the finiteness theorem of Thurston mentioned above that if K is a hyperbolic knot, and some non-trivial resewing of a tubular neighbourhood of K gives S^3, then the new knot is at least not isotopic to K. For, if the resewing in question is the one which 'kills' $\mu\lambda^n$, say, $n \neq 0$, then the new knot's being isotopic to K would imply the existence of a self-homeomorphism h of the exterior X of K taking $\lambda \mapsto \lambda^{\varepsilon}$, $\mu \mapsto \mu^{\varepsilon}\lambda^{\varepsilon n}$, where $\varepsilon = \pm 1$. Since h^r would then take $\mu \mapsto (\mu\lambda^{rn})^{\pm 1}$, the resewing corresponding to $\mu\lambda^{rn}$ would yield S^3, for all r, contradicting the finiteness statement.)

Returning to our diagram of implications, the example of the reef and granny shows that the horizontal implications on the right are not reversible. On the other hand, Johannson [66], [67] and Feustel [36] have shown that if $\pi K_1 \cong \pi K_2$, and X_1 contains no essential annuli, then $X_1 \cong X_2$. Now the only knots whose exteriors contain essential annuli are composite knots and cable knots. The cable knots with unknotted core are just the torus knots, and they are known to be determined by their group [14]. So let K be a non-trivial knot, and let $K_{p,q}$ denote the (p,q)-cable about K, that is, a curve on the boundary ∂N of a tubular neighbourhood N of K, homologous in ∂N to $p[m] + q[\ell]$. (Here, p and q are coprime integers with $|q| \geq 2$, and (ℓ, m) is a longitude-meridian pair on ∂N.) Feustel-Whitten [37] have shown that if $|p| \geq 3$, then $\pi K_{p,q}$ determines $K_{p,q}$. So prime knot complements are known to be determined by their group except possibly for cable knots $K_{p,q}$ with $|p| \leq 2$.

The problem concerning these remaining cable knots turns out to be related to the general question of whether knots are determined by their complement. More precisely, suppose there exist inequivalent knots K_1, K_2 with homeomorphic exteriors X_1, X_2. The homeomorphism $X_1 \to X_2$ must take m_1 to a curve homologous in ∂X_2 to $\pm [m_2] + n[\ell_2]$, for some $n \neq 0$. Then Hempel (unpublished) and Simon [141] show that if there is such a counterexample, with $|n| \neq 1, 2$, or 4, then there exist cable knots of type $(\pm 1, \pm n/2)$ (n even), or $(\pm 2, \pm n)$ (n odd), with isomorphic groups, whose complements are not homeomorphic.

In the other direction, it can be shown (see [37]) that if all knots have Property P, (or even if the answer to Question (2) above is negative), then prime knots are determined by their group.

As regards composite knots, Feustel-Whitten have also shown [37] that if K_1 is composite, and $\pi K_1 \cong \pi K_2$, then the prime factors of K_2 are precisely those of K_1, up to orientations.

To summarize, the question of whether a knot is determined by its group factors naturally into two questions: (A) does the group determine the complement? and (B) does the complement determine the knot? (B) is unsettled, although the expected answer is 'yes'. The answer to (A) is 'no', but may be 'yes' for prime knots; the unsettled cases of this are related to (B). Thus it may be that the failure of knots to be determined by their group is solely due to the phenomenon which arises by changing the (ambient and intrinsic) orientations of the prime factors of a composite knot.

3. Abelian Invariants

The exterior of a knot K has the homology of a circle (as can be seen, for example, by Alexander duality), and as a consequence, once we have chosen orientations for S^3 and K, there is a canonical epimorphism from $\pi_1(X)$ to the cyclic group C_k of order k, for each k, $1 \leq k \leq \infty$. This defines a canonical normal subgroup of index k in $\pi_1(X)$, or, the geometric equivalent, a regular covering space X_k of X with group of covering translations isomorphic to C_k. Although the homology of X is itself uninteresting, this is not always true of these covering spaces, and the derivation of tractable, 'abelian', knot invariants from this point of view has occupied a central place in the development of the subject.

The homology of the X_k can be viewed on at least the following levels (throughout, we shall take coefficients in some commutative Noetherian ring R, with $R = \mathbb{Z}, \mathbb{Z}/p$, or $\mathbb{Q}$ being uppermost in our minds).

(1) the R-module structure of $H_1(X_k;R)$

(2) the module structure of $H_1(X_k;R)$ over the group ring $R[C_k]$.

If R is an integral domain, and $Q(\ )$ denotes field of fractions, we also have

(3) for $k < \infty$, the product structure given by the linking pairing $T_1(X_k;R) \times T_1(X_k;R) \to Q(R)/R$ on the R-torsion subgroup of $H_1(X_k;R)$. ($R = \mathbb{Z}$ is really the only case of interest here.)

(4) the product structure given by the Blanchfield pairing (see §7) $H_1(X_\infty;R) \times H_1(X_\infty;R) \to Q(R[C_\infty])/R[C_\infty]$.

We may remark here that, for $k < \infty$, it is traditional to work with the corresponding branched cyclic covering M_k, rather than with the unbranched covering X_k. Since M_k is a closed 3-manifold, and for other reasons too (see §5), this is perhaps more natural. However, the two are essentially equivalent from the present point of view, as it is not hard to show that $H_1(X_k;R) \cong H_1(M_k;R) \oplus R$, as $R[C_k]$-modules, the module structure on R being induced by the trivial action of C_k.

Apart from the obvious relationships between the above considerations (1)-(4), we have that the $R[C_\infty]$-module $H_1(X_\infty;R)$ determines the $R[C_k]$-module $H_1(X_k;R)$, $1 \leq k < \infty$, (see §5), and the Blanchfield pairing on $H_1(X_\infty;R)$ determines the linking pairing on $T_1(X_k;R)$, $1 \leq k < \infty$.

4. The Infinite Cyclic Cover

Let us first consider the $R[C_\infty]$-module $H_1(X_\infty;R)$. If t denotes the canonical multiplicative generator of C_∞, (determined by the orientations of S^3 and K), we may identify $R[C_\infty]$ with the Laurent polynomial ring $\Pi = R[t,t^{-1}]$. Since R is Noetherian, Π is also, by the Hilbert basis theorem. Furthermore, since X is a finite complex, the chain modules $C_q(X_\infty;R)$ are finitely-generated (free) Π-modules, and hence $H_1(X_\infty;R)$ is a finitely-generated Π-module.

The following argument of Milnor [96] establishes the crucial property that $t-1: H_1(X_\infty;R) \to H_1(X_\infty;R)$ is surjective. (Since $H_1(X_\infty;R)$ is finitely-generated and Π is Noetherian, it follows that $t-1$ is also injective.) The short exact sequence of chain complexes

$$0 \to C_*(X_\infty;R) \xrightarrow{t-1} C_*(X_\infty;R) \to C_*(X;R) \to 0$$

gives rise to a homology exact sequence which ends up with

$$\begin{array}{ccccccccc} H_1(X_\infty;R) & \xrightarrow{t-1} & H_1(X_\infty;R) & \to & H_1(X;R) & \to & H_0(X_\infty;R) & \xrightarrow{t-1} & H_0(X_\infty;R) \\ & & & & \wr\Vert & & \wr\Vert & & \wr\Vert \\ & & & & R & \longrightarrow & R & \xrightarrow{0} & R \end{array}$$

This proves the assertion.

A consequence of this (see [85]) is that $H_1(X_\infty;R)$ is a Π-torsion-module.

Now suppose R is a field. Then Π is a principal ideal domain, and hence $H_1(X_\infty;R)$ decomposes as a direct sum of cyclic Π-modules

$$\Pi/(\pi_1) \oplus \Pi/(\pi_2) \oplus \ldots \oplus \Pi/(\pi_n) ,$$

where the ideals (π_i) satisfy $(\pi_i) \subset (\pi_{i+1})$, $1 \leq i < n$, (and are then uniquely determined). The Π-module $H_1(X_\infty;R)$ is thus completely described by this sequence of ideals $(\pi_1) \subset (\pi_2) \subset \ldots \subset (\pi_n)$. Furthermore, since $H_1(X_\infty;R)$ is a Π-torsion module, no (π_i) is zero. (In the present case, i.e. R a field, the fact that $H_1(X_\infty;R)$ is Π-torsion actually follows immediately from the direct sum decomposition of $H_1(X_\infty;R)$ and the divisibility by $t-1$.)

To determine the R-vector space structure of $H_1(X_\infty;R)$, let $(\Delta) = (\pi_1\pi_2 \ldots \pi_n)$ be the order ideal of $H_1(X_\infty;R)$. We may suppose for convenience that Δ is normalized so that it contains no negative powers of t and has non-zero constant coefficient. Then

$$\dim H_1(X_\infty;R) = \deg \Delta ,$$

and Δ is just the characteristic polynomial of the automorphism t. We shall see later (§7) that $\deg \Delta$ is always even.

Taking $R = \mathbb{Q}$ in particular, we have a complete description of the $\Gamma = \mathbb{Q}[t,t^{-1}]$-module $H_1(X_\infty;\mathbb{Q})$ by a sequence of non-zero ideals $(\gamma_1) \subset (\gamma_2) \subset \ldots \subset (\gamma_n)$. The picture over $\Lambda = \mathbb{Z}[t,t^{-1}]$ is not quite so clear, as Λ is not a principal ideal domain, but one can define some invariants. Thus there are the elementary ideals $E_1 \subset E_2 \subset \ldots$, where E_i is defined to be the ideal in Λ generated by the determinants of all the $(n-i+1) \times (n-i+1)$ submatrices of any $m \times n$ presentation matrix for the module [164, pp. 117-121]. (We may suppose $m \geq n$ without loss of generality, and we put $E_i = \Lambda$ if $i > n$.) Even these are fairly intractable, but since Λ is a unique factorization domain, each E_i is contained in a unique minimal principal ideal (Δ_i). One thus obtains a sequence of elements $\Delta_1, \Delta_2, \ldots, \Delta_n$ of Λ, each determined up to multiplication by a unit (the only units of Λ are $\pm t^r$, $r \in \mathbb{Z}$), such that $\Delta_{i+1} | \Delta_i$, $1 \leq i < n$. Suitably normalized, Δ_i is called the ith Alexander polynomial of the knot, $\Delta_1 = \Delta$ being called simply the Alexander polynomial. Equivalently, one can consider the elements λ_i defined by $\lambda_i = \Delta_i/\Delta_{i+1}$; λ_i is the ith Alexander invariant. These definitions are essentially contained in Alexander's paper [2].

The surjectivity of $t-1: H_1(X_\infty) \to H_1(X_\infty)$ can be expressed by saying that, regarding $\mathbb{Z}$ as a Λ-module via the augmentation homomorphism $\varepsilon: \Lambda \to \mathbb{Z}$, $H_1(X_\infty) \otimes_\Lambda \mathbb{Z} = 0$. It follows that $\varepsilon(E_i) = \mathbb{Z}$, and hence $\varepsilon(\Delta_i) = \Delta_i(1) = \pm 1$. It seems most natural (see §8) to normalize Δ_i so that it is a polynomial in t such that $\Delta_i(0) \neq 0$ and $\Delta_i(1) = 1$. From this it is not too hard to show that if the elements γ_i of Γ which describe the direct sum decomposition of $H_1(X_\infty;\mathbb{Q})$ are normalized so as to be polynomials with integer coefficients with g.c.d. 1, such that $\gamma_i(0) \neq 0$ and $\gamma_i(1) > 0$, then $\lambda_i = \gamma_i$, $1 \leq i \leq n$. It thus transpires that in the presence of the integral information $H_1(X) \cong \mathbb{Z}$, the Alexander polynomials are essentially rational invariants.

In view of the last remark, it is no surprise that the Alexander polynomials do not in general determine the elementary ideals. For example, the knot 9_{46} in the Alexander-Briggs table and the stevedore's knot (6_1) have modules $H_1(X_\infty)$ which are, respectively, $\Lambda/(2-t) \oplus \Lambda/(2t-1)$ and $\Lambda/(2-5t+2t^2)$. In both cases, $H_1(X_\infty;\mathbb{Q})$ is the cyclic Γ-module $\Gamma/(2-5t+2t^2)$. However, for the stevedore's knot, $E_2 = \Lambda$, whereas for 9_{46}, $E_2 = (2-t, 2t-1) \neq \Lambda$ (map Λ onto $\mathbb{Z}$ by $t \mapsto -1$; the image of $(2-t, 2t-1)$ is $3\mathbb{Z}$).

Again, the elementary ideals do not in general determine the Λ-module $H_1(X_\infty)$ (see [47]). Further invariants which have been studied include ideals in certain Dedekind domains, ideal classes, and Hermitian forms over certain rings of algebraic integers [47], [84]. A complete classification has not yet been found.

An important property of the Λ-module $H_1(X_\infty)$ is that it has deficiency 0. (Since $E_1 \neq 0$, any presentation of $H_1(X_\infty)$ must have at least as many relations as generators, so deficiency 0 just means that there is a presentation with the same number of generators and relations.) This may be seen by interpreting $H_1(X_\infty)$ as the abelianized commutator subgroup of the group π of K, and noting that π has a presentation of deficiency 1, for example, either the Wirtinger or Dehn presentation. (Since $H_1(\pi) \cong \mathbb{Z}$, it follows that the deficiency of π is 1.) It is also a consequence of duality (see §7), or, again, follows from the description of $H_1(X_\infty)$ in terms of a Seifert matrix (see §8; this is also related to duality). Deficiency 0 implies that the first elementary ideal E_1 is principal, i.e. $E_1 = (\Delta)$.

Returning briefly to rational coefficients, note that, up to multiplication by a rational unit, γ_1 is the minimal polynomial of the automorphism t of $H_1(X_\infty;\mathbb{Q})$, in other words, the annihilator of $H_1(X_\infty;\mathbb{Q})$ is $(\gamma_1) = (\lambda_1)$. Over Λ, it follows from general considerations, (see [164, p. 123], for example), that E_1 annihilates $H_1(X_\infty)$. Crowell [25] has shown that in fact the annihilator of $H_1(X_\infty)$ is precisely the principal ideal (λ_1) of Λ.

Turning to the abelian group structure of $H_1(X_\infty)$, this seems hard to describe in general, but we do have the result of Crowell [24] that $H_1(X_\infty)$ is always $\mathbb{Z}$-torsion-free. The crucial facts are, firstly, that $H_1(X_\infty)$ has deficiency 0 as a Λ-module, and, secondly, that the Alexander polynomial is primitive (i.e. g.c.d. of coefficients is 1; this follows from $\varepsilon(\Delta) = 1$). Here is the proof. Let A be a square presentation matrix for $H_1(X_\infty)$ over Λ. It must be shown that for any integer q, $A\underline{x} \equiv 0 \pmod q$ implies $\underline{x} \equiv 0 \pmod q$. But $(\text{adj } A)(A\underline{x}) = (\det A)\underline{x} = \Delta\underline{x}$, and since Δ is primitive, this implies the result.

If $x_1, \ldots, x_n$ generate $H_1(X_\infty)$ as a Λ-module, then $\{t^j x_i : 1 \leq i \leq n, -\infty < j < \infty\}$ generate $H_1(X_\infty)$ over $\mathbb{Z}$. Since $\Delta x_i = 0$, $1 \leq i \leq n$, we see that if the constant coefficient of Δ (and hence, by the symmetry of Δ (see §7), the leading coefficient also) is ± 1, then $H_1(X_\infty)$ is finitely-generated over $\mathbb{Z}$, and is therefore free abelian of rank $\deg \Delta$. The converse is also true. For these and other results on the abelian group structure of $H_1(X_\infty)$, see [24], (also [121]).

5. The Finite Cyclic Covers

To relate $H_1(X_k;R)$ to $H_1(X_\infty;R)$, consider the short exact sequence of chain complexes

$$0 \to C_*(X_\infty;R) \xrightarrow{t^k-1} C_*(X_\infty;R) \to C_*(X_k;R) \to 0 .$$

As before, this gives rise to an exact sequence

$$H_1(X_\infty;R) \xrightarrow{t^k-1} H_1(X_\infty;R) \to H_1(X_k;R) \to R \to 0 .$$

If we give R the trivial Π-action, and $H_1(X_k;R)$ the Π-module structure induced by the canonical covering translation, this is an exact sequence of Π-modules. From this and the fact that $H_1(X_k;R) \cong H_1(M_k;R) \oplus R$ (with the trivial Π-action on R), it follows that, as Π- or $R[C_k]$-modules,

$$H_1(M_k;R) \cong \operatorname{coker}(t^k-1)^{(2)} .$$

This relation between $H_1(M_k;R)$ and $H_1(X_\infty;R)$ can be conveniently expressed in matrix terms. Let $B(t)$ be any presentation matrix for $H_1(X_\infty;R)$ over Π, with respect to generators $x_1, \ldots, x_n$, say. Then $\operatorname{coker}(t^k-1)$ is generated

(2) Throughout this section, it is understood that this refers to the action on $H_1(X_\infty;R)$.

over R by the images of $\{t^j x_i : 1 \leq i \leq n,\ 0 \leq j < k\}$, and with respect to these generators, is presented by the matrix $B(T)$ obtained from $B(t)$ by replacing a typical entry $\sum a_r t^r$ by $\sum a_r T^r$, where T is the $k \times k$ matrix

$$\begin{bmatrix} 0 & 1 & 0 & \cdots & 0 \\ 0 & 0 & 1 & \cdots & 0 \\ \vdots & & & & \vdots \\ 1 & 0 & 0 & \cdots & 0 \end{bmatrix}$$

(see [52], [41], [112]).

Over certain coefficient rings, information can be extracted in other ways. For example, over $\mathbb{Z}/p$ (p prime), $(t^{p^r}-1) = (t-1)^{p^r}$ is an automorphism of $H_1(X_\infty;\mathbb{Z}/p)$; hence $H_1(M_{p^r};\mathbb{Z}/p) = 0$. In particular, M_{p^r} is a $\mathbb{Q}$-homology sphere.

Again, if R is any field, from the direct sum decomposition

$$H_1(X_\infty;R) \cong \bigoplus_{i=1}^{n} \Pi/(\pi_i)$$

we obtain a similar decomposition

$$H_1(M_k;R) \cong \bigoplus_{i=1}^{n} \Pi/(\pi_i, t^k-1)\ .$$

Taking $R = \mathbb{C}$, we have the following further simplification pointed out by Sumners [144]. Applying $-\otimes\mathbb{C}$ to the decomposition $\bigoplus_{i=1}^{n} \Gamma/(\lambda_i)$ of $H_1(X_\infty;\mathbb{Q})$, and writing $\Psi = \mathbb{C}[C_\infty]$, we get $H_1(X_\infty;\mathbb{C}) \cong \bigoplus_{i=1}^{n} \Psi/(\lambda_i)$. Over Ψ, however, each $\Psi/(\lambda_i)$ decomposes as a direct sum $\oplus\,\Psi/((t-\alpha)^{e(\alpha)})$ over all distinct roots α of λ_i. Since $((t-\alpha)^{e(\alpha)}, t^k-1) = (t-\alpha)$ or Ψ according as α is or is not a kth root of 1, we see that

$$\dim_{\mathbb{C}} H_1(M_k;\mathbb{C}) = \sum_{i=1}^{n} \ell_i$$

where ℓ_i is the number of distinct roots of λ_i which are k^{th} roots of 1. This result was first obtained by Goeritz [52], by explicitly diagonalizing $B(T)$ over $\mathbb{C}$.

Note that (as was pointed out in [52]), $H_1(M_k;\mathbb{C})$, or equivalently, the first Betti number of M_k, does not just depend on the Alexander polynomial $\Delta=\lambda_1,\ldots,\lambda_n$. The order of $H_1(M_k)$, however, does. Indeed, using Goeritz's diagonalization it may be shown that

$$\text{order } H_1(M_k) = |\det B(T)| = \Big|\prod_{i=1}^{k} \Delta(\omega^i)\Big|, \quad \text{where } \omega = e^{\frac{2\pi i}{k}} .$$

(This was first observed by Fox [41]; the proof given there, however, needs some modification.)

The behaviour of $H_1(M_k)$ as a function of k is sometimes quite interesting. For example, if k is odd, then $H_1(M_k)$ is always of the form $G \oplus G$ [119], [54]. Other results, in particular, necessary and sufficient conditions for $H_1(M_k)$ to be periodic in k, are given in [55].

We shall mention Seifert's work on branched cyclic covers [136], [137] in §8.

6. The Group Again

Let π be the group of a knot K. Since covering spaces of the exterior X of K correspond to subgroups of π, much of the material discussed in §§3-5 can be expressed in purely group-theoretic terms. Thus $\pi_1(X_\infty)$ is just the commutator subgroup π' of π, so $H_1(X_\infty)$ is isomorphic to π'/π''. The Λ-module structure of $H_1(X_\infty)$ can also be described group-theoretically: let $z \in \pi$ be any element which maps to the chosen generator t of C_∞; then the action of t on $H_1(X_\infty)$ corresponds to conjugation by z on π'/π''. Hence, given some presentation of π, it will be possible to derive a Λ-module presentation for π'/π''. If the presentation of π is in turn obtained in some way from a projection of K, we will then have a recipe for computing the Λ-module π'/π'' from a knot diagram. The algorithms described by Alexander [2] and Reidemeister [125, II, §14] are of this kind, based respectively on the Dehn and Wirtinger presentations of the knot group.

Similarly, for $1 \leq k < \infty$, $\pi_1(X_k)$ is isomorphic to the kernel π_k of the canonical epimorphism $\pi \to C_k$, so $H_1(X_k)$ can be identified with π_k/π_k'. Given a presentation of π, a presentation of π_k may be written down (using the Reidemeister-Schreier algorithm, for example), and hence a presentation (over $\mathbb{Z}$) of $H_1(X_k)$. (If one prefers to work with $H_1(M_k)$, then the branching relation must also be added, but as mentioned in §3, the difference between $H_1(X_k)$ and $H_1(M_k)$ is easy to take account of.) Thus again one can give a recipe for writing down a presentation matrix for (say) $H_1(M_k)$ in terms of a projection of the knot. This is done in [3] and [8]. (See also [125].)

Yet another algorithm for writing down a presentation of the Λ-module π'/π'' from a presentation of π is given by the free differential calculus of Fox [39], [40], [41] (see also [23], [26]), which we now briefly describe.

Let $P = (x_1, \ldots, x_n : r_1, \ldots, r_m)$ be a presentation of some group G. Corresponding to P, there is an obvious space X with $\pi_1(X) \cong G$, namely the finite 2-complex consisting of a single 0-cell p, n 1-cells, which we shall call $x_1, \ldots, x_n$, and m 2-cells, $D_1, \ldots, D_m$ (with base-points on their boundaries), the attaching map of D_i being r_i, $1 \leq i \leq m$. Now let H be some quotient of G, and $\tilde{X} \to X$ the regular covering with group of covering translations isomorphic to H. The cell structure of X lifts to a cell structure for $\tilde{X}$; choose a 0-cell $\tilde{p}$ lying over p, let $\tilde{x}_j$ be the unique lift of x_j which starts at $\tilde{p}$, and $\tilde{D}_i$ the unique lift of D_i such that $\partial\tilde{D}_i$ is the lift of r_i which starts at $\tilde{p}$. Then $C_0(\tilde{X})$, $C_1(\tilde{X})$, $C_2(\tilde{X})$ are the free $\mathbb{Z}[H]$-modules on $\{\tilde{p}\}$, $\{\tilde{x}_j : 1 \leq j \leq n\}$, and $\{\tilde{D}_i : 1 \leq i \leq m\}$ respectively.

The free differential calculus is a convenient tool for describing the boundary homomorphism $\partial_2 : C_2(\tilde{X}) \to C_1(\tilde{X})$, and consequently the $\mathbb{Z}[H]$-module $H_1(\tilde{X})$. (Since the latter can be described solely in terms of the group, we could use any space with $\pi_1(X) \cong G$; in particular, the result will be independent of the presentation P.)

Let F be the free group on $x_1, \ldots, x_n$, and $\varphi : \mathbb{Z}[F] \to \mathbb{Z}[G]$ the homomorphism induced by the epimorphism $F \to G$ corresponding to the presentation P. Let $\alpha : \mathbb{Z}[G] \to \mathbb{Z}[H]$ be the quotient homomorphism. For each j, $1 \leq j \leq n$, there is a unique $\mathbb{Z}$-linear function

$$\frac{\partial}{\partial x_j} : \mathbb{Z}[F] \to \mathbb{Z}[F]$$

such that

$$\frac{\partial x_i}{\partial x_j} = \delta_{ij}$$

and
$$\frac{\partial(uv)}{\partial x_j} = \frac{\partial u}{\partial x_j} + u\,\frac{\partial v}{\partial x_j} .$$

If w is any word in the x_j's, regarded as a loop in X based at p, w lifts to a unique path $\tilde{w}$ starting at $\tilde{p}$. It may then be readily verified (for example, by induction on the length of w) that, as a 1-chain in $\tilde{X}$,

$$\tilde{w} = \sum_{j=1}^{n} \alpha\varphi\left(\frac{\partial w}{\partial x_j}\right)\tilde{x}_j .$$

In particular, with respect to the $\mathbb{Z}[H]$-bases $\{\tilde{D}_i : 1 \leq i \leq m\}$, $\{\tilde{x}_j : 1 \leq j \leq n\}$, $\partial_2 : C_2(\tilde{X}) \to C_1(\tilde{X})$ is given by the $m \times n$ matrix

$$\left(\alpha\varphi\left(\frac{\partial r_i}{\partial x_j}\right)\right) .$$

One also sees that $\partial_1 : C_1(\tilde{X}) \to C_0(\tilde{X})$ is given by

$$\partial_1(\tilde{x}_j) = (\alpha\varphi(x_j) - 1)\tilde{p} .$$

The short exact sequence

$$0 \to \ker \partial_1 \to C_1(\tilde{X}) \to \operatorname{im} \partial_1 \to 0$$

gives, after factoring out by $\operatorname{im} \partial_2$, the short exact sequence (of $\mathbb{Z}[H]$-modules)

$$0 \to H_1(\tilde{X}) \to \operatorname{coker} \partial_2 \to \operatorname{im} \partial_1 \to 0 .$$

Since coker ∂_2 is presented by the 'Jacobian' matrix described above, and since we know im ∂_1, we can extract information about $H_1(\tilde{X})$.

In fact, specializing to the knot situation, with $G = \pi$ and $H = C_\infty$, it is not hard to prove that $\operatorname{im} \partial_1 \cong \Lambda$. The above sequence therefore splits, showing that the matrix $(\alpha\varphi(\frac{\partial r_i}{\partial x_j}))$ is a presentation matrix for the Λ-module $(\pi'/\pi'') \oplus \Lambda$.

7. Duality

The modules $H_1(X_\infty;R)$ have additional properties derived from duality. These are somewhat deeper, and the history reflects this. For example, the fact that $\Delta_i(1) = 1$ was proved by Alexander in [2], whereas the symmetry property $\Delta(t) = t^{\deg \Delta}\Delta(t^{-1})$ was first proved by Seifert [136], (the explanation given by Reidemeister in [125, p. 40], in terms of the group, seems to be insufficient), and not fully explained as a duality property until Blanchfield [12]. We now briefly discuss this duality, following Levine [85].

The chain module $C_q = C_q(X_\infty, \partial X_\infty; R)$ is a free Π-module on the q-simplices in $X - \partial X$ of some triangulation of X. Let $C'_q = C'_q(X_\infty;R)$ be the chains on the lifts of the q-simplices of the dual triangulation of X. There is then a non-singular pairing (see [95])

$$\langle\, , \rangle : C_q \times C'_{3-q} \to \Pi$$

defined by

$$\langle c, c' \rangle = \sum_{i=-\infty}^{\infty} (c \cdot t^i c') t^i ,$$

where $\cdot$ denotes ordinary intersection number. This pairing is sesquilinear with respect to the conjugation $\bar{\ }$ of Π induced by $t \longmapsto t^{-1}$. It induces a duality isomorphism

$$\overline{H_q(X_\infty,\partial X_\infty;R)} \cong H^{3-q}(\mathrm{Hom}_\Pi(C'_*,\Pi)) ,$$

where — denotes the conjugate module in which the action of $\pi \in \Pi$ is defined by $a \mapsto \bar{\pi}a$. We are mainly interested in the case $q = 1$. Let us then note that since $H_1(\partial X_\infty;R)$ is generated by the boundary of the lift of a Seifert surface, $H_1(\partial X_\infty;R) \to H_1(X_\infty;R)$ is zero, and hence $H_1(X_\infty;R) \cong H_1(X_\infty,\partial X_\infty;R)$.

Now suppose R is a field, so that Π is a principal ideal domain. Then, by the universal coefficient theorem and the fact that $H_2(X_\infty;R)$ is Π-torsion, (the surjectivity of $t-1$ on $H_2(X_\infty;R)$ follows in the same way as for $H_1(X_\infty;R)$), we get

$$\overline{H_1(X_\infty;R)} \cong \mathrm{Ext}_\Pi(H_1(X_\infty;R),\Pi) .$$

Since $H_1(X_\infty;R)$ is also Π-torsion, we finally obtain the fundamental duality isomorphism

$$\overline{H_1(X_\infty;R)} \cong H_1(X_\infty;R) .$$

In particular, taking $R = \mathbb{Q}$, this implies the familiar duality property of the Alexander polynomials

$$(\Delta_i) = (\bar{\Delta}_i) , \quad \text{i.e.} \quad \Delta_i(t) = t^{\deg \Delta_i}\Delta_i(t^{-1}) .$$

(Note that this, and the fact that $\Delta_i(1) = 1$, implies that $\deg \Delta_i$ is even.)

Now consider the case $R = \mathbb{Z}$. Levine [85] shows that, since Λ has global dimension 2, the universal coefficient spectral sequence still gives us an isomorphism

$$\overline{H_1(X_\infty)} \cong \mathrm{Ext}_\Lambda(H_1(X_\infty),\Lambda) .$$

It follows from this, incidentally, that $H_1(X_\infty)$ is $\mathbb{Z}$-torsion-free. (Here is the argument; see [85, p. 9]. For any positive integer m, the short exact sequence

$$0 \longrightarrow \Lambda \xrightarrow{m} \Lambda \longrightarrow \Lambda/m\Lambda \longrightarrow 0$$

gives rise to an exact sequence

$$\mathrm{Hom}_\Lambda(H_1(X_\infty),\Lambda/m\Lambda) \longrightarrow \mathrm{Ext}_\Lambda(H_1(X_\infty),\Lambda) \xrightarrow{m} \mathrm{Ext}_\Lambda(H_1(X_\infty),\Lambda) \ .$$

But $H_1(X_\infty)$ is annihilated by Δ, which is primitive since $\varepsilon(\Delta) = 1$, and multiplication by a primitive on $\Lambda/m\Lambda$ is injective, by the Gauss lemma. Hence $\mathrm{Hom}_\Lambda(H_1(X_\infty),\Lambda/m\Lambda) = 0$ and multiplication by m on $\mathrm{Ext}_\Lambda(H_1(X_\infty),\Lambda)$ is injective.)

It is interesting to note that over Λ, however, we no longer necessarily have the strong duality statement $\overline{H_1(X_\infty)} \cong H_1(X_\infty)$. Failure of this may sometimes be detected, for example, by the ideal class invariant described in [47].

Returning to arbitrary (Noetherian) coefficients R, here is a slightly different interpretation of duality. Since $H_1(X_\infty;R)$ is Π-torsion, $\langle\ ,\ \rangle$ induces a form

$$\beta_R\colon H_1(X_\infty;R) \times H_1(X_\infty;R) \longrightarrow Q(\Pi)/\Pi \ ,$$

where $Q(\Pi)$ denotes the field of fractions of Π. The definition of β_R is as follows. (Note the analogy with the $\mathbb{Q}/\mathbb{Z}$-valued linking form on the torsion subgroup of the first homology of an oriented 3-manifold.) Let $c \in C_1$, $d \in C_1'$ be representative cycles for elements $x,y \in H_1(X_\infty;R) \cong H_1(X_\infty,\partial X_\infty;R)$. Since $H_1(X_\infty;R)$ is Π-torsion, there exists $c' \in C_2'$ such that $\partial c' = \pi d$ for some nonzero $\pi \in \Pi$. Define

$$\beta_R(x,y) = \frac{\langle c,c'\rangle}{\pi} \ .$$

This form β_R is sesquilinear and Hermitian, and is called the <u>Blanchfield pairing</u> (over R) of the knot. (See [12].)

Consider the case when R is a field. Since the adjoint to β_R is the composition

$$\overline{H_1(X_\infty;R)} \cong \mathrm{Ext}_\Pi(H_1(X_\infty;R),\Pi) \cong \mathrm{Hom}_\Pi(H_1(X_\infty;R),Q(\Pi)/\Pi) ,$$

β_R is non-singular. Here, the first isomorphism comes from duality and universal coefficients, and the second from the short exact sequence

$$0 \longrightarrow \Pi \longrightarrow Q(\Pi) \longrightarrow Q(\Pi)/\Pi \longrightarrow 0 ,$$

using the fact that $H_1(X_\infty;R)$ is Π-torsion.

It turns out that this is also true when $R = \mathbb{Z}$ (see [12], [85]), that is, $\beta = \beta_{\mathbb{Z}}$ induces an isomorphism

$$\overline{H_1(X_\infty)} \cong \mathrm{Hom}_\Lambda(H_1(X_\infty),Q(\Lambda)/\Lambda) .$$

As regards the classification of Blanchfield pairings, the case $R = \mathbb{Q}$ has been done, as follows. In [152], Trotter defines a function $\chi\colon Q(\Gamma)/\Gamma \longrightarrow \mathbb{Q}$ such that

$$\chi\beta_{\mathbb{Q}} : H_1(X_\infty;\mathbb{Q}) \times H_1(X_\infty;\mathbb{Q}) \longrightarrow \mathbb{Q}$$

is non-singular, skew-symmetric, has $t\colon H_1(X_\infty;\mathbb{Q}) \longrightarrow H_1(X_\infty;\mathbb{Q})$ as an isometry, and has the property that the isomorphism class of the pair $(\chi\beta_{\mathbb{Q}},t)$ determines the isometry class of $\beta_{\mathbb{Q}}$. But pairs consisting of a non-singular ε-symmetric ($\varepsilon = \pm 1$) bilinear form on a finite dimensional R-vector space, together with an isometry, have been classified (see [98]). The rational Blanchfield pairings are thereby also classified.

When $R = \mathbb{Z}$, a complete classification has not yet been achieved. (See [84], [152], for partial results; also §9.)

8. The Seifert Form

So far, we have tried to keep the discussion as 'co-ordinate-free' as possible All the cyclic covers of a knot K, however, can be constructed with the aid of any orientable surface spanning K, and it is illuminating to examine the properties discussed above from this point of view. Indeed, this is the way in which many of them were first discovered.

It was an early observation that colouring alternately black and white the regions created by a general position projection of a knot K onto the plane defines two surfaces in S^3 bounded by K. However, it may happen that neither of these is orientable. (It is not hard to show that only the trivial projection of the trivial knot has both surfaces orientable.) The first proof that every knot does bound an orientable surface is given in [48]. Later, Seifert [136] gave another proof, and used such a spanning surface, now called a Seifert surface for K, to derive many important invariants and their properties.

One might proceed as follows. Let the knot K have tubular neighbourhood N and exterior X. Then $N \cong S^1 \times D^2$ and $\partial X = \partial N \cong S^1 \times S^1$. The composition $H_2(N,\partial N) \overset{\text{excision}}{\cong} H_2(S^3,X) \overset{\partial}{\cong} H_1(X)$ shows that $H_1(X) \cong \mathbb{Z}$ is generated by the image of the meridian element $\mu = [* \times S^1] \in H_1(\partial X)$. Let λ be a generator for $\ker(H_1(\partial X) \longrightarrow H_1(X))$. We may choose a trivialization of the normal bundle N so that $\lambda = [S^1 \times *] \in H_1(\partial X)$. Projection $\partial X \longrightarrow S^1$ onto the second factor now extends to $p: X \longrightarrow S^1$. Making p transverse regular, $\text{rel}(p|\partial X)$, to a point in S^1 then gives a bicollared surface $F \subset X$ such that $\partial F (= F \cap \partial X)$ and K together bound an annulus in N. We may assume that F is connected. Cutting X along this Seifert surface F gives a manifold Y whose boundary contains two copies $F^{\pm}$ of F. Taking a countable infinity $\{Y_i\}$ of copies of Y and identifying F^-_{i+1} with F^+_i in the obvious way, for all i, we obtain the infinite cyclic cover X_∞ of X, on which the group C_∞ of covering translations acts by taking each Y_i to Y_{i+1}.

It turns out that all the algebraic information about $H_1(X_\infty)$ discussed above is contained in the Seifert form of F, that is, the bilinear form

$$\alpha: H_1(F) \times H_1(F) \longrightarrow \mathbb{Z}$$

defined by

$$\alpha([w],[z]) = \mathrm{lk}(w^+, z) ,$$

where w, z are 1-cycles in F, w^+ is the cycle obtained by translating w off F in the positive normal direction, and lk denotes linking number in S^3. This form has the property that $\alpha - \alpha^T$ (^{T} denotes transpose) is just the intersection form on $H_1(F)$. In particular, $\det(\alpha-\alpha^T) = 1$. Choosing some basis for $H_1(F)$, we get a $2h \times 2h$ <u>Seifert matrix</u> A representing α, where $h = \text{genus } F$.

A Mayer-Vietoris argument on $X_\infty = \bigcup_{i=-\infty}^{\infty} Y_i$ shows [80] that $H_1(X_\infty)$ is presented as a Λ-module by the matrix $tA - A^T$. In particular, up to a unit of Λ, the Alexander polynomial $\Delta = \det(tA-A^T)$. Since putting $t = 1$ gives the unimodular matrix $A - A^T$ (3), the properties $\varepsilon(E_i) = \mathbb{Z}$, $\varepsilon(\Delta_i) = 1$, of the elementary ideals and Alexander polynomials are immediate.

The consequences of duality are also easily seen in this setting. For example the conjugate module $\overline{H_1(X_\infty)}$ is presented by the matrix $t^{-1}A - A^T$, which is equivalent to $(tA-A^T)^T$. In particular, $\bar{E}_i = E_i$ and $(\bar{\Delta}_i) = (\Delta_i)$ for all i. Since $\det(tA-A^T) \neq 0$, the presentation of $H_1(X_\infty)$ corresponding to $tA - A^T$ is actually a short free resolution

$$0 \longrightarrow F_1 \xrightarrow{\ \varphi\ } F_0 \longrightarrow H_1(X_\infty) \longrightarrow 0 ,$$

where F_0, F_1 are free Λ-modules of rank $2h$. Hence $\mathrm{Ext}_\Lambda(H_1(X_\infty),\Lambda) \cong \mathrm{coker}(\mathrm{Hom}(\varphi, \mathrm{id}))$, and the latter is clearly presented by $(tA-A^T)^T$. So we derive our previous duality statement

---

(3) This is why it is natural, at least for $i = 1$, to normalize so that $\Delta_i(1) = 1$.

$$\overline{H_1(X_\infty)} \cong \mathrm{Ext}_\Lambda(H_1(X_\infty),\Lambda) \ .$$

The Blanchfield pairing $\beta\colon H_1(X_\infty) \times H_1(X_\infty) \longrightarrow Q(\Lambda)/\Lambda$ is also determined by A; it is given by the matrix $(1-t)(tA-A^T)^{-1}$ [70], [85], [152].

Finally, we mention that the matrix M defined by Murasugi [107] in terms of a knot projection can be shown to be a Seifert matrix for a Seifert surface constructed from the knot projection [138].

Turning to the finite cyclic covers, if we write $B(t) = tA - A^T$, then (see §5) B(T) will be a presentation matrix for $H_1(M_k)$. Now B(T) is $2hk \times 2hk$, but Seifert [136] showed how to reduce it, using the permissible matrix operations, to the $2h \times 2h$ matrix $C^k - (C-I)^k$, where $C = A(A-A^T)^{-1}$. He also showed [137] that (and in what sense) the linking form $T_1(M_k) \times T_1(M_k) \longrightarrow \mathbb{Q}/\mathbb{Z}$ is determined by the matrix $(C-I)^k(A-A^T)$. (See [150] for a more general formulation.) This can often be used to detect non-amphicheirality.

9. S-Equivalence

The Seifert form α is clearly an invariant of the pair (S^3,F). Hence, allowing for a change of basis of $H_1(F)$, the equivalence class of A under integral congruence $A \mapsto P^TAP$, P invertible over $\mathbb{Z}$, is an invariant of (S^3,F). (If we choose a symplectic basis for $H_1(F)$, A will satisfy $A - A^T = J = \oplus \begin{pmatrix} 0 & 1 \\ -1 & 0 \end{pmatrix}$. In [150], such an A is called a standard Seifert matrix. Then every Seifert matrix is congruent to a standard one, and two standard Seifert matrices A, B are congruent if and only if they are symplectically congruent, that is, $B = P^TAP$ where P satisfies $P^TJP = J$.)

Since we may always increase the genus of any Seifert surface F for K by adding a 'hollow handle' to it, it is clear that to get an invariant of the knot we must also allow matrix enlargements of the form

$$A \longmapsto \begin{bmatrix} & & & * & 0 \\ & A & & \vdots & \vdots \\ & & & * & 0 \\ 0 & \cdots & 0 & 0 & 1 \\ 0 & \cdots & 0 & 0 & 0 \end{bmatrix} \quad \text{or} \quad \begin{bmatrix} & & & 0 & 0 \\ & A & & \vdots & \vdots \\ & & & 0 & 0 \\ * & \cdots & * & 0 & 0 \\ 0 & \cdots & 0 & 1 & 0 \end{bmatrix}$$

(The *'s record the way the handle links F.) The equivalence relation on Seifert matrices generated by congruence and these enlargements is known as <u>S-equivalence</u>. It will also be convenient to call two knots <u>S-equivalent</u> if they have S-equivalent Seifert matrices.

S-equivalence was first introduced, in an algebraic setting, by Trotter [150]. It also appears in [107]. The following remarks show that it is likely to be an important concept. Firstly, any two Seifert matrices for a given knot K are S-equivalent.(4) (Here is an outline of a proof. Let the matrices be associated with Seifert surfaces F_0, F_1 for K, and let these in turn correspond, via transversality, to maps $p_0, p_1: X \longrightarrow S^1$, such that $p_0|\partial X = p_1|\partial X$, where X is the exterior of K. Then p_0, p_1 extend to $p: X \times I \longrightarrow S^1$, with $p_t|\partial X = p_0|\partial X$ for all $t \in I$. Transverse regularity gives a connected, orientable 3-manifold $M \subset X \times I$ such that $\partial M = F_0 \cup \partial F_i \times I \cup F_1$. Now choose a handle decomposition of M on F_0 with only 1- and 2-handles, such that the former precede the latter, and such that, regarding M as $F_0 \cup \text{collar} \cup \text{handle} \cup \text{collar} \ldots$, each handle is embedded in a level $X \times \{t\}$, and the collars are compatible with the I factor (see [72]). Then in a level between the 1- and 2-handles, M intersects X in a Seifert surface for K which is obtained from each of F_0, F_1 by adding hollow handles.) Secondly, given a Seifert matrix A for K, it is easy to see that any matrix obtained from A by a sequence of enlargements (and congruences) is also a Seifert matrix for K. (But this is not necessarily true for reductions.)

---

(4) In [107], it is noted that by examining the effects of the Reidemeister moves on a knot diagram, the S-equivalence class of the Murasugi matrix can be shown to to be an invariant of K.

Thirdly, in higher (odd) dimensions, S-equivalence completely classifies the so-called simple knots [83].

Probably the most important result concerning S-equivalence relates it to the Blanchfield pairing:

Two knots are S-equivalent if and only if their (integral) Blanchfield pairings are isometric.

A purely algebraic proof of this has been given by Trotter [152]. It is also a consequence of some results of Kearton [70] and Levine [83] on higher-dimensional knots. (In [83] it is shown that, for $n \geq 2$, two simple knots of S^{2n-1} in S^{2n+1} are isotopic if and only if they are S-equivalent, and, in [70], that they are isotopic if and only if their Blanchfield pairings are isometric. Since the algebra only depends on n (mod 2), this implies the stated result.)

In [150] it is shown that every Seifert matrix is S-equivalent to a non-singular one, that is, one with $\det A \neq 0$. Since $\Delta = \det(tA - A^T)$, we see that $\det A = \Delta(0)$ is then an invariant of the knot. Also, the Γ-module $H_1(X_\infty;\mathbb{Q})$ is presented by $tI - A^{-1}A^T$, which shows that $\dim H_1(X_\infty;\mathbb{Q}) = 2h$ (if A is $2h \times 2h$), and that the automorphism t is given by the matrix $A^{-1}A^T$.

It is known that S-equivalence of non-singular Seifert matrices is definitely weaker than integral congruence [83], but there are the following partial results in the other direction. Non-singular Seifert matrices determine isometric rational Blanchfield pairings (recall (§7) that these are classified) if and only if they are congruent over $\mathbb{Q}$ [152]. If A and B are S-equivalent non-singular Seifert matrices, so $\det A = \det B = d$, say, then A and B are congruent over $\mathbb{Z}[d^{-1}]$ [150], [83], [152]. (The converse is false [83].) If $|d|$ is prime, then in fact A and B are congruent over $\mathbb{Z}$ [152]. If d is square-free, then A and B are congruent over the p-adic integers $\mathbb{Z}_p$ for all primes p [152].

10. Characterization

The first realization result concerning the invariants we have been discussing is Seifert's proof [136] that a polynomial Δ is the Alexander polynomial of a knot if and only if it satisfies

(i) $\Delta(1) = 1$, and

(ii) $\Delta(t) = t^{\deg \Delta} \Delta(t^{-1})$.

To do this, Seifert actually shows that any integral matrix A such that $A - A^T = J$ can be realized as a Seifert matrix. This is done by taking an orientable surface of the appropriate genus, regarded as a disc with bands, and embedding it in S^3 by twisting and linking the bands so as to realize A as the matrix (with respect to the basis of $H_1(F)$ represented by the cores of the bands) of the Seifert form. It follows (by changing basis) that any matrix A with $\det(A-A^T) = 1$ is a Seifert matrix.

It turns out that in Seifert's realization of the polynomial, the module which arises, i.e. the module presented by $tA - A^T$, is actually the cyclic Λ-module $\Lambda/(\Delta)$. By taking connected sums, it follows that any sequence of polynomials $\lambda_1, \ldots, \lambda_n$ satisfying the (necessary) conditions

(i) $\lambda_i(1) = 1$, $1 \leq i \leq n$

(ii) $\lambda_i(t) = t^{\deg \lambda_i} \lambda_i(t^{-1})$, $1 \leq i \leq n$, and

(iii) $\lambda_{i+1} | \lambda_i$, $1 \leq i < n$

can occur as the Alexander invariants of a knot. (This can be equivalently expressed in terms of the Alexander polynomials.) A different proof is given in [79].

In particular, the Γ-modules which can occur as $H_1(X_\infty;\mathbb{Q})$ for some knot are completely and simply characterized.

Over the integers, we have the following realization result of Levine [85], which brings in the Blanchfield pairing:

Let H be a finitely-generated Λ-module such that $t-1: H \to H$ is surjective, and let $\beta: H \times H \to Q(\Lambda)/\Lambda$ be a non-singular, sesquilinear, Hermitian pairing. Then β is the Blanchfield pairing of some knot.

To prove this, it is sufficient to show that every such β is given by $(1-t)(tA-A^T)^{-1}$ for some integral matrix A with $\det(A-A^T) = 1$. (Is there a direct algebraic proof of this?) This Levine does by showing that β may be realized as the Blanchfield pairing of some knot of S^{4k+1} in S^{4k+3}, for (any) $k > 0$; a Seifert matrix for this knot is then the desired A.

11. The Quadratic and Other Forms

Because of its historical significance, we shall now make a few remarks about the quadratic form of a knot, although from many points of view this is best discussed in a 4-dimensional setting (see §12).

There are actually two distinct, but related, concepts here. The first is due to Goeritz [51], who associated with a knot diagram an integral quadratic form as follows. Colour the regions of the diagram alternately black and white, the unbounded region being coloured white,[(5)] and number the other white regions $W_1, \ldots, W_n$. At a crossing point c as shown in Figure 1

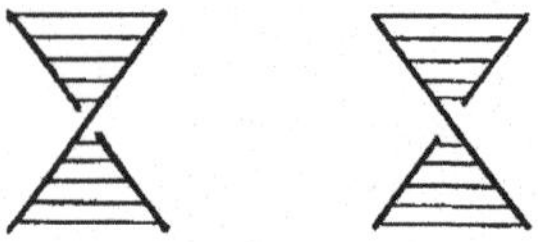

Figure 1

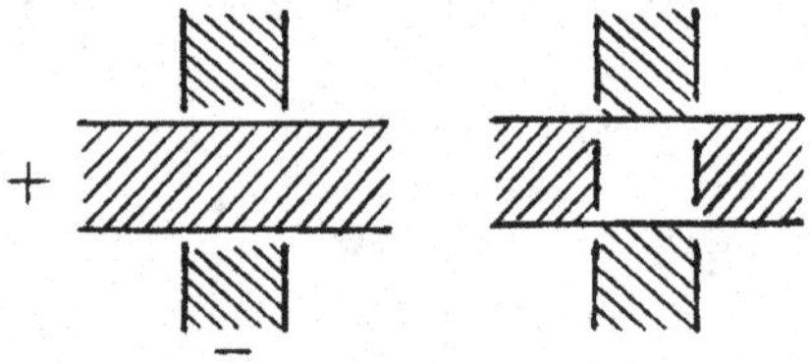

Figure 2

assign 1, -1 respectively if the adjacent white regions are distinct, and 0 otherwise. Call this index $\eta(c)$. Then define the $n \times n$ matrix $G = (g_{ij})$ by

$$g_{ii} = \Sigma\, \eta(c) \quad \text{over crossings adjacent to } W_i ,$$

$$g_{ij} = -\Sigma\, \eta(c) \quad \text{over crossings adjacent to } W_i \text{ and } W_j , \quad i \neq j.$$

It may be verified [51], [76] that the class of G under the equivalence relation generated by (integral) congruence and

$$G \longmapsto \begin{bmatrix} G & 0 \\ 0 & \pm 1 \end{bmatrix}$$

---

(5) Goeritz chose black, but it turns out that this is psychologically confusing.

is invariant under any of the 3 so-called Reidemeister moves [3], [123] on a knot diagram, and is therefore an invariant of the knot K. In particular, the absolute value of the determinant, and the Minkowski units C_p for odd primes p, are invariants of K, (but C_2 and the signature are not) [51].

In [137], Seifert relates G to the 2-fold branched cover M_2 of K, by observing that the latter can be obtained by cutting S^3 along the spanning surface for K corresponding to the shaded regions of the knot projection and gluing together two copies of the resulting manifold in an appropriate fashion. In particular, he shows that G is a presentation matrix for $H_1(M_2)$, and that the linking form $H_1(M_2) \times H_1(M_2) \to \mathbb{Q}/\mathbb{Z}$ is given by $\pm G^{-1}$, the sign depending on the orientation of M_2. (See also §12.) Note that $|\det G| = \text{order } H_1(M_2) = |\Delta(-1)|$ is always odd.

Such linking forms are classified by certain ranks and quadratic characters corresponding to each p-primary component (p an odd prime). See [135], [62]. In [120] (see also [78]) it is shown that these invariants determine the Minkowski units C_p, and, more generally, Kneser-Puppe in [76] show that in fact the linking form completely determines the equivalence class (in the above sense) of the quadratic form.

More recently, Trotter [150] considered the quadratic form given by $A+A^T$, where A is a Seifert matrix for K. (See also [107], which studies $M+M^T$, where M is the Murasugi matrix.) S-equivalence on A induces the equivalence relation on $A+A^T$ generated by congruence and addition of a hyperbolic plane $\begin{bmatrix} 0 & 1 \\ 1 & 0 \end{bmatrix}$. This is a stronger equivalence than the one discussed previously. Also, it may be shown that if the shaded surface F obtained from a knot projection happens to be orientable, then the corresponding Goeritz matrix coincides with $A+A^T$ for some Seifert matrix A associated with F. Finally, for any Seifert matrix A of K, $A+A^T$ is in the equivalence class of Goeritz matrices of K. This may be seen by isotoping the given Seifert surface, regarded as a disc with bands, so that the bands cross over as shown in Figure 2, where + denotes one side of the surface and - the other. The modification shown in Figure 2 produces an orientable surface obtainable from the indicated knot projection by shading; the corresponding Goeritz matrix will then be $B+B^T$ for some Seifert matrix B, which must be S-equivalent to A. (Here, it is easy to see the S-equivalence directly: join the two bands by a 1-handle at each band-crossing.)

From Trotter's form additional invariants may be extracted, notably the signature (therefore referred to as the signature of the knot, $\sigma(K)$), and hence also the Minkowski unit C_2. Also, recall that any non-singular Seifert matrix for K is $2h \times 2h$, where $2h = \deg \Delta$. Hence it follows from Witt's cancellation theorem that over any local ring in which 2 is invertible, the forms $A+A^T$ coming from non-singular Seifert matrices A are all congruent. In particular, this holds for the p-adic integers $\mathbb{Z}_p$, p odd, and $\mathbb{R}$, and hence (since $A+A^T$ is even, see [62]) the genus of $A+A^T$ is an invariant of K [150].

The forms of both Goeritz and Trotter are generalized in [58], where it is shown how a quadratic form may be defined for any spanning surface. The signature of such a form is related to the signature $\sigma(K)$ of the knot. In particular, the correction term needed to obtain $\sigma(K)$ from the signature of a Goeritz matrix can be simply described in terms of the given knot projection.

By symmetrizing A to $A+A^T$, we obtained the signature. Other signatures may be obtained, by Hermitianizing A in other ways. Precisely, let ξ be a complex number, and consider the Hermitian matrix $A(\xi) = (1-\bar{\xi})A + (1-\xi)A^T$. (We may suppose without loss of generality that $\xi \in S^1$, that is, $|\xi| = 1$.) Then S-equivalence on A induces the equivalence relation on $A(\xi)$ generated by congruence (by integral matrices) and addition of $\begin{bmatrix} 0 & 1-\bar{\xi} \\ 1-\xi & 0 \end{bmatrix}$. In particular, the signature of $A(\xi)$ depends only on K, and therefore defines a function $\sigma_K: S^1 \longrightarrow \mathbb{Z}$. Since $A(\xi) = (\xi-1)(\bar{\xi}A - A^T)$, σ_K is continuous away from the roots of the Alexander polynomial $\Delta = \det(tA - A^T)$. These signatures $\sigma_K(\xi)$ are (essentially) those considered by Levine in [81]. For certain roots of unity ξ, they were introduced earlier by Tristram [148]. We shall see later (§12) that for ξ a root of unity, $\sigma_K(\xi)$ actually has a natural geometric interpretation.

Another approach to these signatures is the following. Milnor [96] and Erle [34] show that, over any field R, the (skew-symmetric) cup product pairing

$$H^q(X_\infty, \partial X_\infty; R) \times H^{2-q}(X_\infty, \partial X_\infty; R) \to H^2(X_\infty, \partial X_\infty; R) \cong R$$

is non-singular. Taking $q = 1$ and setting

$$\langle x, y \rangle = x \cup (ty) + y \cup (tx)$$

then defines a non-singular, R-valued, symmetric bilinear form $\langle\ ,\ \rangle$ on $H^1(X_\infty, \partial X_\infty; R)$. With respect to an appropriate basis, $\langle\ ,\ \rangle$ is given by $A + A^T$, where A is a non-singular Seifert matrix, and thus coincides with Trotter's quadratic form (tensored with R). (See [34] for details.)

We remark that the non-singularity of the above cup product pairing can be interpreted as a Poincaré duality in X_∞ of formal dimension 2. However, this non-singularity definitely fails over $\mathbb{Z}$; for example, $H^1(X_\infty, \partial X_\infty)$ $(\cong H^1(X_\infty))$ is often zero.

Taking $R = \mathbb{R}$, let λ be a symmetric, irreducible factor of the Alexander polynomial, so $\lambda = (t-\xi)(t-\bar{\xi})$ where $\xi = e^{i\theta}$, say. Milnor [96] then defines $\sigma_\theta(K)$ to be the signature of the restriction of $\langle\ ,\ \rangle$ to the λ-primary component. The signature of the knot $\sigma(K)$ is the sum of all the $\sigma_\theta(K)$.

These signatures $\sigma_\theta(K)$ turn out to be equivalent to the signature function σ_K; Matumoto has shown [94] that $\sigma_\theta(K)$ is just the jump in σ_K at $e^{i\theta}$.

12. Some 4-Dimensional Aspects

It is enlightening to consider the branched cyclic covers from a 4-dimensional point of view. The basic construction is the following. Pushing the interior of a Seifert surface F for K in S^3 into the interior of the 4-ball B^4 gives a properly embedded surface $\hat{F} \subset B^4$ with $\partial\hat{F} = K$. For $1 \leq k < \infty$, we then have $M_k = \partial V_k$, where M_k, V_k is the k-fold branched cyclic cover of (S^3, K), $(B^4, \hat{F})$ respectively.

Let us first consider the case $k = 2$. In S^3, choose a thickening $F \times [-1,1]$ of $F \times 0$. Then V_2 may be constructed by taking two copies of B^4, and identifying (x,t) in one copy with $(x,-t)$ in the other, for all $x \in F$, $t \in [-1,1]$, (and then smoothing). The canonical covering translation just interchanges the copies of B^4. A Mayer-Vietoris argument shows that $H_2(V_2) \cong H_1(F)$, and that if A is the Seifert matrix associated with some basis of $H_1(F)$, then the intersection form on $H_2(V_2)$, with respect to the corresponding basis, is given by the matrix $A+A^T$ (see, for example, [69]).

Actually this works even if F is non-orientable. A thickening of F will now be a twisted $[-1,1]$-bundle over F, but we may still carry out the above construction using the local product structure. The intersection form on $H_2(V_2)$ can again be described in terms of F; in particular, if F arises from the shaded regions of a knot diagram, then the intersection form is given by the Goeritz matrix G [58].

By duality we have the exact sequence

$$H_2(V_2) \xrightarrow{\varphi} \mathrm{Hom}(H_2(V_2), \mathbb{Z}) \longrightarrow H_1(M_2) \longrightarrow 0$$

where φ is adjoint to the intersection form. Thus, if the latter is given by a matrix B, say, then φ will be represented by B with respect to dual bases. It is then clear that B is a presentation matrix for $H_1(M_2)$. It also follows, using $\det B \neq 0$, that the linking form on $H_1(M_2)$ is given by $-B^{-1}$. This recovers the results of Seifert [137] on the 2-fold branched cover.

Now let us consider the higher order branched covers; here, F must be orientable. As before, the intersection form on $H_2(V_k)$ may be described in terms of the Seifert form of F. In particular, one may write down a presentation matrix B for $H_1(M_k)$ in terms of a Seifert matrix, and again, if $H_1(M_k)$ is finite (as will always be the case if k is a prime-power, for example), the linking form on $H_1(M_k)$ will be given by $-B^{-1}$.

Using the cyclic group action, one may derive finer information. The intersection form on $H_2(V_k)$ extends naturally to a Hermitian form on $H_2(V_k;\mathbb{C})$, with respect to which the automorphism τ of $H_2(V_k;\mathbb{C})$, induced by the canonical covering translation, is an isometry. Let $\omega = e^{\frac{2\pi i}{k}}$. Then $H_2(V_k;\mathbb{C})$ decomposes as an orthogonal direct sum $E_0 \oplus E_1 \oplus \ldots \oplus E_{k-1}$, where E_r is the ω^r-eigenspace of τ. Let $\sigma_r(V_k)$ be the signature of the restriction of our Hermitian form to E_r. It then turns out that

$$\sigma_r(V_k) = \mathrm{sign}((1-\omega^{-r})A + (1-\omega^r)A^T), \quad 0 \leq r < k$$

where A is a Seifert matrix for F. (See [32], [154], [18]). These signatures $\sigma_r(V_k) = \sigma_K(\omega^r)$, $0 < r < k$, are the <u>k-signatures</u> of the knot K. In particular, $\sigma_1(V_2)$ is just the signature of V_2.

We saw earlier that $\sigma_K(\xi)$ depends only on K. Here, rather more is true. We could construct V_k with $\partial V_k = M_k$ using any (orientable) surface $F \subset B^4$ with $\partial(B^4, F) = (S^3, K)$. Then $\sigma_r(V_k)$ is independent of F. To see this, we shall use the G-signature theorem [6]; (for an elementary proof for semi-free actions in dimension 4, which is all that is needed here, see [57]). Recall that the τ^s-signatures $\mathrm{sign}(\tau^s, V_k)$ are defined as follows. We have $H_2(V_k; \mathbb{C}) = H^+ \oplus H^- \oplus H^0$, where the Hermitian 'intersection' form is $\pm$-definite on $H^{\pm}$ and zero on H^0. Then

$$\mathrm{sign}(\tau^s, V_k) = \mathrm{trace}(\tau^s | H^+) - \mathrm{trace}(\tau^s | H^-) .$$

By similarly decomposing each eigenspace $E_r = E_r^+ \oplus E_r^- \oplus E_r^0$, we may take $H^+ = E_0^+ \oplus E_1^+ \oplus \ldots \oplus E_{k-1}^+$, etc., which (recalling that $\sigma_0(V_k) = 0$) shows that

$$\mathrm{sign}(\tau^s, V_k) = \sum_{r=1}^{k-1} \omega^{rs} \sigma_r(V_k) , \qquad 0 < s < k .$$

Inverting, we obtain

$$\sigma_r(V_k) = \frac{1}{k} \sum_{s=1}^{k-1} (\omega^{-rs} - 1)\mathrm{sign}(\tau^s, V_k) , \qquad 0 < r < k .$$

Now suppose V_k, V'_k arise from surfaces F, $F' \subset B^4$ with $\partial F = \partial F' = K$. Let $W = V_k \cup -V'_k$, identified along M_k. Since the fixed-point set $F \cup -F'$ of the C_k-action on W has trivial normal bundle, the G-signature theorem gives $\operatorname{sign}(\tau^s, W) = 0$, $0 < s < k$, and therefore, by Novikov additivity,

$$\operatorname{sign}(\tau^s, V_k) = \operatorname{sign}(\tau^s, V'_k), \qquad 0 < s < k .$$

Hence

$$\sigma_r(V_k) = \sigma_r(V'_k), \qquad 0 < r < k ,$$

as required.

A variation of the above proof in the case $k = 2$ allows one to compute the signature of a knot K from an arbitrary (not necessarily orientable) surface F in B^4 with $\partial F = K$ [58]. In particular, this leads to the relation between the signature of K and the signature of any Goeritz matrix for K which was alluded to in §11.

13. Concordance

Two knots K_0, K_1 in S^3 (everything oriented) are said to be concordant if there is a smooth, oriented, submanifold T of $S^3 \times I$, homeomorphic to $S^1 \times I$, such that $T \cap S^3 \times 0 = K_0$, $T \cap S^3 \times 1 = -K_1$. This concept was introduced by Fox and Milnor [46]. Concordance is an equivalence relation, and the equivalence classes form an abelian group C_1 under connected sum, the zero element 0 being represented by the unknot, and the inverse of $[K]$ being represented by the inverted mirror-image of K. A knot represents 0 in C_1 if and only if it is slice, that is, bounds a smooth 2-disc in B^4. This knot concordance group C_1 has not yet been computed; indeed our comparative lack of knowledge about its structure is a central example of our present ignorance concerning 4-manifolds in general.

Historically, the first necessary condition to be established for a knot to be slice (see [46]) was that the Alexander polynomial must satisfy $\Delta \sim \lambda\bar{\lambda}$, for some $\lambda \in \Lambda$. This is enough to show that C_1 is not finitely-generated. Later, the concordance invariance of the signature was proved [107]. (Murasugi works entirely with his matrix M, but as we remarked earlier, this is a particular Seifert matrix.) This implies the existence of elements of infinite order in C_1. The Miwkowski units are also concordance invariants [106], as are the p-signatures [148] and the signatures $\sigma_\theta(K)$ [96].

This information is all subsumed under the invariance of the 'Witt class' of the Seifert form, which we shall discuss soon, but we pause briefly to consider the signature function $\sigma_K: S^1 \longrightarrow \mathbb{Z}$ of §11, as a direct approach to this is possible via branched covering spaces.

Recall (§12) that for $\xi = e^{\frac{2\pi r i}{k}}$ a kth root of 1, $\sigma_K(\xi) = \sigma_r(V_k)$, the signature of the restriction to the ξ-eigenspace of the intersection form on the k-fold branched cyclic cover V_k. Now suppose $(S^3 \times I, T)$ is a concordance, between knots K_0 and K_1, say, and let W_k be its k-fold branched cyclic cover. If k is a prime-power, then (as in §5), $H_*(W_k;\mathbb{Q}) \cong H_*(S^3 \times I;\mathbb{Q})$; in particular, $H_2(W_k;\mathbb{Q}) = 0$. Hence, by Novikov additivity of the eigenspace signatures, $\sigma_{K_0}(\xi) = \sigma_{K_1}(\xi)$. (In particular, the p-signatures of Tristram are concordance invariants.) Since the roots of 1 of prime-power order are certainly dense in S^1, and since σ_K is continuous except at finitely many points in S^1, it follows that $\sigma_{K_0} = \sigma_{K_1}$ almost everywhere. Hence if we define $\tau_K: S^1 \longrightarrow \mathbb{Z}$ by taking the average of the one-sided limits of σ_K at each point, we see that τ_K is a concordance invariant. This is equivalent to the concordance invariance of the $\sigma_\theta(K)$'s, proved in [96] (see §11). Compare also [81, p. 242]. (Note that τ_K takes values in $\mathbb{Z}$, since if ξ is not a root of the Alexander polynomial of K, $A(\xi)$ (see §11) is non-singular; hence $\sigma_K(\xi) \equiv \operatorname{rank} A(\xi) \pmod 2$ is even. Also, Matumoto has shown [94] that if the 1st Alexander invariant (or minimal polynomial) λ_1 has no repeated roots, then $\tau_K = \sigma_K$.)

We now turn to the Seifert form. (The treatment which follows is that of Levine [81], [82].) Let K be a slice knot, so $(S^3,K) = \partial(B^4,D)$ for some smooth 2-disc D. A tubular neighbourhood of D in B^4 may be identified with $D \times D^2$; let $V = B^4 - D \times \operatorname{int} D^2$ be the exterior of D. Then $\partial V = X \cup D \times S^1$, where X is exterior of K, and projection $D \times S^1 \longrightarrow S^1$ extends to a map $p: V \longrightarrow S^1$. Transverse regularity gives a bicollared 3-manifold $M \subset B^4$ with $\partial M = F \cup D$ where F is a Seifert surface for K in S^3. By duality and universal coefficients, the homology exact sequence of $(M, \partial M)$ gives an exact sequence

$$H_1(M,\partial M)^* \xrightarrow{j^*} H_1(M)^* \longrightarrow H_1(\partial M) \xrightarrow{i} H_1(M) \xrightarrow{j} H_1(M,\partial M) \;,$$

where $*$ denotes $\mathrm{Hom}(\;,\mathbb{Z})$. Thus $\operatorname{coker} j^* \cong \ker i$, and $\ker j = \operatorname{im} i$. But $\ker j$ and $\operatorname{coker} j^*$ have the same rank; hence $\operatorname{rank}(\ker i) = \frac{1}{2} \operatorname{rank} H_1(\partial M)$. Moreover the Seifert form α on $H_1(F) \cong H_1(\partial M)$ vanishes on $\ker i$. (If w, z are 1-cycles in F representing elements in $\ker i$, there are 2-chains u, v in M with $\partial u = w$, $\partial v = z$. Then u^+, obtained by pushing u off M in the positive normal direction, has $\partial u^+ = w^+$ and is disjoint from v. Hence $\mathrm{lk}(w^+, z) = 0$.)

Recall that Seifert forms can be characterized algebraically as just those bilinear forms $\alpha: H \times H \longrightarrow \mathbb{Z}$, H a finitely-generated, free abelian group, such that $\det(\alpha - \alpha^T) = 1$. (This implies that H has even rank.) Write $\alpha \sim \beta$ if the orthogonal sum $\alpha \oplus (-\beta)$ vanishes on a subgroup (and hence on a direct summand) of half the total rank. This is an equivalence relation on Seifert forms, and the equivalence classes form a 'Witt group' $W_S(\mathbb{Z})$ under $\oplus$. Since connected sum of knots induces orthogonal sum of Seifert forms, the discussion in the previous paragraph shows that there is an epimorphism

$$\psi: C_1 \longrightarrow W_S(\mathbb{Z}) \;.$$

The first step in the computation of $W_S(\mathbb{Z})$ is to pass to the rationals. Thus one defines $W_S(\mathbb{Q})$ to be the analogous Witt group of finite-dimensional bilinear forms α over $\mathbb{Q}$ with $\det((\alpha - \alpha^T)(\alpha + \alpha^T)) \neq 0$. The natural map $W_S(\mathbb{Z}) \longrightarrow W_S(\mathbb{Q})$ is injective.

The problem of computing $W_S(\mathbb{Q})$ can be translated into a more standard one by symmetrizing, as follows. Consider pairs $(\langle\;,\;\rangle, t)$ consisting of a finite-dimensional non-singular symmetric bilinear form $\langle\;,\;\rangle$ over $\mathbb{Q}$ together with an isometry t, such that ± 1 is not an eigenvalue of t. The classes of such <u>isometric structures</u>, under the equivalence relation obtained by factoring out forms with a t-invariant subspace of half the total dimension on which $\langle\ ,\ \rangle$ vanishes, form a Witt group $W_0(C_\infty, \mathbb{Q})$ under $\oplus$. An isomorphism

$$W_S(\mathbb{Q}) \longrightarrow W_0(C_\infty, \mathbb{Q})$$

is induced by sending a non-singular matrix A representing a class in $W_S(\mathbb{Q})$ to the class in $W(C_\infty, \mathbb{Q})$ with matrix representatives $(A + A^T, A^{-1}A^T)$. (Every class in $W_S(\mathbb{Q})$ has a non-singular representative.) Note that if A is a non-singular Seifert matrix for a knot K, then $A + A^T$ represents the quadratic form of K, and $A^{-1}A^T$ represents the automorphism $t\colon H_1(X_\infty;\mathbb{Q}) \longrightarrow H_1(X_\infty;\mathbb{Q})$.

A complete set of invariants for $W_0(C_\infty, \mathbb{Q})$ has been given by Levine [82], using results of Milnor [98]. These are defined for each λ-primary component V_λ, where λ is a symmetric, irreducible factor of the characteristic polynomial of t, and are: the exponent mod 2 of λ in the characteristic polynomial, the signature of the restriction of $\langle\ ,\ \rangle$ to V_λ (this is the σ_θ of §11), and a Witt class invariant version (analogous to a Minkowski unit) of the Hasse invariant of the restriction of $\langle\ ,\ \rangle$ to V_λ. In particular, $W_0(C_\infty, \mathbb{Q}) \cong \mathbb{Z}^\infty \oplus (\mathbb{Z}/4)^\infty \oplus (\mathbb{Z}/2)^\infty$. The image of the injection $W_S(\mathbb{Z}) \longrightarrow W_0(C_\infty, \mathbb{Q})$ is also isomorphic to $\mathbb{Z}^\infty \oplus (\mathbb{Z}/4)^\infty \oplus (\mathbb{Z}/2)^\infty$.

A different but related approach to the computation of $W_S(\mathbb{Z})$ is described by Kervaire in [73]. For further results on the structure of $W_S(\mathbb{Z})$, see [143].

Similar definitions and results hold for knots of S^{4n+1} in S^{4n+3} for $n > 0$; in particular, there is a knot concordance group C_{4n+1} and a homomorphism

$$\psi_{4n+1}\colon C_{4n+1} \longrightarrow W_S(\mathbb{Z}) \ .$$

Levine has shown that, if $n > 0$, ψ_{4n+1} is an isomorphism [81]. According to Casson-Gordon [17], [18], however, this is not the case for $n = 0$. We shall briefly summarize their argument.

Let K be a knot in S^3, and N the closed 3-manifold obtained from S^3 by 0-framed surgery along K. Write N_k for the k-fold cyclic cover of N, $1 \leq k \leq \infty$. Recall that M_k denotes the k-fold branched cyclic cover of K, $1 \leq k < \infty$. Suppose that, for some k, $\chi: H_1(M_k) \longrightarrow \mathbb{C}^*$ is a character of order m (that is, the image of χ is C_m, the group of mth roots of 1). Composing χ with the canonical epimorphism $H_1(N_k) \longrightarrow H_1(M_k)$ gives a character χ' of order m on $H_1(N_k)$, inducing an m-fold cyclic covering $\tilde{N}_k \longrightarrow N_k$. Similarly χ induces an m-fold cyclic covering $\tilde{N}_\infty \longrightarrow N_\infty$. Then $\tilde{N}_\infty$ is a regular cover of N_k with group of covering translations $C_m \times C_\infty$. Since $\Omega_3(K(C_m \times C_\infty, 1))$ is finite, there is a regular $C_m \times C_\infty$-covering $\tilde{V}_\infty \longrightarrow V_k$ of compact, oriented 4-manifolds such that

$$\partial \begin{pmatrix} \tilde{V}_\infty & \longrightarrow & \tilde{V}_k \\ \downarrow & \searrow & \downarrow \\ V_\infty & \longrightarrow & V_k \end{pmatrix} = r \begin{pmatrix} \tilde{N}_\infty & \longrightarrow & \tilde{N}_k \\ \downarrow & \searrow & \downarrow \\ N_\infty & \longrightarrow & N_k \end{pmatrix}$$

for some integer $r \neq 0$.

Let $\mathbb{C}(t)$ be the field of rational functions in t with coefficients in $\mathbb{C}$; $\mathbb{C}(t)$ is a $\mathbb{Z}[C_m \times C_\infty] = \mathbb{Z}[C_m][t,t^{-1}]$-module. Write $H_*^t(V_k;\mathbb{C}(t))$ for the twisted homology $H_*(C_*(\tilde{V}_\infty) \otimes_{\mathbb{Z}[C_m \times C_\infty]} \mathbb{C}(t))$. The intersection pairing on the chains of $\tilde{V}_\infty$ (compare §7) induces a form

$$H_2^t(V_k;\mathbb{C}(t)) \times H_2^t(V_k;\mathbb{C}(t)) \longrightarrow \mathbb{C}(t)$$

which is Hermitian with respect to the involution J on $\mathbb{C}(t)$ given by $t \mapsto t^{-1}$ and complex conjugation. This form therefore defines an element $w(V_k) \in W(\mathbb{C}(t),J)$, the Witt group of finite-dimensional Hermitian forms over $\mathbb{C}(t)$. The ordinary intersection form on $H_2(V_k;\mathbb{Q})$ represents an element of $W(\mathbb{Q})$; let $w_0(V_k)$ be the image of this element in $W(\mathbb{C}(t),J)$. Then define

$$\tau(K,\chi) = \frac{1}{r}\,(w(V_k) - w_0(V_k)) \in W(\mathbb{C}(t),J) \otimes_{\mathbb{Z}} \mathbb{Q}\ .$$

It can be shown that $\tau(K,\chi)$ is independent of r and V_k.

Now suppose that K is a slice knot, so $(S^3,K) = \partial(B^4,D)$, say. Let W_k be the k-fold branched cyclic cover of (B^4,D), and take k to be a prime-power. Then $\widetilde{H}_*(W_k;\mathbb{Q}) = 0$ (see §5), so, by duality, $H_1(M_k)$ has order ℓ^2, where $G = \ker(H_1(M_k) \longrightarrow H_1(W_k))$ has order ℓ. Note that G has the property, intrinsic to M_k, that the linking form $H_1(M_k) \times H_1(M_k) \longrightarrow \mathbb{Q}/\mathbb{Z}$ vanishes on G. Let V be the closure of the complement of a tubular neighbourhood of D in B^4, and write V_k for the k-fold cyclic cover of V, $1 \leq k \leq \infty$. Then $\partial V_k = N_k$.

Let χ be a character of prime-power order m on $H_1(M_k)$, such that $\chi(G) = 1$. There is then a character $\overline{\chi}$ on $H_1(W_k)$ such that

$$\begin{array}{ccc} H_1(M_k) & \longrightarrow & H_1(W_k) \\ & {\scriptstyle \chi'}\searrow \quad \swarrow{\scriptstyle \overline{\chi}} & \\ & \mathbb{C}^* & \end{array}$$

commutes. Suppose (but only to simplify the exposition) that $\overline{\chi}$ also has order m. Composing with the canonical epimorphism $H_1(V_k) \longrightarrow H_1(W_k)$, we get a character $\overline{\chi}'$ on $H_1(V_k)$ such that

$$\begin{array}{ccc} H_1(N_k) & \longrightarrow & H_1(V_k) \\ & {\scriptstyle \chi'}\searrow \quad \swarrow{\scriptstyle \overline{\chi}'} & \\ & C_m & \end{array}$$

commutes. We can therefore use V_k to compute $\tau(K,\chi)$. But it can be shown that since V is a homology circle and m is a prime-power, $H_*(\widetilde{V}_\infty;\mathbb{Q})$ is finite-dimensional. In particular, $H_2(\widetilde{V}_\infty)$ is $\mathbb{Z}[C_\infty]$-torsion. Since $\mathbb{C}(t)$ is flat over $\mathbb{Z}[C_m \times C_\infty]$, it follows that $H_2^t(V_k;\mathbb{C}(t)) = H_2(\widetilde{V}_\infty) \otimes_{\mathbb{Z}[C_m \times C_\infty]} \mathbb{C}(t) = 0$, and therefore $w(V_k) = 0$. Again, since V is a homology circle and k is a prime-power, $H_2(V_k;\mathbb{Q}) = 0$ (see §5). Hence $w_0(V_k) = 0$ also, giving $\tau(K,\chi) = 0$.

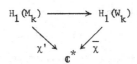

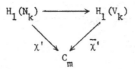

The vanishing of $\tau(K,\chi)$ for certain characters χ is therefore a necessary condition for K to be slice. To utilize this condition, we first define a signature homomorphism

$$\sigma_1 : W(\mathbb{C}(t),J) \otimes_{\mathbb{Z}} \mathbb{Q} \longrightarrow \mathbb{Q} .$$

It suffices to consider $\varphi \in W(\mathbb{C}(t),J)$; suppose φ has a representative which is given with respect to some basis by the matrix $B(t)$. Then set

$$\sigma_1(\varphi) = \frac{1}{2}\left(\lim_{\theta\to 0^+} \text{sign}\, B(e^{i\theta}) + \lim_{\theta\to 0^-} \text{sign}\, B(e^{i\theta})\right) .$$

It turns out that $\sigma_1(\tau(K,\chi))$ is sometimes related to another invariant, analogous to, but simpler than, τ. The general definition of this goes as follows. Let M be a closed, oriented 3-manifold and χ a character of order m on $H_1(M)$ inducing an m-fold cyclic covering $\tilde{M} \longrightarrow M$. Since $\Omega_3(K(C_m,1))$ is finite, there exists an m-fold cyclic covering $\tilde{W} \longrightarrow W$ of compact, oriented 4-manifolds with $\partial(\tilde{W} \longrightarrow W) = r(\tilde{M} \longrightarrow M)$ for some integer $r \neq 0$. Writing $H_*^t(W;\mathbb{C})$ for the twisted homology $H_*(C_*(\tilde{W}) \otimes_{\mathbb{Z}[C_m]} \mathbb{C})$, we have a Hermitian intersection form

$$H_2^t(W;\mathbb{C}) \times H_2^t(W;\mathbb{C}) \longrightarrow \mathbb{C} .$$

Let $s(W)$ be the signature of this form, $s_0(W)$ the ordinary signature of W, and define

$$\sigma(M,\chi) = \frac{1}{r}(s(W) - s_0(W)) \in \mathbb{Q} .$$

This is independent of r and W.

Returning to the knot situation, recall our original character χ on $H_1(M_k)$, inducing $\tilde{M}_k \longrightarrow M_k$. It can be shown that if $H_1(\tilde{M}_k;\mathbb{Q}) = 0$, then

$$|\sigma_1(\tau(K,\chi)) - \sigma(M_k,\chi)| \leq 1$$

If, in addition, K is slice, and χ satisfies the conditions described earlier which then imply that $\tau(K,\chi) = 0$, we obtain

$$|\sigma(M_k,\chi)| \leq 1 .$$

Since the invariant $\sigma(M_k,\chi)$ can often be calculated, this is a workable condition. For example, if K is a 2-bridge (or rational) knot, and $k = 2$, then M_k is a lens space, and $\sigma(M_k,\chi)$ can be calculated fairly easily using the G-signature theorem. Also, in this case, $\tilde{M}_k$ will always be a rational homology sphere, so K can be slice only if (for suitable χ) $|\sigma(M_k,\chi)| \leq 1$. From this it can be shown that a large number of 2-bridge knots K have $\psi([K]) = 0$ in $W_S(\mathbb{Z})$, but are not slice knots.

14. 3-Manifolds and Knots

In this section and the next we shall discuss some of the functions

$$\{\text{knots}\} \longrightarrow \{\text{3-manifolds}\}$$

which may be defined. Such a function relates knot theory to the general theory of 3-manifolds, and hence by means of it any development in one theory will have consequences for the other. Here, among other things, we shall look at some of the ways in which general results about 3-manifolds have had implications for knot theory. Possible influences in the other direction will be considered in §15.

Probably the most obvious function of the above type is the one which simply associates to a knot its exterior. (This is not known to be injective, but the odds seem good that it is.) Here, as we have already mentioned, Dehn's lemma implied that $\pi K \cong \mathbb{Z}$ only if K is trivial, the sphere theorem implies the asphericity of knots, and Waldhausen's work implies that knots are classified by the triples $(\pi K, \lambda, \mu)$.

We might also mention the fibration theorem of Stallings [142], (see also [111]) which, when applied to knot exteriors, implies that many knots K (in fact, precisely those such that the commutator subgroup of πK is finitely-generated, see [142], [110], [112]), correspond to a 'singular' fibring of S^3 over S^1, in the following sense: $S^3 = \bigcup_{\theta=0}^{2\pi} F_\theta$, where each F_θ is homeomorphic to some compact surface F, $\partial F_\theta = K$ for all $\theta \in S^1$, and $S^3 - K = \bigcup_{\theta=0}^{2\pi} \operatorname{int} F_\theta$ is a fibre bundle over S^1, the fibres being the $\operatorname{int} F_\theta$. In other words, K is the binding of an open book structure on S^3. The fibred knots with finite bundle group are precisely the torus knots; see [165] for a nice description of the fibration in this case.

Thurston's recent (unpublished) work on 3-manifolds implies that a knot K which has no companions and is not a torus knot has an exterior which supports a 'hyperbolic structure'. Also, the decomposition theorem of Johannson [66], [67] and Jaco-Shalen [64] applies to knot exteriors. In particular, using this together with his own work, Thurston has shown that knot groups are residually finite. One hopes and expects that in the near future knot theory will be further enriched by these ideas from hyperbolic geometry.

Another advance in the theory of 3-manifolds which has striking consequences for knot theory is discussed in [158]. There it is indicated how Haken's results on hierarchies of incompressible surfaces in irreducible 3-manifolds, and Hemion's recent solution of the conjugacy problem for the group of isotopy classes of homeomorphisms of a compact, bounded, surface, together imply that the knot problem is algorithmically solvable, or, equivalently, that knots can be classified (i.e. listed, without repetition). Again, the connection is via the exterior of the knot.

Branched covering spaces provide examples of functions {knots} $\longrightarrow$ {closed 3-manifolds}, and, as mentioned in §1, invariants of these covers have been used to distinguish knots. Also, by means of such a function, bridge decompositions of the knot are related to Heegaard splittings of the 3-manifold (see [13]), as follows. Let K be a b-bridge knot. Then $(S^3,K) = (B^3_+,A_+) \cup_\partial (B^3_-,A_-)$, where $A_\pm$ is a set of b arcs properly embedded in $B^3_\pm$, and $(B^3_\pm, A_\pm) \cong (B^2,P) \times I$, where P is a set of b points in $\operatorname{int} B^2$. Now let M be some cover of S^3 branched over K. From the bridge decomposition of (S^3,K) one obtains

$M = H_+ \cup_\partial H_-$, say, with $\partial H_\pm$ connected, and $H_\pm \cong \tilde{B} \times I$, where $\tilde{B}$ is the corresponding branched cover of (B^2, P). If the projection $M \longrightarrow S^3$ is k-sheeted away from the branch set, and m-sheeted over K, then

$$\chi(\tilde{B}) = k\,\chi(B^2) - (k-m)\chi(P) = mb - (b-1)k \ .$$

It follows that $H_\pm$ is a solid handlebody of genus $(b-1)k - mb + 1$, giving a Heegaard splitting of M of that genus. For the k-fold branched cyclic cover, the genus is $(k-1)(b-1)$. In particular, for the 2-fold branched cover, we just get $b-1$. In this way, knots of increasing complexity are mapped to 3-manifold decompositions of increasing complexity.

Now it is known that the 2-fold branched covering function is not injective; many examples of pairs of prime knots with the same 2-fold branched cover are described in [11]. It is injective, however, on the set of 2-bridge knots. There, the 2-fold branched cover has genus 1, and is therefore a lens space, and Schubert has proved [132] that this lens space determines the knot. This injectivity already fails for 3-bridge knots [11]. It has been shown by Birman-Hilden [10], however, as a consequence of a rather special feature of the group of isotopy classes of homeomorphisms of a closed surface of genus 2, that if we regard the 2-fold branched covering function as a function $\{\text{knots}\} \longrightarrow \{$equivalence classes of Heegaard splittings of 3-manifolds$\}$, then it is injective on the set of 3-bridge knots.

Finally, in this context we might mention the result of Waldhausen [157], which says that only the unknot has S^3 as its 2-fold branched cover.

15. Knots and 3- and 4-Manifolds

Continuing in the general framework of §14, let us now consider the possibility of using knowledge about knots to give information about 3-manifolds. In particular, functions $\{\text{knots}\} \longrightarrow \{\text{3-manifolds}\}$ which are surjective, or at least have a sizeable image, will be of interest.

Returning to branched covers, Alexander showed [1] that every closed, orientable 3-manifold is a cover of S^3 branched over some link. This has recently been refined (independently) by Hilden [60], Hirsch, and Montesinos [100], who show that every closed, orientable 3-manifold is actually a 3-fold (irregular dihedral) cover of S^3 branched over a knot. This result is best possible in the sense that there are 3-manifolds which are not 2-fold branched covers of S^3. To utilize this function to get invariants of 3-manifolds, it would be helpful to have a purely knot-theoretic description of the equivalence relation on knots which corresponds to homeomorphism of the associated branched covers. Some moves on the knot which leave the branched cover unchanged are known (see [99], [100]), but it has not yet been established whether or not these suffice. (In the same vein, even though the 2-fold branched covering function is not surjective, it would still be interesting to have an intrinsic description of the appropriate equivalence relation on knots.)

Cappell-Shaneson [15] have obtained a formula for the Rohlin μ-invariant of a $\mathbb{Z}/2$-homology sphere M, given as a 3-fold dihedral branched cover of a knot K, which involves (among other things) the classical invariants of K given by the linking numbers of the lifts of K in M [124], [116].

As a concrete example of an application to 3-manifolds of the branched covering space point of view we cite [61], which proves a sharpening of the Hilden-Hirsch-Montesinos theorem, and obtains as a consequence the (known) result that closed, orientable 3-manifolds are parallellizable.

Other interesting ways of constructing 3-manifolds from knots are provided by what is now referred to as Dehn surgery. More precisely, given a knot K and a pair of coprime integers α,β, one can consider the closed, orientable 3-manifold $M(K;\beta/\alpha)$ (we use the 'rational surgery coefficient' notation of [128]), obtained by removing from S^3 a tubular neighbourhood of K and sewing it back so as to identify a meridian on the boundary of the solid torus with a curve on the boundary of the exterior of K homologous to $\alpha[\ell]+\beta[m]$, where (ℓ,m) is a longitude-meridian pair for K. Note that $H_1(M(K;\beta/\alpha)) \cong \mathbb{Z}/|\beta|$. With $|\beta|=1$, this construction first appeared in [27], where Dehn showed that many non-simply-connected homology spheres, in particular, the dodecahedral space discovered earlier by Poincaré, could be obtained in this way from torus knots. Indeed, the Property P conjecture (see §2) is that if K is non-trivial and $\alpha \neq 0$, then $M(K;1/\alpha)$ is never simply-connected.

It seems likely that the function $M(\_;\beta/\alpha)$ is never injective, although this has only been verified for certain β/α [53], [87]. However, it may not be unreasonable to conjecture that, denoting the unknot by 0, and excluding the trivial case $\alpha = 0$, $M(K;\beta/\alpha) \cong M(0;\beta/\alpha)$ only if $K = 0$. The case $|\beta| = 1$ is just a weakened form of the Property P conjecture, and the case $\beta = 0$ has also received some attention (under the name 'Property R').

Turning to the question of surjectivity, clearly the most one could hope to obtain in this way is the set of all closed, orientable 3-manifolds M with $H_1(M)$ cyclic. This seems highly unlikely. In particular, it is surely not true that all homology spheres can be obtained by Dehn's original method, although this is apparently rather difficult to prove.

The a priori restriction on the homology disappears if one allows, instead of knots, links with arbitrarily many components, and it is indeed the case that one can now obtain all closed, orientable 3-manifolds. Actually a stronger statement is possible. If L is a framed link in S^3, then (ordinary) framed surgery on S^3 along L gives a 3-manifold $M(L)$, say. Wallace [159] and Lickorish [86] have shown that this function $\{\text{framed links}\} \longrightarrow \{\text{closed, orientable 3-manifolds}\}$ is surjective. Wallace's proof is essentially 4-dimensional; it uses the theorem of Rohlin [126] that 3-dimensional oriented cobordism $\Omega_3 = 0$, together with handlebody techniques. (The argument is: given M, there exists W such that $M = \partial W$; W has a handle decomposition with one 0-handle and no 4-handles. Replace the 1- and 3-handles by 2-handles ('handle trading'), giving W'. The attaching maps of the 2-handles in W' now define a framed link L with $M \cong M(L)$.) Lickorish's proof, on the other hand, is 2-dimensional, in the sense that it is based on the fact that the group of isotopy classes of orientation-preserving homeomorphisms of a closed surface is generated by 'twists'. (This was first proved by Dehn [29].)

Since the trace of the surgery is a 4-manifold bounded by the given 3-manifold, this approach gives another proof that $\Omega_3 = 0$.

The equivalence relation on framed links (in the oriented 3-sphere) which corresponds to (orientation-preserving) homeomorphism of the associated 3-manifolds M(L) has been identified by Kirby [74], in the sense that it is shown to be generated by certain moves on the link. (Craggs (unpublished) has also obtained results along these lines.) Kirby uses two moves; Fenn-Rourke [35] show that these can be incorporated into a single move. Also, Rolfsen (private communication) has provided the modification necessary to describe the equivalence relation appropriate to the more general process of Dehn surgery on a link.

Armed with these results, it is clear, in theory, how one might go about getting new invariants of 3-manifolds. For instance, with respect to any kind of complexity of a framed link, every 3-manifold will be obtained by surgery on some class of links of minimal complexity, so invariants of this class will be invariants of the 3-manifold. However, this point of view has not yet had much effect on the theory of 3-manifolds, mainly because, although the above-mentioned equivalence relation on links is easy to describe, it seems hard to decide in practice whether two given links are equivalent, or to find link-theoretic invariants of the equivalence relation. It is clear that more remains to be done in this direction.

The work described above also relates knot and link theory to 4-manifolds, and offers the prospect of obtaining, perhaps first, known results (Rohlin's $\Omega_4 \cong \mathbb{Z}$ theorem [127] is an obvious example), but ultimately, new results, about 4-manifolds via link theory. In this spirit, Kaplan has shown [68] that, given a framed link, it may be modified by Kirby's moves so as to make all the framings even. This, together with the Wallace-Lickorish theorem, implies the (known) result that every closed, orientable 3-manifold bounds a parallellizable 4-manifold.

We might also mention here the Rohlin Theorem, that the signature of a smooth closed, oriented, almost parallellizable 4-manifold is divisible by 16. Elementary proofs of this (assuming $\Omega_4 \cong \mathbb{Z}$) have been given by Casson and (independently) Matsumoto (both unpublished), and link-theoretic ideas are involved in these proofs.

Many questions concerning the existence of certain surfaces in 4-manifolds are equivalent, or closely related, to questions about knot and link concordance. Thus Tristram [148] used his p-signatures to show that a class $ax+by$ in $H_2(S^2 \times S^2) \cong \mathbb{Z} \oplus \mathbb{Z}$ can be represented by a smoothly embedded 2-sphere only if a and b are coprime. (It is still unknown whether this condition is sufficient, except for the cases $|a| \leq 1$ or $|b| \leq 1$.) As we have seen in §12, signatures of knots (and links) are probably best studied from a 4-dimensional point of view anyway, so this kind of connection is not surprising.

Perhaps more surprising is the result of Casson (unpublished) that simply-connected surgery is possible in dimension 4 if each of a certain explicit set of infinite sequences of links contains a slice link. On the other hand, failure of the latter condition implies the existence of some kind of pathology in dimension 4. For example, if the sequence of (untwisted) doubles of the Whitehead link contains no slice link, then there is a 4-manifold proper homotopy equivalent to $S^2 \times S^2$-point whose end is not diffeomorphic to $S^3 \times R$, and a 4-dimensional counterexample to the McMillan cellularity criterion. (These results are also due to Casson.)

16. Knots and the 3-Sphere

All the abelian algebra discussed so far is valid for knots in homology 3-spheres. Similarly, all known knot concordance invariants are actually homology-cobordism invariants. The group of a knot in S^3, of course, has weight 1 (being generated by the conjugates of any meridian element), but again this is true of a knot in any homotopy 3-sphere. Still, it is clear that the theory of knots in the 3-sphere, having the concreteness and immediacy of the physical world, is of prime importance. Moreover, even properties which hold in more general settings might be more easily observed in the 3-sphere. This has certainly been the case historically. For example, the property $\Delta(1) = 1$ of the Alexander polynomial was first proved by means of knot projections [2]. (In fact the purely combinatorial view of knot theory, in which invariants are defined in terms of knot diagrams and then shown to be unchanged under the Reidemeister moves, dominated the subject for a long time [2], [3], [123], [125], [51].) Also, the symmetry property of link polynomials has been obtained [147] as a consequence of the existence of 'dual' Wirtinger presentations of the group. (See [95] for a 'co-ordinate-free' proof in the more general setting of a homology sphere.) Again, in trying to relate knot and link theory to 3- and 4-manifolds, presumably the hope is that one might be able to 'see' new information about the manifolds precisely because one is working with visualizable objects in ordinary space.

In dealing with the 3-sphere, however, there is more involved than just convenience, for it is known that different 3-manifolds have different knot theories. More precisely, it is known that if $\mathcal{K}(M)$ denotes the set of isomorphism classes of groups of knots in the closed 3-manifold M, then $M \cong N$ if and only if $\mathcal{K}(M) = \mathcal{K}(N)$. The result in this generality is due to Jaco-Myers [63] (for orientable manifolds) and, independently, Row (unpublished). The fact that the 3-sphere is determined by its knot groups was apparently proved earlier by Connor (unpublished). The idea of trying to classify 3-manifolds by their knot theories goes back to Fox, who used it to recover the (known) classification of lens spaces.

This suggests the problem of trying to characterize the groups of knots in the 3-sphere. A characterization was given, several years ago, by Artin [5], but this is in terms of the existence of a particular kind of presentation, and whether this can be expressed more intrinsically is still unknown.

A good example of a problem which specifically concerns knots in the 3-sphere is the Smith conjecture that no non-trivial knot is the fixed-point set of a $\mathbb{Z}/p$-action on S^3 (clearly it is enough to consider p prime). This is false for knots in homology spheres. On the other hand, most of the partial results on the conjecture are essentially homological in nature. A notable exception is Waldhausen's proof [157] for the case $p = 2$, which uses (as it must) the geometry of the 3-sphere (in particular, the uniqueness of Heegaard splittings [156]) in an essential way.

17. Other Topics

Here we briefly mention one or two topics which we shall not be able to discuss in detail.

First, there is the whole question of symmetries of knots, For S^1-actions, the answer is known: only the unknot can be the fixed-point set of an S^1-action on S^3, and the only knots which are invariant under (effective) S^1-actions are the torus knots. (This follows from the theory of Seifert fibre spaces [134]; see [65].) For the case of $\mathbb{Z}/p$-actions on S^3 fixing a knot K, we of course have the Smith conjecture that K must be trivial. This is surely one of the major unsolved problems in knot theory. It is known to be true for $p=2$ [157], and there exist various other partial results, including [16], [42], [45], [50], [56], [109]. Necessary conditions are given in [149] and [108], for a knot K to have a symmetry of order n in the sense that there is a homeomorphism h of S^3 of period n, with fixed-point set a circle disjoint from K, such that $h(K)=K$.

Given an unoriented knot K in oriented S^3, one can ask whether or not there exists an orientation-reversing homeomorphism of S^3 taking K to itself, (or equivalently, an orientation-preserving homeomorphism of S^3 taking K to its mirror-image). If there is, K is amphicheiral. If K is now oriented, one can ask whether there is an orientation-preserving homeomorphism of S^3 taking K onto K but reversing its orientation. If so, K is invertible. If K is amphicheiral, then, for example, all its branched covers will support orientation-reversing homeomorphisms. Because of this, amphicheirality is often relatively easy to detect [135]. Since many knot invariants are independent of the orientation of the knot, however, it is harder to establish non-invertibility. This was first done in [151], by analysing automorphisms of the group. See [71], [161] for further results. Two interesting conjectures relating these concepts to symmetries (see [75]) are: K is amphicheiral if and only if K is invariant under reflection through the origin (van Buskirk); and: K is invertible if and only if there is an orientation-preserving involution of S^3 taking K to itself, reversing its orientation (Montesinos). Apparently these are true for knots with small crossing number.

Alternating knots have always occupied a special place in the subject; for instance, their asphericity was proved [7] before the sphere theorem. Othere interesting results on alternating knots are contained in [21], [22], [101], [102], [103], [104], [105].

The important work of Schubert on unique factorization [129] and companionship [130], [131] should be mentioned.

For results on the genus of a knot see [136], [110], [59], [133], [22], [101].

The question of the uniqueness of Seifert surfaces of minimal genus has received considerable attention [4], [153], [162].

Finally, there is an extensive literature on the knots which arise as links of complex algebraic plane curve singularities. (These are certain iterated cables of the unknot.) See [97] and references therein.

References

[1] J.W. Alexander, Note on Riemann spaces, Bull. Amer. Math. Soc. 26(1919), 370-372.

[2] ________, Topological invariants of knots and links, Trans. Amer. Math. Soc. 30(1928), 275-306.

[3] ________ and G.B. Briggs, On types of knotted curves, Ann. of Math. 28(1927), 562-586.

[4] W.R. Alford, Complements of minimal spanning surfaces of knots are not unique, Ann. of Math. 91(1970), 419-424.

[5] E. Artin, Theorie der Zöpfe, Abh. Math. Sem. Univ. Hamburg 4(1925), 47-52.

[6] M.F. Atiyah and I.M. Singer, The index of elliptic operators; III, Ann. of Math. 87(1968), 546-604.

[7] R.J. Aumann, Asphericity of alternating knots, Ann. of Math. 64(1956), 374-392.

[8] C. Bankwitz, Über die Torsionszahlen der zyklischen Überlagerungsräume des Knotenaussenraumes, Ann. of Math. 31(1930), 131-133.

[9] R.H. Bing and J.M. Martin, Cubes with knotted holes, Trans. Amer. Math. Soc. 155(1971), 217-231.

[10] J.S. Birman and H.M. Hilden, The homeomorphism problem for S^3, Bull. Amer. Math. Soc. 79(1973), 1006-1010.

[11] ________, F. González-Acuña, and J.M. Montesinos, Heegaard splittings of prime 3-manifolds are not unique, Michigan Math. J. 23(1976), 97-103.

[12] R.C. Blanchfield, Intersection theory of manifolds with operators with applications to knot theory, Ann. of Math. 65(1957), 340-356.

[13] G. Burde, On branched coverings of S^3, Can. J. Math. 23(1971), 84-89.

[14] _________ and H. Zieschang, Eine Kennzeichnung der Torusknoten, Math. Ann. 167(1966), 169-175.

[15] S.E. Cappell and J.L. Shaneson, Invariants of 3-manifolds, Bull. Amer. Math. Soc. 81(1975), 559-562.

[16] _________, A note on the Smith conjecture, to appear.

[17] A.J. Casson and C. McA. Gordon, Cobordism of classical knots, mimeographed notes, Orsay, 1975.

[18] _________, On slice knots in dimension three, to appear in Proceedings AMS Summer Institute in Topology, Stanford, 1976.

[19] J.H. Conway, An enumeration of knots and links, and some of their algebraic properties, Computational Problems in Abstract Algebra, Pergamon Press, Oxford and New York, 1969, 329-358.

[20] _________ and C. McA. Gordon, A group to classify knots, Bull. London Math. Soc. 7(1975), 84-86.

[21] R.H. Crowell, Non-alternating links, Illinois J. Math. 3(1959), 101-120.

[22] _________, Genus of alternating link types, Ann of Math. 69(1959), 258-275.

[23] _________, Corresponding group and module sequences, Nagoya Math. J. 19(1961), 27-40.

[24] _________, The group G'/G'' of a knot group G, Duke Math. J. 30(1963), 349-354.

[25] _________, On the annihilator of a knot module, Proc. Amer. Math. Soc. 15(1964), 696-700.

[26] _________ and R.H. Fox, An Introduction to Knot Theory, Ginn and Co., Boston, Mass., 1963.

[27] M. Dehn, Über die Topologie des dreidimensionalen Raumes, Math. Ann. 69(1910), 137-168.

[28] _________, Die beiden Kleeblattschlingen, Math. Ann. 75(1914), 402-413.

[29] _________, Die Gruppe der Abbildungsklassen, Acta Math. 69(1938), 135-206.

[30] _________ and P. Heegaard, Analysis Situs, Encyklopädie der Mathematischen Wissenschaften, Band III, Heft 6, B.G. Teubner, Leipzig, 1920, 153-216.

[31] G.W. Dunnington, Carl Friedrich Gauss, Titan of Science, Hafner Publishing Co., New York, 1955.

[32] A. Durfee and L. Kauffman, Periodicity of branched cyclic covers, Math. Ann. 218(1975), 157-174.

[33] C.H. Edwards, Concentricity in 3-manifolds, Trans. Amer. Math. Soc. 113(1964), 406-423.

[34] D. Erle, Die quadratische Form eines Knotens und ein Satz über Knotenmannigfaltigkeiten, J. Reine Angew. Math. 236(1969), 174-218.

[35] R. Fenn and C. Rourke, On Kirby's calculus of links, to appear.

[36] C.D. Feustel, On the torus theorem and its applications, Trans. Amer. Math. Soc. 217(1976), 1-43.

[37] ________ and W. Whitten, Groups and complements of knots, to appear.

[38] R.H. Fox, On the complementary domains of a certain pair of inequivalent knots, Nederl. Akad. Wetensch. Proc. Ser. A 55 = Indagationes Math. 14(1952), 37-40.

[39] ________, Free differential calculus I, Ann. of Math. 57(1953), 547-560.

[40] ________, Free differential calculus II, Ann. of Math. 59(1954), 196-210.

[41] ________, Free differential calculus III, Ann. of Math. 64(1956), 407-419

[42] ________, On knots whose points are fixed under a periodic transformation of the 3-sphere, Osaka Math. J. 10(1958), 31-35.

[43] ________, A quick trip through knot theory, Topology of 3-Manifolds and Related Topics (Proc. The Univ. of Georgia Inst., 1961), Prentice-Hall, Englewood Cliffs, New Jersey, 1962, 120-167.

[44] ________, Some problems in knot theory, ibid, 168-176.

[45] ________, Two theorems about periodic transformations of the 3-sphere, Michigan Math. J. 14(1967), 331-334.

[46] ________ and J.W. Milnor, Singularities of 2-spheres in 4-space and cobordism of knots, Osaka J. Math. 3(1966), 257-267.

[47] ________ and N. Smythe, An ideal class invariant of knots, Proc. Amer. Math. Soc. 15(1964), 707-709.

[48] F. Frankl and L. Pontrjagin, Ein Knotensatz mit Anwendung auf die Dimensionstheorie, Math. Ann. 102(1930), 785-789.

[49] C.F. Gauss, Werke, ed. Königliche Gesellschaft für Wissenschaften, Göttingen, Leipzig and Berlin, 1863-1929.

[50] C.H. Giffen, Cyclic branched coverings of doubled curves in 3-manifolds, Illinois J. Math. 11(1967), 644-646.

[51] L. Goeritz, Knoten und quadratische Formen, Math. Zeit. 36(1933), 647-654.

[52] ________, Die Betti'schen Zahlen der zyklischen Überlagerungsräume der Knotenaussenräume, Amer. J. Math. 56(1934), 194-198.

[53] F. González-Acuña, Dehn's construction on knots, Bol. Soc. Mat. Mex. 15(1970), 58-79.

[54] C. McA. Gordon, A short proof of a theorem of Plans on the homology of the branched cyclic coverings of a knot, Bull. Amer. Math. Soc. 77(1971), 85-87.

[55] ________, Knots whose branched cyclic coverings have periodic homology, Trans. Amer. Math. Soc. 168(1972), 357-370.

[56] ________, Uncountably many stably trivial strings in codimension two, to appear in Quart. J. Math. Oxford.

[57] ________, On the G-signature theorem in dimension 4, to appear in Proceedings Oklahoma Topology Conference, 1978.

[58] ________ and R.A. Litherland, On the signature of a link, to appear.

[59] W. Haken, Theorie der Normalflächen, Acta Math. 105(1961), 245-375.

[60] H.M. Hilden, Three-fold branched coverings of S^3, Amer. J. Math. 98(1976), 989-997.

[61] ________, J.M. Montesinos and T. Thickstun, Closed oriented 3-manifolds as 3-fold branched coverings of S^3 of special type, Pacific J. Math. 65(1976), 65-76.

[62] F. Hirzebruch, W.D. Neumann and S.S. Koh, Differentiable Manifolds and Quadratic Forms, Marcel Dekker, Inc., New York, 1971.

[63] W. Jaco and R. Myers, An algebraic characterization of closed 3-manifolds, to appear.

[64] ________ and P.B. Shalen, Seifert fibered spaces in 3-manifolds, I; the mapping theorem, to appear.

[65] R. Jacoby, One parameter transformation groups on the 3-sphere, Proc. Amer. Math. Soc. 7(1956), 131-140.

[66] K. Johannson, Équivalences d'homotopie des variétés de dimension 3, C.R. Acad. Sc. Paris 281(1975), 1009-1010.

[67] ________, Homotopy equivalences of knot spaces, to appear.

[68] S. Kaplan, Constructing framed 4-manifolds with given almost framed boundaries, Dissertation, Berkeley, CA., 1976.

[69] L. Kauffman and L. Taylor, Signature of links, Trans. Amer. Math. Soc. 216(1976), 351-365.

[70] C. Kearton, Blanchfield duality and simple knots, Trans. Amer. Math. Soc. 202(1975), 141-160.

[71] ________, Noninvertible knots of codimension 2, Proc. Amer. Math. Soc. 40(1973), 274-276.

[72] ________ and W.B.R. Lickorish, Piecewise linear critical levels and collapsing, Trans. Amer. Math. Soc. 170(1972), 415-424.

[73] M.A. Kervaire, Knot cobordism in codimension two, Manifolds-Amsterdam 1970, Lecture Notes in Mathematics 197, Springer-Verlag, Berlin, Heidelberg, New York, 1971, 83-105.

[74] R. Kirby, A calculus for framed links in S^3, to appear.

[75] ________, ed., Problems in low dimensional manifold theory, to appear in Proceedings of AMS Summer Institute in Topology, Stanford, 1976.

[76] M. Kneser and D. Puppe, Quadratische Formen und Verschlingungsinvarianten von Knoten, Math. Zeit. 58(1953), 376-384.

[77] C.G. Knott, Life and Scientific Work of P.G. Tait, Cambridge University Press, 1911.

[78] R.H. Kyle, Branched covering spaces and the quadratic forms of links, Ann. of Math. 59(1954), 539-548.

[79] J. Levine, A characterization of knot polynomials, Topology 4(1965), 135-141.

[80] ________, Polynomial invariants of knots of codimension two, Ann. of Math. 84(1966), 537-554.

[81] ________, Knot cobordism groups in codimension two, Comment. Math. Helv. 44(1969), 229-244.

[82] ________, Invariants of knot cobordism, Inventiones Math. 8(1969), 98-110, and 355.

[83] ________, An algebraic classification of some knots of codimension two, Comment. Math. Helv. 45(1970), 185-198.

[84] ________, Knot modules, Knots, Groups, and 3-Manifolds, Ann. of Math. Studies 84, Princeton University Press, Princeton, N.J., 1975, 25-34.

[85] ________, Knot modules I, Trans. Amer. Math. Soc. 229(1977), 1-50.

[86] W.B.R. Lickorish, A representation of orientable combinatorial 3-manifolds, Ann. of Math. 76(1962), 531-540.

[87] ________, Surgery on knots, Proc. Amer. Math. Soc. 60(1976), 296-298.

[88] J.B. Listing, Vorstudien zur Topologie, Göttingen Studien, 1847.

[89] R.A. Litherland, Surgery on knots in solid tori, to appear.

[90] C.N. Little, On knots, with a census for order ten, Trans. Conn. Acad. Sci. 18(1885), 374-378.

[91] ________, Non-alternate $\pm$ knots, of orders eight and nine, Trans. Roy. Soc. Edinburgh 35(1889), 663-664.

[92] ________, Alternate $\pm$ knots of order eleven, Trans. Roy. Soc. Edinburgh 36(1890), 253-255.

[93] ________, Non-alternate $\pm$ knots, Trans. Roy. Soc. Edinburgh 39(1900), 771-778.

[94] T. Matumoto, On the signature invariants of a non-singular complex sesquilinear form, to appear.

[95] J. Milnor, A duality theorem for Reidemeister torsion, Ann. of Math. 76(1962), 137-147.

[96] ________, Infinite cyclic coverings, Conference on the Topology of Manifolds, Prindle, Weber, and Schmidt, Boston, Mass., 1968, 115-133.

[97] ________, Singular Points of Complex Hypersurfaces, Ann. of Math. Studies 61, Princeton University Press, Princeton, N.J., 1968.

[98] ________, On isometries of inner product spaces, Inventiones Math. 8(1969), 83-97.

[99] J. Montesinos, Reducción de la conjetura de Poincaré a otras conjeturas geométricas, Rev. Mat. Hisp.-Amer. 32(1972), 33-51.

[100] ________, Three-manifolds as 3-fold branched covers of S^3, Quart. J. Math. Oxford 27(1976), 85-94.

[101] K. Murasugi, On the genus of the alternating knot, I and II, J. Math. Soc. Japan 10(1958), 94-105 and 235-248.

[102] ________, On the Alexander polynomial of the alternating knot, Osaka Math. J. 10(1958), 181-189.

[103] ________, On alternating knots, Osaka Math. J. 12(1960), 277-303.

[104] ________, Non-amphicheirality of the special alternating links, Proc. Amer. Math. Soc. 13(1962), 771-776.

[105] ________, On a certain subgroup of the group of an alternating link, Amer. J. Math. 85(1963), 544-550.

[106] ________, On the Minkowski unit of slice links, Trans. Amer. Math. Soc. 114(1965), 377-383.

[107] ________, On a certain numerical invariant of link types, Trans. Amer. Math. Soc. 117(1965), 387-422.

[108] ________, On periodic knots, Comment. Math. Helv. 46(1971), 162-174.

[109] R. Myers, Companionship of knots and the Smith conjecture, Dissertation, Rice University, Houston, Tx., 1977.

[110] L. Neuwirth, The algebraic determination of the genus of knots, Amer. J. Math. 82(1960), 791-798.

[111] ________, On Stallings fibrations, Proc. Amer. Math. Soc. 14(1963), 380-381.

[112] ________, Knot Groups, Ann. of Math. Studies 56, Princeton University Press, Princeton, N.J., 1965.

[113] ________, The status of some problems related to knot groups, Topology Conference VPISU, 1973, Lecture Notes in Mathematics 375, Springer-Verlag, Berlin, Heidelberg, New York, 1974, 209-230.

[114] C.D. Papakyriakopoulos, On Dehn's lemma and the asphericity of knots, Ann. of Math. 66(1957), 1-26.

[115] K.A. Perko, On the classification of knots, Proc. Amer. Math. Soc. 45(1974), 262-266.

[116] ________, On dihedral covering spaces of knots, Inventiones Math. 34(1976), 77-82.

[117] ________, On ninth order knottiness, to appear.

[118] ________, On 10-crossing knots, to appear.

[119] A. Plans, Aportación al estudio de los grupos de homología de los recubrimientos ciclicos ramificados correspondientes a un nudo, Rev. Acad. Ci. Madrid 47(1953), 161-193.

[120] S.D. Puppe, Minkowskische Einheiten und Verschlingungsinvarianten von Knoten, Math. Zeit. 56(1952), 33-48.

[121] E.S. Rapaport, On the commutator subgroup of a knot group, Ann. of Math. 71(1960), 157-162.

[122] K. Reidemeister, Knoten und Gruppen, Abh. Math. Sem. Univ. Hamburg, 5(1927), 7-23.

[123] ________, Elementare Begründung der Knotentheorie, Abh. Math. Sem. Univ. Hamburg 5(1927), 24-32.

[124] ________, Knoten und Verkettungen, Math. Zeit. 29(1929), 713-729.

[125] ________, Knotentheorie, Ergebnisse der Mathematik, 1, Springer-Verlag, Berlin, 1932; reprint, Chelsea, New York, 1948.

[126] V.A. Rohlin, A 3-dimensional manifold is the boundary of a 4-dimensional one, Dokl. Akad. Nauk. SSSR 81(1951), 355-357.

[127] ________, New results in the theory of 4-dimensional manifolds, Dokl. Akad. Nauk. SSSR 84(1952), 221-224.

[128] D. Rolfsen, Knots and Links, Mathematics Lecture Series 7, Publish or Perish Inc., Berkeley, Ca., 1976.

[129] H. Schubert, Die eindeutige Zerlegbarkeit eines Knotens in Primknoten, S.-B. Heidelberger Akad. Wiss. Math.-Natur Kl., 3(1949), 57-104.

[130] ________, Knoten und Vollringe, Acta Math. 90(1953), 131-286.

[131] ________, Über eine numerische Knoteninvariante, Math. Zeit. 61(1954), 245-288.

[132] ________, Knoten mit zwei Brücken, Math. Zeit. 65(1956), 133-170.

[133] ________, Bestimmung der Primfaktorzerlegung von Verkettungen, Math. Zeit. 76(1961), 116-148.

[134] H. Seifert, Topologie dreidimensionaler gefaserter Räume, Acta Math. 60(1933), 147-238.

[135] ________, Verschlingungsinvarianten, S.-B. Preuss. Akad. Wiss. 26(1933), 811-828.

[136] ________, Über das Geschlecht von Knoten, Math. Ann. 110(1934), 571-592

[137] ________, Die Verschlingungsinvarianten der zyklischen Knotenüberlagerungen, Abh. Math. Sem. Univ. Hamburg 11(1936), 84-101.

[138] Y. Shinohara, On the signature of knots and links, Dissertation, Florida State University, Tallahassee, Fla., 1969.

[139] J. Simon, Some classes of knots with Property P, Topology of Manifolds, (Proc. Univ. of Georgia Inst. 1969), Markham, Chicago, 1970, 195-199.

[140] ________, An algebraic classification of knots in S^3, Ann. of Math. 97(1973), 1-13.

[141] ________, On the problems of determining knots by their complements and knot complements by their groups, Proc. Amer. Math. Soc. 57(1976), 140-142.

[142] J. Stallings, On fibering certain 3-manifolds, Topology of 3-Manifolds and Related Topics (Proc. The Univ. of Georgia Inst., 1961), Prentice-Hall, Englewood Cliffs, N.J., 1962, 95-100.

[143] N.W. Stoltzfus, Unraveling the integral knot concordance group, AMS Memoir 192, 1977.

[144] D.W. Sumners, On the homology of finite cyclic coverings of higher-dimensional links, Proc. Amer. Math. Soc. 46(1974), 143-149.

[145] P.G. Tait, On knots, I, II, and III, Scientific Papers, Vol. I, Cambridge University Press, London, 1898, 273-347.

[146] H. Tietze, Über die topologischen Invarianten mehrdimensionaler Mannigfaltigkeiten, Monatsh. Math. Phys. 19(1908), 1-118.

[147] G. Torres and R.H. Fox, Dual presentations of the group of a knot, Ann. of Math. 59(1954), 211-218.

[148] A.G. Tristram, Some cobordism invariants for links, Proc. Cambridge Philos. Soc, 66(1969), 251-264.

[149] H.F. Trotter, Periodic automorphisms of groups and knots, Duke Math. J. 28(1961), 553-558.

[150] ________, Homology of group systems with applications to knot theory, Ann. of Math. 76(1962), 464-498.

[151] ________, Noninvertible knots exist, Topology 2(1964), 275-280.

[152] ________, On S-equivalence of Seifert matrices, Inventiones Math. 20(1973), 173-207.

[153] ________, Some knots spanned by more than one unknotted surface of minimal genus, Knots, Groups, and 3-Manifolds, Ann. of Math. Studies 84, Princeton University Press, Princeton, N.J., 1975, 51-62.

[154] O. Ja. Viro, Branched coverings of manifolds with boundary and link invariants. I, Math. USSR Izvestija 7(1973), 1239-1256.

[155] F. Waldhausen, On irreducible 3-manifolds which are sufficiently large, Ann. of Math. 87(1968), 56-88.

[156] ________, Heegaard-Zerlegungen der 3-Sphäre, Topology 7(1968), 195-203.

[157] ________, Über Involutionen der 3-Sphäre, Topology 8(1969), 81-91.

[158] ________, Recent results on sufficiently large 3-manifolds, to appear in Proceedings of AMS Summer Institute in Topology, Stanford, 1976.

[159] A.H. Wallace, Modifications and cobounding manifolds, Can. J. Math. 12(1960), 503-528.

[160] H. Wendt, Die gordische Auflösung von Knoten, Math. Zeit. 42(1937), 680-696.

[161] W. Whitten, Surgically transforming links into noninvertible knots, Amer. J. Math. 94(1972), 1269-1281.

[162] ________, Isotopy types of knot spanning surfaces, Topology 12(1973), 373-380.

[163] ________, Algebraic and geometric characterizations of knots, Inventiones Math. 26(1974), 259-270.

[164] H.J. Zassenhaus, The Theory of Groups, 2nd. edition, Chelsea Publishing Co., New York, 1958.

[165] E.C. Zeeman, Twisting spun knots, Trans. Amer. Math. Soc. 115(1965), 471-495.

Department of Mathematics
The University of Texas
Austin, Texas 78712

A SURVEY OF MULTIDIMENSIONAL KNOTS

by

M. KERVAIRE and C. WEBER

CHAPTER I : INTRODUCTION

§ 1. Some historical landmarks.

Knotted n-spheres $K = f(S^n) \subset S^{n+2}$ with $n \geq 2$ make what seems to be their first appearance in a famous paper by E. Artin published in 1925, where he describes a construction which produces examples of non-trivial n-knots for arbitrary $n \geq 2$. (Detailed reference data are provided at the end of the survey). In today's terminology, introduced by E.C. Zeeman (1959), the construction is called spinning and it goes as follows.

Let $K \subset S^{n+2}$ be an n-knot, i.e. a smoothly embedded n-sphere K in S^{n+2}. Take the associated knotted disk pair $(B, bB) \subset (D^{n+2}, S^{n+1})$ obtained by removing from S^{n+2} a small open disk U centered at a point of K. Here, $D^{n+2} = S^{n+2} - U$ and $B = K - K \cap U$. The subset $D = \{ \Sigma_{i=1}^{n+4} x_i^2 = 1, \ x_{n+3} \geq 0, \ x_{n+4} = 0 \}$ in $S^{n+3} \subset R^{n+4}$ is an (n+2)-dimensional disk which we identify with D^{n+2}. Thus, $B \subset D$. Now, the sphere S^{n+3} can be obtained by rotating this disk D in R^{n+4} around the (n+2)-plane $P = \{x_{n+3} = 0, \ x_{n+4} = 0\}$. Note that P contains the unknotted boundary sphere $bD = S^{n+1} \subset S^{n+3}$ which thus remains pointwise fixed during the rotation. In the process, the set $B \subset D$ will sweep out a smooth (n+1)-dimensional sphere embedded in S^{n+3}. This is the spun knot $\Sigma_K \subset S^{n+3}$ of the knot $K \subset S^{n+2}$.

E. Artin observed in his paper that

$$\pi_1(S^{n+3} - \Sigma_K) \cong \pi_1(S^{n+2} - K) .$$

Thus, $\Sigma_K \subset S^{n+3}$ is certainly knotted if $\pi_1(S^{n+2} - K) \not\cong \mathbb{Z}$.

Starting with a non-trivial "classical" knot (i.e. n = 1) and iterating the construction, one gets non-trivial n-knots for all n .

A similar construction can be performed on linked spheres and it also leaves unchanged the fundamental group of the complement. See van Kampen (1928) and Zeemann (1959) for details.

The objective of multidimensional knot theory is, as for classical knots, to perform classification, ultimately (and ideally) with respect to isotopy, and meanwhile with respect to weaker equivalence relations. There is however with higher dimensional knots the additional difficulty that the construction of a knot cannot merely be described by the simple-minded drawing up of a knot projection. Thus, efforts at classification (i.e. finding invariants) now have to be complemented by construction methods (i.e. showing that the invariants are realizable). This is why Artin's paper is so significant. It gives the first construction showing that the groups of classical knots are all realizable as fundamental groups of the complement of n-knots for arbitrary n.

After Artin's paper, multidimendional knot theory went into a long sleep. Strangely enough, the theory awoke subsequently to Papa's proof of the sphere theorem. One of the consequences of this famous result is that classical knots have aspherical complements, i.e. : $\pi_i(S^3 - f(S^1)) = 0$ for i > 1. Hence a natural question : What about multidimensional knots ? The answer came quickly : In 1959, J.J. Andrews and M. L. Curtis showed that the complement of the spun trefoil has a non-vanishing second homotopy group. In fact their result is more general and also better : there is an embedded 2-sphere which represents a non-zero element.

This paper was followed less than a month later by D.Epstein (1959) who gave a formula expressing π_2, the second homotopy group of the complement of any spun 2-knot. A corollary of Epstein's result is that the complement of a non-trivial spun 2-knot has a π_2 which is not finitely generated as an abelian group.

The question was then raised by R.H.Fox (1961) to describe π_2 as a π_1-module. This gave the impulse for the subsequent research in that direction. (See for example S.J. Lomonaco Jr (1968)).

One thus began to suspect that multidimensional knots would behave quite differently from the classical ones. The major breakthrough came from the development of surgery techniques which made it possible to get a general method of constructing knots with prescribed properties of their complements. In a perhaps subtler way, surgery techniques were also decisive in classification problems. See our chapters III and IV.

Here is an illustration of the power of surgery techniques. A common feature to the examples (all based on spinning) known in 1960 was that π_2 was $\neq 0$ because $\pi_1 \neq Z$. It was thus natural to ask : Can one produce an n-knot with $\pi_2 \neq 0$ but $\pi_1 = Z$? Clearly such an example cannot be obtained by spinning a classical knot. However, J. Stallings (see M. Kervaire's paper (1963), p. 115) and C.T.C. Wall (see his book : Surgery on compact manifolds, p.18) proved in 1963 that for all $n \geq 3$, there exist many knots $K \subset S^{n+2}$ with $\pi_1(S^{n+2}-K)=Z$ but $\pi_2(S^{n+2} - K) \neq 0$. The construction is an easy exercise in surgery.

At the same time, another construction method was invented by E.C. Zeeman (1963). It is a deep generalization of Artin's spinning called twist-spinning. We shall talk about it in chap. V § 4.

To close this short historical survey, we ought to mention Kinoshita's paper (1960). It gives a construction of 2-knots by pasting together discs in 4-space which is probably the unique knot construction prior to 1963 not based on spinning. There is also the somewhat related method used by R. Fox (1961a), where a 2-knot is described slice by slice, by the moving picture of its intersection with a 3-dimensional hyperplace sliding across $\mathbb{R}^4$.

Hovewer, one cannot expect Kinoshita's nor Fox's level curve methods to be applicable in higher dimensions because they still rely on drawings and intuitive descriptions in the next lower-dimensional 3-space.

As a conclusion, let us make a few remarks :

1) The use of surgery techniques showed that multidimensional knot theory could do well without direct appeal to 3-dimensional goemetric intuition nor immediate computability. There resulted a useful kickback for classical knot theory which benefited much, since 1965, from the use of geometrical tools borrowed from higher dimensional topology and from a partial relinquisment of computational methods.

2) Around 1964, it became generally accepted that the theory of imbeddings in codimensions ≥ 3 was well understood. Piecewise linear imbeddings $S^n \longrightarrow S^{n+q}$ with $q \geq 3$ are all unknotted by a theorem of E.C. Zeeman of 1962, published in "Unknotting combinatorial balls", Ann. of Math. 78 (1963)p.501-526. The differentiable theory was in good shape with the works, both in 1964, of J. Levine, "A classification of Differentiable Knots", Ann. of Math. 82 (1965), 15-50 on the one hand, and A. Haefliger, "Differentiable Embeddings of S^n in S^{n+q} for $q > 2$", Ann. of Math. 83(1966)p.402-436 on the other.

These impressive pieces of work provided a decisive encouragement to take up the certainly less tractable codimension 2 case. A lot of effort went into it and since then the growth of the subject has been so important that we cannot follow a chronological presentation. We have chosen instead to talk about articles published after 1964 in the chapters corresponding to their subject as listed in the table of contents below. Of course, at some points, whenever convenient, we did go back again to papers which appeared before this date.

For the same reason we had to delete from this survey the mention of many beautiful papers. In particular, we have mostly disregarded the papers centering around a discussion of the equivalence (or non-equivalence) of various possible definitions. We have rather tried to emphasize the moving aspect of the subject.

§ 2 . Some definitions and notations.

Do we now have to tell the reader what a knot is ?

Usually an n-knot is a codimension 2 submanifold K in S^{n+2}. Most of the time S^{n+2} will be the standard (n+2)-dimensional smooth sphere. However, in some cases, one is forced to relax this condition. (For instance, when n+2 = 4, in order to get the realization theorems for π_1. (See Chap.II, § 3).

What K should be is a little harder to make definite. For us, it will be a locally flat, oriented, PL-submanifold of S^{n+2}, PL-homeomorphic to the standard n-sphere or a differential submanifold homeomorphic (or diffeomorphic) to the standard n-sphere.

The reason for such hesitations can easily be explained. The proof of the algebraic properties of the various knot invariants usually does not require a very restrictive definition of a knot. In some cases, S^{n+2} could as well be replaced by a homotopy sphere and K by a homology sphere, or even less (see chap. V, § 5), sometimes not even locally flat.

On the other hand, to be able to perform geometrical constructions we usually need more restrictions. For instance the proof of the existence of a Seifert surface requires local flatness in order to get a normal bundle (which will be trivial).

Moreover, when one wants to prove realization theorems for the algebraic invariants, the stronger the restrictions on the knot definition, the better the theorems.

So we decided to let a little haze about the definition of a knot, leaving to the reader the task to get to the original papers whenever needed and see what is really required (or used).

The dimension of a knot is n if it is an n-dimensional sphere K in S^{n+2}. We also say an n-knot.

We refer to 1-knots as being "classical" ; n-knots with $n \geq 2$ are "multidimensional".

NOTATIONS :

$X_o = S^{n+2} - K$ is the complement of the knot.

X is the exterior of the knot. (See beginning of Chap. II for the definition).

bX is the boundary of X.

C denotes an infinite cyclic group, written multiplicatively.

t is a generator of C. When $C = H_1(X_o)$, t is usually chosen according to orientation conventions.

$\Lambda = \mathbb{Z}C$ is the integral group ring of C. If t has been chosen, Λ is canonically isomorphic to the ring $\mathbb{Z}[t,t^{-1}]$.

This paper is mainly intended to topologists not working in multidimensional knot theory. As the standard joke goes : the specialist will find here nothing new, except mistakes.

Therefore, in this spirit,

1) We have often written up in some detail elementary arguments which are well known to people working in the field, but perhaps not so easy to find in the literature.

2) We did not attempt to talk about everything in the subject, but rather tried to emphasize what seems to be its most exciting aspects.

3) The latest news is often not here. Other parts of this book should fill this gap and provide references.

On the other hand, we have assumed that the reader knows some algebraic and geometric topology, and even sometimes that he is moderately familiar with classical knot theory.

TABLE OF CONTENTS

CHAPTER II : THE COMPLEMENT OF A KNOT AS AN INVARIANT

§ 1 : Completeness theorems

The idea of distinguishing knots by the topology of their complements goes back at least to M. Dehn (Ueber die Topologie des dreidimensionalen Raumes, Math. Annalen 69 (1910), 137-168).

However, the question to decide just how close the complement comes to be a complete invariant of the knot does not seem to have occured (for higher knots) before the paper of H. Gluck in 1962.

Actually it is technically advantageous to replace the complement $X_0 = S^{n+2} - K$ by the so-called exterior , that is the complement $X = S^{n+2} - N$ of an open tubular neighborhood N of K. Observe that K has trivial normal bundle ν so that N is diffeomorphic to $S^n \times D^2$ and a trivialization of ν will give an identification $N \cong S^n \times D^2$. Observe also that X_0 is diffeomorphic to the interior of the compact manifold X and that $bX = b\overline{N} \cong S^n \times S^1$. Thus X determines X_0. The converse is true at least for $n \geq 3$. Since this point seems left in the dark in the printed literature, the following explanations may perhaps be helpfu Suppose X and X' are knot exteriors and let $F_0 : X_0 \longrightarrow X_0{}'$ be a diffeomorphism. Take a neighborhood U of bX of the form $U \cong bX \times [0,1]$, i.e. a collar. Look at the submanifold $M = X' - F_0(X-U)$.

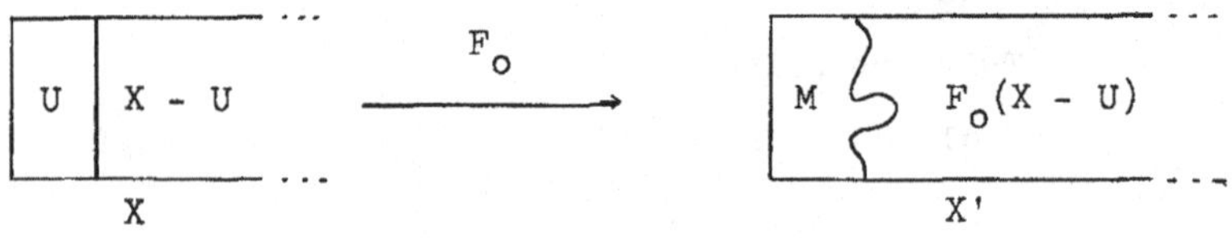

If U has been taken narrow enough, M is contained in a collar around bX' and it is easy to construct continuous retractions of M onto

each of its two boundary components $M_o = bX'$ and $M_1 = bF_o(X-U)$. Thus M is an h-cobordism between M_o and M_1. Now, M_o and M_1 are both diffeomorphic to $S^n \times S^1$, and $\pi_1 M_o = \mathbb{Z}$. A basic theorem of differential topology, the s-cobordism theorem, now states that under these conditions tions and if dim $M \geq 6$, then the diffeomorphism $F_o : b(X-U) \longrightarrow M_1$ can be extended to a diffeomorphism $F : X \longrightarrow X'$. (For a proof of the s-cobordism theorem see M. Kervaire, Comm. Math. Helv. 40 (1965),31-42. Here one also needs the fact that the Whitehead group of the infinite cyclic group is trivial. For this fact, see H. Bass, A. Heller and R.G. Swan, Publications mathématiques, I.H.E.S. No 22. In the case $n = 3$, the s-cobordims theorem does not apply and one needs Theorem 16.1 in C.T.C. Wall's book, Surgery on compact manifolds, p. 232).

Now, if K and K' are two knots and a diffeomorphism $F : X \longrightarrow X'$ is given between their exteriors, then after choosing identifications $N \cong S^n \times D^2 \cong N'$, F will restrict on boundaries to a diffeomorphism $f : S^n \times S^1 \longrightarrow S^n \times S^1$. The equivalence of K and K' thus reduces to a question of extendability of f to a (core preserving) diffeomorphism $S^n \times D^2 \longrightarrow S^n \times D^2$.

One is then led to study the group $\mathcal{D}(S^n \times S^1)$ of concordance classes of diffeomorphisms of $S^n \times S^1$ onto itself. Two diffeomorphisms h_o, $h_1 : M \longrightarrow M$ are concordant if there exists a diffeomorphism $h : M \times [0,1] \longrightarrow M \times [0,1]$ such that $h(x, 0) = (h_o(x), 0)$ and $h(x, 1) = (h_1(x), 1)$.

It is clear that indeed, only the concordance class of $f : S^n \times S^1 \longrightarrow S^n \times S^1$ in $\mathcal{D}(S^n \times S^1)$ matters for the extension problem at hand.

The final result is then

<u>THEOREM</u> : <u>For</u> n > 1, <u>there exist at most two</u> n-<u>knots with a given exterior</u>.

Sketch of proof. The group $\mathcal{D}(S^n \times S^1)$ projects onto the group of concordance classes of homeomorphisms $\mathcal{H}(S^n \times S^1)$ and it turns out that the extendability question for the above $f : S^n \times S^1 \longrightarrow S^n \times S^1$ depends only on its image in $\mathcal{H}(S^n \times S^1)$.

H. Gluck (1961) calculated $\mathcal{H}(S^2 \times S^1)$ and proved that $\mathcal{H}(S^2 \times S^1) \cong Z/2Z \times Z/2Z \times Z/2Z$, which means that there are at most eight 2-knots with a given exterior.

This number can however be cut down to two, as H. Gluck observed, since $\mathcal{H}(S^2 \times S^1)$ has a subgroup of order 4 generated by

a reflection on S^2 $\times$ the identity on S^1, and
the identity on S^2 $\times$ a reflection on S^1,

which both obviously extend to core preserving diffeomorphisms $S^2 \times D^2 \longrightarrow S^2 \times D^2$.

The calculation of $\mathcal{H}(S^n \times S^1)$ for $n \geqslant 5$ was achieved by W.Browder (1966) and finally completed to include the cases $n = 3$ and $n = 4$ by R.K. Lashof and J. Shaneson (1969). In all cases $\mathcal{H}(S^n \times S^1) \cong Z/2Z \times Z/2Z \times Z/2Z$ with generators which are the obvious generalizations of those for $n = 2$.

It still remained the question whether inequivalent knots with diffeomorphic complements do actually exist.

Examples of such knots were more recently produced by S. Cappell and J. Shaneson (1975) in dimensions $n = 3, 4$, (and possibly 5) and by C. Gordon (1975) for $n = 2$.

The method of S. Cappell and J. Shaneson is general and should yield examples of non-equivalent knots with diffeomorphic complements for all $n \geqslant 3$. It stumbles for $n \geqslant 6$ on the following purely algebraic open problem : does there exist for all n an automorphism A of Z^{n+1} without any real-negative eigenvalue and with determinant + 1 such that for all exterior powers $\lambda^i A$, $i = 1,\ldots,n$, the endomorphism $\lambda^i A - 1 : \lambda^i Z^{n+1} \longrightarrow \lambda^i Z^{n+1}$ is again an automorphism ?

Such an A can be concocted fairly easily for n = 3,4 and if one finds other values of n for which A exists with the required properties, it can be fed into the machinery of S. Cappell and J. Shaneson to produce new examples of inequivalent n-knots with diffeomorphic complements.

§ 2 . Unknotting theorems

There is one case where one would certainly like the complement $X_0 = S^{n+2} - K$ to determine the knot. That is the case where X_0 has the homotopy type of S^1, i.e. the homotopy type of the complement of the trivial, unknotted imbedding $K_0 = S^n \subset S^{n+2}$. Is it then true that K is isotopic to K_0 ?

In 1957 this was known to hold in the classical case (n = 1) as a consequence of the so-called Dehn's lemma proved by C.Papakyriakopoulos. (See Ann. of Math., 66 (1957)p.1-26).

For n = 2, this problem is still unsolved today, as far as we know.

For $n \geq 3$, it was solved by J. Stallings in 1962 for topological knots. If $K \subset S^{n+2}$ is a locally flat, topologically imbedded n-sphere with $n \geq 3$ and if $S^{n+2} - K$ has the homotopy type of S^1, then there exists a homeomorphism $h : S^{n+2} \longrightarrow S^{n+2}$ such that $hK = K_0$.

From the point of view of differential topology however, the major problem is whether a smooth knot $K \subset S^{n+2}$ with $S^{n+2} - K \simeq S^1$ is smoothly unknotted, i.e. whether there exists a diffeomorphism $h : S^{n+2} \longrightarrow S^{n+2}$ such that $hK = K_0$.

J. Levine's paper proving this and a little more in 1964 certainly played a decisive role in getting multidimensional knot theory off the ground.

His precise result is as follows.

LEVINE'S UNKNOTTING THEOREM : - Let $K \subset S^{n+2}$ be a smooth n-knot with $n \geq 4$ and let $X_0 = S^{n+2} - K$. Suppose that $\pi_i(X_0) \cong \pi_i(S^1)$ for $i \leq \frac{1}{2}(n+1)$ Then there is a diffeomorphism h of S^{n+2} onto itself such that hK is

<u>the standard</u> n-<u>sphere</u> S^n <u>in</u> S^{n+2} .

The proof shows in fact that under the stated hypotheses, K is the boundary of a contractible (n+1)-manifold V smoothly imbedded in S^{n+2}. We come back on this in the section on Seifert surfaces. See Chap. III, § 2.

By a theorem of S. Smale, the manifold V is then diffeomorphic to a disc. (See Ann. of Math. 74 (1961) p. 391-406). Thus, under the stated hypotheses, K bounds an (n+1)-disc smoothly imbedded in S^{n+2}.

The remainder of the proof is then relatively easy and has to do with the equivalence of various definitions of isotopy.

There remained the case of a smooth 3-knot $K^3 \subset S^5$. It was solved by C.T.C. Wall (1965) and independently by J. Shaneson (1968). (Note that these two references are only announcements of results. For a complete proof see C.T.C. Wall's book : Surgery on compact manifolds, § 16, p. 232).

<u>Remark</u>. The reader has perhaps noticed that we have slided from the homeomorphism type of the complement to its homotopy type, in the beginning of this paragraph. The invariants we are going to talk about in the next paragraph are invariants of the homotopy type of the complement. So, the question arises whether the homotopy type determines the topology of the complement. There are several results in this direction. See S. Cappell (1969) for a discussion. Here are some striking results :

Let us treat the exterior as a pair (X, bX). Then, the homotopy type of (X, bX) determines the homeomorphism type :

1) For classical knots. This is a beautiful result due to F. Waldhausen : "On irreductible 3-manifolds which are sufficiently large"

Annals of Math. 87 (1968) p. 56-88.

2) When $n \geq 4$ and $\pi_1(X) = \mathbf{Z}$. See R.K. Lashof and J.L. Shaneson (1968).

§ 3 . Invariants derived from the complement.

In view of the importance of the complement $X_o = S^{n+2} - K$, or the exterior X, as an invariant of the knot, it is desirable to extract from X weaker but calculable invariants such as for example the Alexander polynomial in the case of classical knots.

The homology of X is uninteresting. By Alexander duality, $H_*(X) \cong H_*(S^1)$, and thus $H_*(X)$ is in fact independent of the knot.

It was then natural to turn attention to the homotopy groups $\pi_i(X)$ of X .

$\pi_1(X)$ was easy to understand once surgery techniques were available to perform the necessary knot constructions. (See M. Kervaire (1963)). The fundamental group π of the complement of an n-knot, $n \geq 3$, is characterized by the following properties :

(1) π is finitely presented.

(2) $H_1(\pi) = \mathbf{Z}$, $H_2(\pi) = 0$,

(3) There is an element in π whose set of conjugates generates π.

Surgery techniques (for instance) enable one to construct an (n+2) dimensional oriented manifold M with $\pi_1(M) \cong \pi$, and $H_i(M) = 0$ for $i \neq 0, 1, n+1, n+2$. (For this the properties (1) and (2) of π are used. Surgery is not essential here) .

Then one takes an imbedding $\varphi : S^1 \times D^{n+1} \longrightarrow M$ representing an element $\alpha \in \pi$ whose conjugates generate π. One constructs a new manifold Σ by removing from M the interior of the image $\varphi(S^1 \times D^{n+1})$, say $X = M - \operatorname{int} \varphi(S^1 \times D^{n+1})$, and replacing it by $D^2 \times S^n$. Since $D^2 \times S^n$

and $S^1 \times D^{n+1}$ have the same boundary $S^1 \times S^n$, it follows that $D^2 \times S^n$ can be glued to X along $S^1 \times S^n$ by the map φ. The resulting manifold $\Sigma = X \cup \varphi(D^2 \times S^n)$ has the homotopy type of S^{n+2}, and for $n \geqslant 3$ is therefore homeomorphic to S^{n+2} by the theorems of S. Smale (Annals of Math. 74(1961)p.391-406). Actually, with some patching up one can even assume that Σ is diffeomorphic to S^{n+2}. By construction Σ contains a beautifully imbedded n-sphere, namely the core $K = \{0\} \times S^n$ in the subspace $N = D^2 \times S^n \subset \Sigma$. The subspace $X = \Sigma - N$ is just the exterior of the obtained n-knot $K \subset \Sigma$ and $\pi_1(X) \cong \pi_1(M) \cong \pi$.

The construction of Σ from M is one of the simplest examples of surgery.

For a discussion of the case $n = 2$, see M. Kervaire (1963) as well as J. Levine's article : "Some results on higher- dimensional knot groups" in this volume.

These references also contain some analysis of the above algebraic conditions (1), (2), (3) on a group. For further work in this direction see J.-Cl. Hausmann et M. Kervaire : "Sous-groupes dérivés des groupes de noeuds", l'Enseignement Mathématique XXIV (1978), pp. 111-123.

As to the higher π_i, $i > 1$, we have already mentionned in the introduction the papers of J.J. Andrews and M.L. Curtis (1959) and D.B.A. Epstein (1959).

More recently the subject has been taken up again. See E. Dyer and A. Vasquez (1972) and B. Eckmann (1975). Their result is that for $n > 1$, the space $X_o = S^{n+2} - K$ is never aspherical unless the knot is trivial.

Nevertheless, a complete understanding of the higher homotopy groups of knot complements seems out of reach today.

The most gratifying invariants at present are the homology modules of coverings of X and in particular those of the maximal abelian cover $\tilde{X}$ corresponding to the kernel of the surjection $\pi_1(X) \longrightarrow H_1(X)$.

These are simple enough to be tractable and yet non-trivial enough to provide a beautiful theory.

The homology modules $A_q(K) = H_q(\tilde{X})$ are modules over the integral group ring Λ of $H_1(X)$ which operates on $\tilde{X}$ as the group of covering transformations. The group $H_1(X)$ is infinite cyclic and if we denote by t a generator of $H_1(X)$, then Λ is the ring $\mathbb{Z}[t, t^{-1}]$ of Laurent polynomials in t. Observe that $H_1(X)$ is generated by a fibre of the normal circle-bundle over $K \subset S^{n+2}$ and thus a choice of generator t is provided by the orientations of K and S^{n+2}.

We shall follow M. Hirsch and L. Neuwirth (1964) in calling $A_q(K)$ $q \geq 1$, the Alexander modules of the knot or simply, following J. Levine (1974), the <u>knot modules</u>.

The general problem is : What sequences of Λ-modules $A_1, \ldots, A_n$ are modules of n-knots ? (It turns out that $A_q = 0$ for $q > n$).

Observe that $\pi_1(\tilde{X})$ is just the commutator subgroup $G = [\pi, \pi]$ of the knot group $\pi = \pi_1(X)$. Therefore $H_1(\tilde{X})$ is G/G' viewed as a group with operators from $H_1(X)$ via the extension $1 \longrightarrow G/G' \longrightarrow \pi/G' \longrightarrow H_1(X) \longrightarrow 1$. Thus $A_1(K)$ is determined by the knot group π .

In the classical case, $A_1(K)$ is the only (non-zero) Alexander module. It possesses a square presentation matrix (over Λ) whose determinant is the familiar Alexander polynomial.

The fundamental group $\pi = \pi_1(X)$ influences $A_2(K)$ also . Since $G = \pi_1(\tilde{X})$, there is an exact sequence

$$\pi_2(\tilde{X}) \longrightarrow H_2(\tilde{X}) \longrightarrow H_2(G) \longrightarrow 0$$

by a celebrated theorem of H. Hopf (Fundamentalgruppe und zweite Betti'sche Gruppe, Comm. Math. Helv. 14 (1941), 257-309) and thus $A_2(K)$ must surject onto $H_2(G)$.

It may then perhaps be more appropriate to ask : what set $\{\pi, A_1, A_2, \ldots, A_n\}$ with $A_1 = H_1(G)$ and surjection $A_2 \longrightarrow H_2(G)$, $G = [\pi,\pi]$, is realizable with π the knot group and A_q the knot modules for $q = 1, \ldots, n$?

A start on this question with π infinite cyclic was made by M. Kervaire (1964). The formulation (in terms of the homotopy modules of the knot complement) was however very ackward. The decisive break-through was accomplished by J. Levine (1974) which we now follow.

Let X again be the exterior of a knot $K \subset S^{n+2}$. Assume X is triangulated as a finite complex and let $\tilde{X}$ be the infinite cyclic covering of X with the natural triangulation (such that $\tilde{X} \longrightarrow X$ is a simplicial map). We denote by C the multiplicative infinite cyclic group with generator t . C operates on $\tilde{X}$ without fixed point and the chain groups $C_q(\tilde{X})$ are finitely generated free $\mathbb{Z}C$-modules.

Since Λ has no divisors of zero, the multiplication by 1-t induces an injection $1-t : C_*(\tilde{X}) \longrightarrow C_*(\tilde{X})$. The quotient module is (canonically) isomorphic to the chain group of X (regarded as Λ-module with trivial action) and we get an exact sequence of complexes :

$$0 \longrightarrow C_*(\tilde{X}) \xrightarrow{1-t} C_*(\tilde{X}) \longrightarrow C_*(X) \longrightarrow 0 .$$

Passing to the associated long homology sequence

$$\cdots \longrightarrow H_{q+1}(X) \longrightarrow H_q(\tilde{X}) \xrightarrow{1-t} H_q(\tilde{X}) \longrightarrow H_q(X) \longrightarrow \cdots$$

in which $H_q(X) = 0$ for $q > 1$ by Alexander duality, one obtains that

$$1 - t : H_q(\tilde{X}) \longrightarrow H_q(\tilde{X})$$

is an isomorphism for $q \geqslant 2$. Inspection of the sequence near $q = 1$, i.e.

$$0 \rightarrow H_1(\tilde{X}) \xrightarrow{1-t} H_1(\tilde{X}) \longrightarrow H_1(X) \longrightarrow H_0(\tilde{X}) \xrightarrow{1-t} H_0(\tilde{X})$$

reveals that $1\text{-}t : H_1(\tilde{X}) \longrightarrow H_1(X)$ is also an isomorphism.

Following J. Levine (1974), we shall say that a Λ-module A is <u>of type</u> K if

(1) A is finitely generated (over Λ), and

(2) $1\text{-}t : A \longrightarrow A$ is an isomorphism.

We have just seen that all knot modules are of type K.

Of course, one cannot expect this property to characterize the Alexander modules of knots.

It is a remarkable theorem of J. Levine that there is however just one property missing : Blanchfield duality. (Except perhaps for a condition on the Z-torsion submoduble of A_1).

In order to understand Blanchfield duality, recall that an oriented, triangulated, m-dimensional manifold M possesses an intersection pairing

$$I : C_q(M, bM) \otimes C_{m-q}(M^*) \longrightarrow Z ,$$

where $C.(M^*)$ is the chain complex of the dual cellular subdivision M^* of M. If M is compact, this gives rise to Poincaré duality. Here, we shall take $M = \tilde{X}$, the infinite cyclic cover of the exterior of a knot $K \subset S^{n+2}$. Of course, $\tilde{X}$ is non-compact but $C \cong H_1(X)$ operates on

$\tilde{X}$ simplicially with compact quotient X.

One first uses the action of C on $\tilde{X}$ to construct a Λ-valued intersection pairing on $\tilde{X}$

$$C_q(\tilde{X}, b\tilde{X}) \otimes C_{n+2-q}(\tilde{X}^*) \longrightarrow \Lambda = \mathbb{Z}C$$

defined by

$$(x,y^*) = \Sigma_{s\in C}\ I(x, sy^*)s .$$

This construction actually goes back to K. Reidemeister (Durchschnitt und Schnitt von Homotopieketten, Monathefte Math. 48(1939), 226-239).

The above pairing has nice algebraic properties and because $X=\tilde{X}/C$ is a finite complex, it is a completely orthogonal pairing and one gets an isomorphism

$$H_q(\tilde{X}, b\tilde{X}) \cong H^{n+2-q}(\tilde{X},\Lambda).$$

The left hand side is the ordinary homology of the pair $(\tilde{X}, b\tilde{X})$ with integral coefficients. The right hand side is the cohomology of the complex $\mathrm{Hom}_\Lambda(C.(\tilde{X}^*),\Lambda)$. The isomorphism is an isomorphism of Λ-modules provided that $H^{n+2-q}(\tilde{X},\Lambda)$ is given its natural right-module structure and $H_q(\tilde{X}, b\tilde{X})$ is turned into a right module by the usual formula $x \cdot \lambda = \bar{\lambda} \cdot x$, $\lambda \in \Lambda$, where $\lambda \longmapsto \bar{\lambda}$ is the obvious involution on Λ sending the elements of C to their inverses.

An elegant reformulation due to J. Levine (1974) using $H_q(\tilde{X}, b\tilde{X}) = H_q(\tilde{X})$ for $0 < q \leq n$ and some non-trivial homological algebra yields the following statements.

Recall $A_q = H_q(\tilde{X})$. Let T_q be the $\mathbb{Z}$-torsion submodule of A_q and $F_q = A_q/T_q$. Then,

(1) There is a $(-1)^{q(n-q)}$-hermitian completely orthogonal pairing

$$F_q \otimes F_{n-q+1} \longrightarrow Q(\Lambda)/\Lambda$$

over Λ, where $Q(\Lambda)$ is the field of fractions of Λ and $Q(\Lambda)/\Lambda$ is the quotient Λ-module. (Note that $Q(\Lambda)$ is merely the field of rational fractions $\mathbf{Q}(t)$. The hermitian property of the pairing is of course with respect to the involution of Λ defined above).

(2) There is a $(-1)^{q(n-q)}$-symmetric completely orthogonal pairing

$$[\ ,\ ] : T_q \otimes T_{n-q} \longrightarrow \mathbf{Q}/\mathbf{Z}$$

with respect to which C operates by isometries, i.e.

$$[t\alpha,\ t\beta] = [\alpha,\beta] .$$

This second pairing has also been discovered by M.Š. Farber(1974).

Now, J. Levine's realization theorem reads as follows.

THEOREM. - Given a sequence $A_1, \ldots, A_n$ of Λ-modules of type K. Let T_q be the torsion submodule of A_q and $F_q = A_q/T_q$. Suppose that $T_1 = 0$ and that the families F_q and T_q are provided with pairing as in (1) and (2) above. Then, there exists an n-knot K such that $A_1, \ldots, A_n$ is the sequence of Alexander modules of K.

Hopefully the unfortunate assumption $T_1 = 0$ will turn out to be removable. It is known that this assumption is not a necessary condition on T_1.

CHAP. III : TOWARDS A CLASSIFICATION UP TO ISOTOPY .

§ 1 . Seifert surfaces.

A basic concept for any attempt at classification is that of a Seifert surface.

A Seifert surface for an n-knot K is a compact, orientable submanifold $V \subset S^{n+2}$, such that $bV = K$.

The fact that V should be orientable is important and was first emphasized by H. Seifert (1934) who introduced the concept and proved existence in the classical case.

For multidimensional knots, the existence of a Seifert surface seems to have become public knowledge during the Morse Symposium at Princeton in 1963. (However, H. Gluck had proved it earlier for 2-knots. See H. Gluck (1961)). It appears in print in M. Kervaire (1963) and E.C. Zeeman (1963).

Here is a sketch of proof. Recall that a trivialization of the normal bundle of the knot K provides an identification $bX \cong S^n \times S^1$, and thus a projection $bX \longrightarrow S^1$.

The first step consists in showing that with a proper choice of trivialization above, the projection $bX \longrightarrow S^1$ extends to a map $X \longrightarrow S^1$. This is not difficult. The homotopy classes of maps into S^1 are classified by the first cohomology group H^1 with integral coefficients and one has enough control on both $H^1(X) \cong Z$ and the restriction homomorphism $i^* : H^1(X) \longrightarrow H^1(bX)$.

The existence of a Seifert surface now follows by transversality. One chooses an extension $X \longrightarrow S^1$ which is transverse regular to the point $1 \in S^1$. The inverse image of 1 is then a codimension one submanifold W in S^{n+2} equipped with a non-vanishing normal vector field (pulled back from a tangent vector to S^1 at 1). Hence, W is orientable. The boundary of W is precisely $S^n \times \{1\} \subset bX$. We can then add a collar to W, joining bW to K along the radii of the normal bundle to K and get the submanifold V we are looking for.

Many constructions in knot theory depend on a Seifert surface. We collect in this section some of the notions derived from a Seifert surface which we shall need in the subsequent chapters of this survey (even though they may not pertain directly to the subject of the present chapter).

First, a Seifert surface enables one to perform a paste and scissors construction of the infinite cyclic cover of a knot.

Let V be a Seifert surface for the knot K. Let N be an open tubular neighborhood of K and set $X = S^{n+2} - N$. We assume V to be radial inside N and set $W = V \cap X$.

Let Y be the manifold with boundary obtained by cutting X along W. Equivalently, Y is obtained from X by removing a small tubular neighborhood of W, homeomorphic to $W \times [-1, +1]$. Notice that it is here that the orientability of V comes in.

The boundary of Y is the union of two copies of W, i.e. $W \times \{-1\} = W_-$ and $W \times \{+1\} = W_+$ together with $bW \times I$, where $I = [-1, +1]$. These pieces are glued together to form bY in the obvious way.

Notice also that there is a natural projection map π from Y onto X which sends the two copies of W onto W and is otherwise injective.

(Glue again what you had cut !).

Now, let $\{Y_i\}_{i\in Z}$ be a collection of copies of Y, indexed by the integers $\mathbf{Z}$. Let $\tilde{X}$ be the quotient of the disjoint union $\coprod_{i\in Z} Y_i$ by the obvious identification of $(W_-)_i$ with $(W_+)_{i+1}$ for all $i \in \mathbf{Z}$. The maps $\pi_i : Y_i \longrightarrow X$ are compatible with these identifications and provide a map $p : \tilde{X} \longrightarrow X$.

It is not hard to verify that p is a covering map. The covering is regular and its Galois group is C. (We denote by C the group of integers written multiplicatively).

Hence, $p : \tilde{X} \longrightarrow X$ is "the" infinite cyclic covering of X.

This construction has been used by L.P. Neuwirth (1959) to give a description of the knot group. It is also the first step in proving the Neuwirth-Stallings fibration theorem. (We come back on this in the chapter on fibered knots, Chapter V, § 3).

The above description of the infinite cyclic cover leads of course to a computation of the homology of this covering by a Mayer-Vietoris sequence. (See M. Hirsch and L. Neuwirth (1964)).

Indeed, let $\tilde{X}_{odd}$ be the subspace of $\tilde{X}$ which is equal to the canonical image of $\coprod_{i \text{ odd}} Y_i$ in $\tilde{X}$, and let $\tilde{X}_{even}$ be the analogous subspace for i even. Obviously $\tilde{X}_{odd} \cup \tilde{X}_{even} = \tilde{X}$ and $\tilde{X}_{odd} \cap \tilde{X}_{even} = \coprod_{i\in Z} W_i$, W_i being identified with $(W_-)_i$, say.

Let now H_j denote homology with some fixed coefficient group and let $\Lambda = \mathbf{Z}\, C$ be the integral group ring of C. One has

$$H_j(\tilde{X}_{odd}) \oplus H_j(\tilde{X}_{even}) = H_j(Y) \otimes \Lambda ,$$

$$H_j(\coprod_{i\in Z} W_i) = H_j(W) \otimes \Lambda ,$$

the isomorphisms being Λ-isomorphisms, C acting on the left hand side via the Galois operations.

The Mayer-Vietoris sequence for the decomposition $\tilde{X} = \tilde{X}_{odd} \cup \tilde{X}_{even}$ produces the following exact sequence :

$$(*) \quad \dots \longrightarrow H_j(W) \otimes \Lambda \xrightarrow{\alpha} H_j(Y) \otimes \Lambda \xrightarrow{\beta} H_j(\tilde{X}) \longrightarrow \dots .$$

The homomorphisms are all Λ-modules homomorphisms.

Moreover, if we denote by i_+ the homomorphisms $H_j(W) \longrightarrow H_j(Y)$ induced by the inclusion $W_+ \subset Y$, and similarly with i_-, then

$$\alpha(x \otimes \lambda) = i_+(x) \otimes t\lambda - i_-(x) \otimes \lambda ,$$

the minus sign coming from the Mayer-Vietoris sequence. Here, t is a correctly chosen generator for C .

Caution. Different identifications in the construction may lead to slightly different formulas.

A useful fact, due to J. Levine, is that this sequence always breaks up into short exact sequences

$$0 \longrightarrow H_j(W) \otimes \Lambda \xrightarrow{\alpha} H_j(Y) \otimes \Lambda \xrightarrow{\beta} H_j(\tilde{X}) \longrightarrow 0 .$$

In some circumstances, we may thus be on the way to get a free resolution of the module $H_j(\tilde{X})$. See J. Levine (1976).

Remarks.

1. For a very nice application of this sequence to the symmetry properties of the Alexander polynomials, see also J. Levine (1966).

2. A variant of this process gives a description of the g-th cyclic covering X_g of X, g an integer > 1. Alternatively, X_g can be obtained as a quotient of X via the automorphism t^g, where t is a generator of

the Galois group. One then gets for the homology of X_g a sequence analogous to the one described above in chap. II, § 3, p. 21.

A notion of paramount importance for all classification problems of knots is that of the Seifert pairing associated with a Seifert surface for an odd dimensional knot.

This notion was introduced in the classical case by H. Seifert (1934). We proceed to describe it in general.

Let $K \subset S^{2m+1}$ be a (2m-1)-knot. Choose a trivialization of the normal bundle of a (truncated) Seifert surface W for the knot K. The trivialization determines a map

$$i_+ : W \longrightarrow Y ,$$

where Y, as above, is the complement of a neighborhood of V.

There is a pairing

$$L : H_m(W) \times H_m(Y) \longrightarrow \mathbf{Z}$$

defined by the linking number in S^{2m+1}. Now, define

$$A : H_m(W) \times H_m(W) \longrightarrow \mathbf{Z}$$

by the formula $A(x, y) = L(x, i_+(y))$.

Observe that A is bilinear and thus vanishes on the torsion subgroup of $H_m(W)$.

We note F_m the free part of the integral homology H_m, i.e. $F_m = H_m/\text{Torsion}$.

Since $H_m(V) = H_m(W)$, we have obtained a bilinear pairing

$$A : F_m(V) \times F_m(V) \longrightarrow Z \quad .$$

By definition, A is called the Seifert pairing associated with the Seifert surface V .

In general, there is no symmetry nor non-degeneracy properties satisfied by A itself. However, let A^T denote the transpose of A. One has

$$\begin{aligned}(A + (-1)^m A^T)(x, y) &= L(x, i_+y) + (-1)^m L(y, i_+x) \\ &= L(x, i_+y) - L(x, i_-y) \\ &= L(x, i_+y - i_-y) \ ,\end{aligned}$$

and this is equal to the intersection number of x and y in V .

So, $A + (-1)^m A^T = I$ is the intersection pairing on $F_m(V) =$ $= H_m(V)/\text{Torsion}$. Since bV is a sphere, Poincaré duality on V implies that $A + (-1)^m A^T$ is unimodular.

We shall come back to the study of the Seifert pairing in § 3 below in the case of simple knots, and in Chap. IV again, where we talk about knot cobordism.

§ 2 . Improving a Seifert surface.

For a given knot, there are many possible Seifert surfaces. The surfaces may be abstractly different (non homeomorphic), or abstractly the same but imbedded differently. (However, the existence proof shows that they are all cobordant).

It is hence natural to look for Seifert surfaces which are "minimal in some sense. For classical knots, it is clear what "minimal" should mean : V should be connected and its genus as small as possible. But, for multidimensional knots, the notion is not so clear, except under special circumstances (such as for the odd dimensional simple knots which we discuss in § 3 below).

We shall now review some cases in which one can "improve" or "simplify" a Seifert surface. The main point is that there is a strong connection between the connectivity of $\tilde{X}$ and the best possible connectivity of a potential Seifert surface.

a) For all $n \geq 1$, if a knot has a 1-connected Seifert surface, then $\tilde{X}$ is 1-connected. The first proof of this fact is due to M. Hirsch and L. Neuwirth (1964) and it goes as follows : if V is 1-connected, then by van Kampen, $\pi_1(X) \cong \pi_1(Y) * \mathbf{Z}$ and a generator z of $\mathbf{Z}$ represents a meridian of the knot. It follows that the normal closure of z in $\pi_1(X)$ should be the entire group. (Compare the characterization of knot groups in Chap. I, § 3). We see immediately that this is possible only if $\pi_1(Y) = \{1\}$, and thus $\pi_1(X) = \mathbf{Z}$.

Caution . It is essential in this proof to be able to identify a generator of the factor $\mathbf{Z}$ as a meridian of the knot. The question whether in general a free product $G * \mathbb{Z}$ with $G \neq \{1\}$, may or may not contain an element whose normal closure is the whole group is still an unsolved problem.

b) The converse of a) is almost true. In fact, M. Hirsch and L. Neuwirth (1964) proved by an argument of exchange of handles that if $\pi_1(\tilde{X}) = \{1\}$ and if $n \geq 3$, then there exists a 1-connected Seifert surface for the knot.

The case $n = 1$ is also true. (Dehn's lemma). So there remains only the case $n = 2$ which is still open.

c) By the above case a), Alexander duality and the homology exact sequence (*) of the preceeding paragraph one sees immediately that if there exists a k-connected Seifert surface for a knot, then $\tilde{X}$ is also k-connected.

d) Now, again the converse is almost true. But this is the content of a deep theorem of J. Levine (1964). For clarity we separate the statements in two parts :

Part. 1 : Let $n \geq 2k+1$ and suppose that $\tilde{X}$ is k-connected. Then, there exists a k-connected Seifert surface for the knot.

Part 2 : Let $n = 2m$ or $n = 2m-1$ and suppose that $\tilde{X}$ is m-connected. Then, if $n \geq 4$, there exists a m-connected Seifert surface V for the knot.

Observe that by Blanchfield duality the condition on $\tilde{X}$ in Part 2 is equivalent to $\tilde{X}$ being contractible. Similarly, Poincaré duality and the Hurewicz theorem imply that the Seifert surface V in Part 2 must be contractible.

These statements constitute the essential part of J. Levine's unknotting theorem. Suppose that X has the homotopy type of S^1 and that $n \geq 4$. Then $\widetilde{X}$ is contractible and so K bounds a contractible Seifert surface V. By S. Smale, V is a P.L. disk and so K is P.L. unknotted. If $n \geq 5$ and K is differentiable, then K is differentiably unknotted and so has the standard differential structure.

§ 3. Simple knots.

In view of Levine's unknotting theorem, it is natural to study the n-knots which are "almost" trivial ; that is those for which $\pi_i(\tilde{X}) = 0$ for $i < m$ with $n = 2m$ or $n = 2m-1$. These knots have been called simple by J. Levine. Their study breaks up into two cases, depending upon the parity of n .

First case : n odd.

This case has been much studied by J. Levine (1969). We describe now the content of his paper.

By the statement under d), Part. 1, in the preceeding paragraph, one can find for any simple knot $K \subset S^{2m+1}$ a (m-1)-connected Seifert surface V. As dim V = 2m the only non-trivial homology group of V is $H_m(V)$, where we use integer coefficients.

It is not difficult, using Poincaré duality and the parallelizability of V in the case m even, to see that $H_m(V)$ is a free abelian group of even rank. Moreover, for $m \neq 2$, the conditions we have on V imply that V is obtained from a 2-dimensional disk by attachning handles of type m. (See C.T.C. Wall : "Classification of (n-1)-connected 2n-manifolds" in Annals of Math. 75 (1962), p. 163-198).

So, odd dimensional simple knots have a tendency to look like classical knots. For instance, it is obvious how to define a minimal Seifert surface V for them : V should be (m-1)-connected and the rank of $H_m(V)$ as small as possible.

In order to classify odd dimensional simple knots, J. Levine undertakes to classify all (m-1)-connected Seifert surfaces whether minimal or not, which are associated to such a knot.

It turns out that the Seifert pairing does the job. Let K be a simple (2m-1)-knot and let $V = V^{2m}$ be a (m-1)-connected Seifert surface for K. Since $H_m(V)$ is torsion free, the Seifert pairing is a bilinear map

$$A : H_m(V) \times H_m(V) \longrightarrow Z$$

such that $A + (-1)^m A^T$ is $(-1)^m$-symmetric and unimodular.

THEOREM . *For* $m \geq 3$, *the isotopy class of an* (m-1)-*connected Seifert surface* V *for a simple* (2m-1)-*knot is determined by its associated* Seifert *pairing*.

For a proof, see J. Levine (1969), p. 191, sections 14 to 16.

Furthermore, using the fact that two Seifert surfaces for the same knot are cobordant, J. Levine shows :

Fact 1 : For $m \geq 1$, any two Seifert pairings for a given knot are S-equivalent.

S-equivalence is the equivalence relation generated by isomorphisms and by the following elementary operations : replace the underlying Z-module H by $H \times Z \times Z$ and A by A' or A", where A', A" are expressed matricially by

$$A' = \left(\begin{array}{c|cc} & 0 & 0 \\ A & \vdots & \vdots \\ & 0 & 0 \\ \hline * \;\ldots\; * & 0 & 0 \\ 0 \;\ldots\; 0 & 1 & 0 \end{array}\right) \qquad A'' = \left(\begin{array}{c|cc} & * & 0 \\ A & \vdots & \vdots \\ & * & 0 \\ \hline 0 \ldots 0 & 0 & 1 \\ 0 \ldots 0 & 0 & 0 \end{array}\right)$$

Fact. 2 : Suppose $m \geq 2$. Let K and K' be two simple (2m-1)-knots, each equipped with a (m-1)-connected Seifert surface. Suppose that the two corresponding Seifert pairings are S-equivalent. Then the two knots are isotopic.

This is of course the most difficult part of the theory. It relies heavily on the classification of Seifert surfaces described in the above theorem.

Definition : Given an integer $m \neq 2$, define a Seifert form (for m) to be a bilinear form

$$A \quad : \quad E \times E \longrightarrow \mathbf{Z}$$

on a finitely generated free $\mathbf{Z}$-module E such that $A + (-1)^m A^T$ is unimodular.

For m = 2, observe that the Seifert surface is a smooth, parallelizable 4-manifold, with boundary a sphere, and therefore, by V.Rochlin's theorem its intersection pairing has a signature divisible by 16. (For V. Rochlin's theorem, see J. Milnor and M. Kervaire, Bernoulli numbers, Homotopy groups and a theorem of Rochlin, Proc. of the Int. Congress of Math., 1958, p. 454-458). Thus, for m = 2, a Seifert form will be defined as a bilinear map A as above subject to the additional condition that signature $(A + A^T) \equiv 0 \bmod 16$.

We can now state the last needed fact.

Fact 3 : Given a Seifert form A for m . Then, if $m \neq 2$, there exists a (m-1)-connected (orientable) submanifold $V^{2m} \subset S^{2m+1}$ such that bV is homeomorphic to the (2m-1)-sphere and A is the associated Seifert pairing. For m = 2, the same statement holds, except that now A is only S-equivalent to the Seifert pairing of the constructed Seifert surface V.

In the classical case ($m = 1$), this fact is due to H. Seifert himself. For multidimensional knots, see M. Kervaire (1964) in the case $m \neq 2$, and J. Levine (1969) in general.

Putting all these facts together, J. Levine obtains the theorem :

For $m \geq 2$, the isotopy classes of simple $(2m-1)$-knots are in one-to-one correspondence with the S-equivalence classes of Seifert forms.

In the classical case ($m = 1$), the isotopy classes of 1-knots are mapped onto the set of S-equivalence classes of Seifert forms. This fact was known already to H. Trotter (1960) and to K. Murasugi (1963). But the mapping is not injective. For instance, knots with trivial Alexander polynomial are mapped into the trivial S-equivalence class.

Remarks : From J. Levine's theorem, the set of simple 5-knots is isomorphic to the set of 9-knots, to the set of 13-knots, etc... The bijection is well defined. So, it is natural to ask whether one can define this bijection directly. In the case of fibered knots, such a construction is provided by L. Kauffman and W. Neumann (1976).

Let K be a simple $(2m-1)$-knot. Look at the set of all its minimal Seifert surfaces. Question : Are all these surfaces isotopic ? If they are, we would say that the minimal Seifert surface for K is (essentially unique.

By J. Levine's theorem this question can now be attacked algebraically. Look at the (minimal) Seifert pairing associated with the minimal surfaces. We know that they are all S-equivalent. But if the answer to the question is "yes", they should all be isomorphic ($m \geq 1$). Conversely, for $m \geq 2$, if they are isomorphic, the Seifert surfaces are isotopic. Thus the problem is to determine the isomorphism classes of of Seifert forms within a given S-equivalence class. This algebraic problem has been attacked by H. Trotter in several papers (1960), (1970) and (1972). Sometimes the S-equivalence class determines the isomorphism class, sometimes it does not. Sometimes the answer is unknown. The problem involves the determinant of a minimal Seifert pairing (which is an invariant of the knot and therefore of the S-equivalence class of the Seifert form). As an example, there is only one isomorphism class in the given S-equivalence class if this determinant is ± 1, a result which can be interpreted (and proved) geometrically, using fibered knots.

H. Trotter's papers give also nice answers to other old questions. For instance, it is easy to see that if we change the orientation of the knot, we must also change the orientation of the Seifert surface because K and V are given orientations which correspond each other via the homology exact sequence, and then, the normal vectors to V have to change direction. It is then easy to see that the initial Seifert form is changed into its transposed (up to a sign which seems to be $(-1)^{m+1}$). H. Trotter then gives examples of Seifert forms which are not S-equivalent to their transpose, showing thus that non-invertible knots exist for $m \geqslant 1$. For $m = 1$ this is the famous result first proved by H.Trotter (1963).

For $m \geqslant 2$, it is also rather nice, because it is not based on the non-symmetry of some Alexander invariant. Related reference : C. Kearton (1974).

<u>Second case</u> : n even . The case of even dimensional simple knots is much harder than the first case because there is no such simple algebraic invariant as the Seifert pairing. There is only a complicated invariant consisting of a composite algebraic object. However, the classification has almost been completed by C. Kearton (1975). The problem has also been taken up by S. Kojima (1977) and A. Ranicki(1977).

§ 4 . Seifert pairings and the infinite cyclic covering .

As may perhaps be expected, the Seifert pairing is related to the calculation of $H_*(\tilde{X})$ using a Seifert surface as explained in § 1.

Let $K \subset S^{2m+1}$ be an odd dimensional simple knot and V a (m-1)-connected Seifert surface for K .

Recall that the Alexander duality gives an isomorphism

$$d \;:\; H_m(Y) \longrightarrow H^m(W) \;,$$

where we keep the notations of § 1. (W is the truncated Seifert surface and Y is the exterior X of the knot cut along W).

Because W is (m-1)-connected, the evaluation map

$$e \;:\; H^m(W) \longrightarrow \mathrm{Hom}(H_m(W),\, \mathbf{Z})$$

is also an isomorphism.

Now, let $a \in H_m(V)$ and $b \in H_m(Y)$ be given. Then L(a,b) is by definition the integer obtained by evaluating on $a \in H_m(V)$ the homomorphism ed(b). (Recall $H_m(V) = H_m(W)$) .

In other words, the (right) adjoint to $L : H_mV \times H_mY \longrightarrow \mathbf{Z}$ is precisely ed.

So, the adjoint to $A : H_mV \times H_mV \longrightarrow \mathbf{Z}$ is $edi_+ : H_m(V) \longrightarrow \mathrm{Hom}(H_m(V),\mathbf{Z})$.

As e and d are (canonical) isomorphisms, we see that the algebraic properties of i_+ will be reflected by those of A.

Now, if we start from a $\mathbf{Z}$- basis of $H_m(V)$, we can take the dual basis for $\mathrm{Hom}(H_m(V),\, \mathbf{Z})$ and get via d and e a basis for $H_m(Y)$.

With respect to these bases, the matrix expressing the bilinear form A will be precisely the matrix for the map $i_+ : H_m(V) \longrightarrow H_m(Y)$.

Returning to the short exact sequence

$$(*) \quad 0 \longrightarrow H_m(V) \otimes \Lambda \xrightarrow{\alpha} H_m(Y) \otimes \Lambda \xrightarrow{\beta} H_m(\widetilde{X}) \longrightarrow 0$$

of § 1, with integer coefficients, the $\mathbb{Z}$-bases for $H_m(V)$ and $H_m(Y)$ give Λ-bases for the tensor products.

It now follows from the formula

$$\alpha\,(\chi \otimes \lambda) = i_+(x) \otimes t\lambda - i_-(x) \otimes \lambda$$

of § 1, that the matrix expressing α is $At + (-1)^m A^T$.

So, from a Seifert matrix for K (i.e. the matrix of a Seifert pairing for K), one can get a presentation matrix for $H_m(\widetilde{X})$. For classical knots this result is due to H. Seifert.

We now consider the Blanchfield pairing on $H_m(\widetilde{X})$, still assuming that K is a simple (2m-1)-knot. A study of the above exact sequence (*) with various coefficients reveals that for simple knots, $H_m(\widetilde{X})$ is $\mathbb{Z}$-torsion free. See, for instance, the thorough study made by J. Levine (1976), § 14.

So, the Blanchfield pairing reduces to a pairing

$$H_m(\widetilde{X}) \times H_m(\widetilde{X}) \longrightarrow Q(\Lambda)/\Lambda$$

(Compare § 3 in Chap. I).

Now, H. Trotter (1972) and C. Kearton (1973) have shown that this Blanchfield pairing is determined by the Seifert form in the following way. Let us take as generators for $H_m(\widetilde{X})$ the images by $\beta : H_m(Y) \otimes \Lambda \longrightarrow H_m(\widetilde{X})$ of the basis elements chosen for $H_m(Y) \otimes \Lambda$. Of course, they do not form a basis for $H_m(\widetilde{X})$, but there still is a matrix representative of the Blanchfield pairing with respect to this set of generators, and it is

$$(1-t)(At + (-1)^m A^T)^{-1} .$$

(See the exposition in J. Levine (1976), prop. 14.3).

Again, different conventions will lead to slightly different formulas.

This result is the starting point of H. Trotter's paper (1972). More precisely, to every free abelian group equipped with a Seifert form A, H. Trotter associates a $\mathbb{Z}C$-module with presentation matrix $At + (-1)^m A^T$ and equipped with a Blanchfield pairing represented by the matrix $(1 - t)(At + (-1)^m A^T)^{-1}$.

He then goes on to prove that

(1) S-equivalent Seifert forms give rise to isomorphic Blanchfield pairings, and the deep result :

(2) If two Seifert forms give rise to isomorphic Blanchfield pairings, then they are S-equivalent.

A nice geometric consequence of this result is that simple (2m-1)-knots (for $m \geq 2$) are classified by their Blanchfield duality. This furnishes an intrinsic classification for these knots. The same result has also been proved by C. Kearton (1973).

An interesting question, asked by C. Kearton, and which provides our conclusion to this chapter, is whether the same is true for simple even dimensional knots. Possibly, A. Ranicki will tell you the answer.

CHAP. IV : KNOT COBORDISM.

§ 1 : Prehistory .

The notion of knot cobordism was invented in the context of classical knots around 1954 by R. Fox and J. Milnor.

An announcement appeared in 1957 but the paper itself (with simplified proofs) was only published in 1966.

Knot cobordism is a weaker equivalence relation between knots than isotopy and part of the motivation for introducing it certainly is the discouraging difficulties involved in the classification up to isotopy. But there is another motivation. The idea of knot cobordism is also related to the topological study of isolated codimension two singularities.

Suppose that $M^{n+1} \subset N^{n+3}$ is an embedded submanifold which is locally flat except at one point $x_0 \in M$. Intersecting M with the boundary of a small disk neighborhood U of x_0 in N will yield a (knotted)sphere K of dimension n in $bU = S^{n+2}$. Thus an n-knot.

Definition : A knotted n - sphere $K \subset S^{n+2}$ is null-cobordant if K is the boundary of a locally flat embedded disk $B \subset D^{n+3}$.

The requirement of local flatness for B is of course essential, or else the cone over K from the center of D^{n+3} would trivially do the job.

It has been believed that at least for $n = 1$, the singularity at the vertex of the cone may be removable, yielding a null-cobordism.

This is definitely not the case. In fact we shall see that it is even worse than its higher dimensional analogues.

Going back to the embedded $M^{n+1} \subset N^{n+3}$ with $K = M \cap bU$ as defined above, it is clear that if the knot $K \subset S^{n+2}$ is null-cobordant, then the embedding $M \subset N$ can be replaced near x_o by a locally flat embedding of M.

Conversely, if the embedding $M \subset N$ can be changed near x_o within some neighborhood of x_o in N, to produce a locally flat embedding, the above knot was null-cobordant.

Thus in some sense, the local singularity of M at x_o is described by the knot cobordism class of K.

An additional pleasant feature is that the set of cobordism classes has nicer algebraic properties than the set of isotopy classes. The set K_n of isotopy classes of n-knots forms a commutative monoid under ambiant connected sum (joining the knotted spheres by a tube). It turns out that modulo null-cobordism, the quotient monoid actually is an abelian group C_n. (Incidentally, it does not seem to have attracted attention to investigate whether or not C_n is in any sense the largest quotient group of K_n).

R. Fox and J. Milnor looked at C_1 and after proving that the Alexander polynomial of a null-cobordant knot must be of the form $t^g f(t)f(1/t)$ for some polynomial $f \in Z[t]$, they recognized that C_1 could not be finitely generated.

A good surprise came with the simple result

$$C_{2m} = 0 \qquad \text{for all} \quad m \geq 1 ,$$

proved in M. Kervaire (1964). But it soon appeared that (contrary to tempting dreams) the groups C_{2m-1} are indeed non-finitely generated for all $m \geqslant 1$.

Much effort was then devoted to the rather formidable task of computing C_{2m-1}.

§ 2. The algebraization of the problem.

By analysing the obstructions which arise if one tries to apply to the odd dimensional case the surgery methods used to calculate C_{2m}, J. Levine (1969) extracted a purely algebraic description of C_{2m-1} which again hinges on the concept of a Seifert form.

Recall from Chap. III, § 1, that a Seifert form for m is a bilinear pairing $A : E \times E \longrightarrow \mathbf{Z}$ on some finitely generated free $\mathbf{Z}$-module E such that $A + (-1)^m A^T$ is unimodular. (If m = 2, there is also a condition on the signature).

If $K \subset S^{2m+1}$ is a (2m-1)-knot and V is a Seifert surface for K, then the Seifert pairing

$$A \;:\; F_m(V) \times F_m(V) \longrightarrow \mathbf{Z}$$

on the torsion free part $F_m(V)$ of $H_m(V)$ is a Seifert form.

Moreover, by our discussion of simple knots in Chap. III, § 3, Fact 3, every Seifert form is (essentially) the Seifert pairing of a (simple) knot.

The first step is to carry over to Seifert forms the notion of cobordism.

Definition : A Seifert form $A : E \times E \longrightarrow \mathbf{Z}$ is said to be null-cobordant (or split) if there exists a totally isotopic subspace $E_0 \subset E$ such that $E_0 = E_0^{\perp}$, where

$$E_0^{\perp} = \{ x \in E \mid A(y,x) = A(x,y) = 0 \text{ for all } y \in E_0 \} .$$

It turns out that the monoid of Seifert forms (for a given m) modulo null-cobordant forms becomes a group under the operation of direct sum.

Of course the definition of this group resembles much the definition of the Witt group of $\mathbb{Z}$. But here the forms A are not assumed to be symmetric, and the resulting group is tremendously more complicated than $W(\mathbb{Z})$.

J. Levine's theorem says that for a given $m \geq 2$, the group of Seifert forms modulo split forms is isomorphic to the knot cobordism group C_{2m-1} of (2m-1)-knots.

For $m = 1$, there is a surjection of C_1 onto the cobordism group of Seifert forms. But it is known that the kernel is non-zero. (Compare C. Gordon's survey of classical knot theory in this volume).

A corollary of J. Levine's theorem is that C_n is periodic of period 4 for $n \geq 4$. Again, it was natural to try and explain the periodicity by direct geometric arguments. This was done by S. Cappell and J. Shaneson (1972) and by G. Bredon (1972).

In order to prove that $C_{2m} = 0$, one takes a Seifert surface V for the given knot $K \subset S^{2m+2}$. Thus, $\dim V = 2m+1$. The method consists in performing surgery on V, increasing its connectivity by attaching handles which are imbedded in D^{2m+3}. Thus, the effect of surgery is to increase the connectivity of V at the cost of pushing it into D^{2m+3}. As the dimension of the core of the handles does not exceed $m + 1$ (because it suffices to make V m-connected), there is no obstruction to imbedding problems in D^{2m+3}. It follows that the given knot $K \subset S^{2m+2}$ bounds a contractible submanifold of D^{2m+3}, and thus is null-cobordant.

In contrast, for a (2m-1)-knot $K \subset S^{2m+1}$, the Seifert surface V has dimension 2m . The above method will still enable one to replace V by $M \subset D^{2m+2}$ with M (m-1)-connected . In fact, one proves in this way that K is cobordant to a simple knot $K' \subset S^{2m+1}$, where K' bounds an (m-1)-connected Seifert surface $V' \subset S^{2m+1}$.

But at the last step, i.e. in the attempt to make V' m-connected, one hits obstructions. They arise from the problem of extending to the interior of a bunch of (m+1)-dimensional discs a given embedding of their boundaries into S^{2m+1}.

The cobordism group of Seifert forms measures precisely these obstructions to constructing a null knot cobordism.

§ 3. Unraveling the integral knot-cobordism group.

We borrow the title of this section from the paper of N.Stoltzfus (1976) containing a major part of the calculation of C_{2m-1}. Here is a summary of some of this work:

The reader will have guessed that together with the purely algebraic definition of C_{2m-1} comes the possibility of defining cobordism groups of bilinear forms over other coefficient domains than $\mathbb{Z}$. It was J. Levine (1969) who recognized this possibility as an essential tool in calculating C_{2m-1} by algebraic methods.

We begin by recasting accordingly the definition of a Seifert form.

First note that our previous definition of a Seifert form for m depended on m via the sign $(-1)^m$ only. The condition on the signature for $m = 2$ need not really be dragged along as it is easily recaptured at the end of the calculation of C_3.

Definition 1 : Let R be a commutative ring and M an R-module. Let $\varepsilon = \pm 1$. An ε-form is an R-bilinear form

$$A \; : \; E \times E \longrightarrow M$$

on a finitely generated R-module E, satisfying the condition that $S = A + \varepsilon A^T$ is unimodular, i.e.

$$\mathrm{ad}\,(S) \; : \; E \longrightarrow \mathrm{Hom}_R(E, M)$$

is an isomorphism, where as usual

$$\mathrm{ad}\,(S)\,(x)\,(y) \; = \; S(x, y).$$

As before, A itself is not assumed to possess any symmetry nor non-degeneracy property.

Basic examples are the cases $R = M = \mathbb{Z}$, and E is then $\mathbb{Z}$-free of finite rank, or $R = M = k$, a field, or $R = \mathbb{Z}$, $M = \mathbb{Q}/\mathbb{Z}$ and E is a finite abelian group.

Definition 2 : An ε-form $A : E \times E \longrightarrow M$ is split (or metabolic) if there exists an R-direct summand $E_0 \subset E$ such that $E_0 = E_0^{\perp}$, where $E_0^{\perp} = \{x \in E \mid A(x,y) = A(y,x) = 0 \text{ for all } y \in E_0\}$.

Note that two ε-forms $A : E \times E \longrightarrow M$ and $B : F \times F \longrightarrow M$ can be added by direct sum $A \oplus B : (E \oplus F) \times (E \oplus F) \longrightarrow M$, where $(A \oplus B)(x \oplus y, x' \oplus y') = A(x,x') + B(y,y')$, and $A \oplus B$ is again an ε-form. Obviously, if A and B are split forms, so is $A \oplus B$.

Given a commutative ring R and an R-module M, one can then define the group $C_R^{\varepsilon}(M)$ of cobordism classes of M-valued ε-forms $A : E \times E \longrightarrow M$ modulo split forms. Two ε-forms A and A' represent the same element in $C_R^{\varepsilon}(M)$ if there exist split forms H and H' such that $A \oplus H \cong A' \oplus H'$. The addition in $C_R^{\varepsilon}(M)$ is induced by the direct sum of ε-forms and the inverse of a class represented by the form $A : E \times E \longrightarrow M$ is represented by the form $-A$.

We shall abbreviate $C_R^{\varepsilon}(R)$ to $C^{\varepsilon}(R)$.

By J. Levine's theorem in the preceeding paragraph, $C_{2m-1} = C^{(-1)^m}(\mathbb{Z})$ for $m \geq 3$. For $m = 2$, C_3 is the subgroup of $C^{+1}(\mathbb{Z})$ generated by the (+1)-forms A such that $S = A + A^T$ has signature divisible by 16. For $m = 1$, the group C_1 surjects onto $C^{-1}(\mathbb{Z})$ and the kernel is definitely non-zero. (Compare C. Gordon's survey in this volume).

Caution : Unfortunately, $C_R(M)$ is not a functor in M. On the other hand, in the special cases which one needs to consider in order to calculate $C^{\varepsilon}(\mathbb{Z})$, it turns out that if A is equivalent to a split form, i.e. $A \oplus H \cong H'$, where H and H' are split, then A itself is a split form. However, we do not know how much of this remains true under reasonably general conditions for R and M.

The starting point for the study of $C^{\varepsilon}(\mathbf{Z})$ is the inclusion

$$C^{\varepsilon}(\mathbf{Z}) \subset C^{\varepsilon}(\mathbf{Q})$$

and the calculation, due to J. Levine (1969), of $C^{\varepsilon}(\mathbf{Q})$ which yields a complete system of algebraic invariants detecting the elements of $C^{\varepsilon}(\mathbf{Z})$.

Here is a summary of the method, extended by N. Stoltzfus (1976) to any perfect field k.

Given an ε-form $A : E \times E \longrightarrow M$, let $s : E \longrightarrow E$ be the endomorphism defined by

$$S(sx,y) = A(x,y)$$

for all $x,y \in E$, where $S = A + \varepsilon A^T$.

We have $S(sx,y) + S(x,sy) = A(x,y) + A(y,x) = S(x,y)$, or
$S(Sx,y) = S(x, (1-s)y)$.

We propose to call s an additive isometry.

Suppose now that $R = M = k$, a perfect field.

Let $f = f_s$ be the (monic) minimal polynomial of $s : E \longrightarrow E$. It is easily verified, using the isometric property of s, that f is self-dual, i.e.

$$f(1-X) = (-1)^{\deg f} . f(X).$$

Suppose that $f \in k[X]$ is irreducible. Then E becomes a vector space over the extension field $K = k[X]/(f)$. Denoting by σ the element corresponding to X in K, the action of σ on E is defined by $\sigma.x = s(x)$ for all $x \in E$,

Observe also that K possesses an involution $a \mapsto \bar{a}$ determined by $\bar{\sigma} = 1 - \sigma$.

The form S is then lifted to an ε-hermitian, K-valued form

$$(\ , \ ) : E \times E \longrightarrow K$$

on E, where $(x,y) \in K$ is defined by the formula

$$\mathrm{trace}_{K/k} \{ a.(x,y) \} = S(a.x, y)$$

for all $a \in K$.

This construction is due to J. Milnor and it plays a decisive role in the calculation of knot cobordism groups. (See J. Milnor, On isometries of inner product spaces, Inventiones Math. 8 (1969), 83-97).

Thus, if the minimal polynomial f_s of s is irreducible, there is associated with the ε-form $A : E \times E \longrightarrow k$ an ε-hermitian form over $K = k[X]/(f_s)$. Conversely, the above trace formula redefines S and A if a non-singular ε-hermitian form is given on some K-space E.

It turns out that every ε-form $A : E \times E \longrightarrow k$ over a field k is cobordant to a direct sum of ε-forms A_i whose associated endomorphisms $s_i : E_i \times E_i \longrightarrow k$ have irreducible minimal polynomials.

Denoting by $H^\varepsilon(K)$ the Witt group of ε-hermitian forms over the field $K = k[X]/(f)$ with involution induced by $X \longmapsto 1-X$, the result of the calculation of $C^\varepsilon(k)$, due to J. Levine (1969) in a somewhat different formulation, is that

$$C^\varepsilon(k) = \oplus_{f \in P} H^\varepsilon(K_f) ,$$

where P is the set of self-dual irreducible polynomials over k, and $K_f = k[X]/(f)$.

Note that $f = X - \frac{1}{2}$ if and only if $K_f = k$ with trivial involution. In that case

$$H^{\varepsilon}(K_f) = W^{\varepsilon}(k)$$

is the ordinary Witt group of ε-symmetric forms. If $f \neq X - \frac{1}{2}$, then K_f/k has a non-trivial involution. In this case also the Witt group $H^{\varepsilon}(K_f)$ is well known by the work of W. Landherr (Abh. Math. Sem. Hamburg Univ. 11 (1935)p.245). It is not hard to derive a presentation of $H^{\varepsilon}(K_f)$ by generators and relations similar to the one for Witt groups of symmetric forms. (See the book of J. Milnor and D. Husemoller, Symmetric bilinear forms, Springer Verlag, 1974). More precisely, let F be the fixed field of the involution on K. If $a \in F^{\cdot}$, let $< a >$ denote the hermitian form

$$(\, , \,) : K \times K \longrightarrow K \text{ of rank 1 given by } (x,y) = a x\bar{y}$$

Then, there is a surjection $Z[F^{\cdot}] \longrightarrow H(K)$ given by $[a] \longrightarrow < a >$. Here $H(K) = H^{+1}(K)$. The kernel is the ideal of $Z[F^{\cdot}]$ generated by the elements of one of the forms

$$[a] - [a.x.\bar{x}]$$

$$[a] + [-a]$$

$$[a] + [b] - [a + b] - [ab.(a + b)]$$

for $a, b, a + b \in F^{\cdot}$, $x \in K^{\cdot}$.

For $\varepsilon = -1$, observe that if the involution on K is non-trivial then $H^{-1}(K) \cong H^{+1}(K)$ under the map $(\, , \,) \longrightarrow \sqrt{d}.(\, , \,)$, where $K = F(\sqrt{d})$.

The above argument yields in particular

$$C^{\varepsilon}(Q) = \oplus_{f \in P} H^{\varepsilon}(K_f) ,$$

where P is the set of irreducible polynomials which are self-dual, i.e. $f(1-X) = (-1)^{\deg f} f(X)$.

Actually, what is needed for the calculation of $C^{\varepsilon}(Z)$ is the group

$$C_0^{\varepsilon}(Q) = \oplus_{f \in I} H^{\varepsilon}(K_f) ,$$

where I is the set of irreducible <u>integral</u> self-dual polynomials .

N. Stoltzfus (1976)observes that there is an exact sequence

$$0 \longrightarrow C^{\varepsilon}(Z) \longrightarrow C_0^{\varepsilon}(Q) \xrightarrow{\partial} C_Z^{\varepsilon}(Q/Z) \longrightarrow 0$$

where $\partial : C_0^{\varepsilon}(Q) \longrightarrow C_Z^{\varepsilon}(Q/Z)$ is defined as follows.

Let $A : E \times E \longrightarrow Q$ be an ε-form representing some element in $C_0^{\varepsilon}(Q)$. Because $s = S^{-1}A$ has integral characteristic polynomial, there is an integral lattice $L \subset E$ on which S is integral valued and which is invariant by S. Define

$$L' = \{ x \in E \mid S(x,y) \in Z \text{ for all } y \in L \} .$$

Then $L \subset L'$ and $E^* = L' / L$ is a Z-module with a Q/Z -valued form $S^* : E^* \times E^* \to Q/Z$ defined by $S^*(x^*,y^*) = S(x,y) \bmod Z$ for $x,y \in L'$ representing $x^*, y^* \in E^*$.

The dual L' is also invariant by s and therefore there is an additive isometry $s^* : E^* \longrightarrow E^*$ induced by s and satisfying

$$S^*(s^*x,y) = S^*(x,(1-s^*)y)$$

for all $x,y \in E^*$. It is not hard to verify that

$$\text{ad } S^* : E^* \longrightarrow \text{Hom}(E^*, Q/Z)$$

is an isomorphism.

By definition

$$\partial [E,S,s] = [E^*, S^*, s^*] .$$

At this point, one has to calculate $C^{\varepsilon}_Z(Q/Z)$. A localization argument gives first

$$C^{\varepsilon}_{\mathbf{Z}}(\mathbb{Q}/\mathbf{Z}) = \oplus_p \; C^{\varepsilon}(\mathbb{F}_p) ,$$

where p runs over all rational primes. Now, $C^{\varepsilon}(\mathbb{F}_p)$ has been evaluated above since $\mathbb{F}_p$ is a perfect field, and it can be calculated explicitly. We quote Corollary 2.9. of N. Stoltzfus' paper (1976) :

$$C^{\varepsilon}(\mathbb{F}_p) = \oplus \mathbf{Z}/2\mathbf{Z} \oplus W^{\varepsilon}(\mathbb{F}_p) ,$$

where $W^{\varepsilon}(\mathbb{F}_p)$ is the Witt group of the prime field and the first direct sum is taken over all irreducible, self-dual, monic polynomials except $X - \frac{1}{2}$. This is for p an odd prime, if $p = 2$ only the first summand is present.

In the remainder of his paper N. Stoltzfus uses algebraic number theory to make the above results more explicit. We cannot enter into the details here.

CHAP. V : FIBERED KNOTS

§ 1 : General properties .

In this chapter we study the very important special case of fibered knots. At least two reasons make this special case worth of study :

1) Knots which appear as local singularities of complex hypersurfaces are fibered knots.

2) The geometry of the complement of fibered knots can be made quite explicit and thus many knot invariants get a very nice geometrical interpretation.

Let us start with the definition. Racall that $H^1(X;\mathbb{Z}) \approx \mathbb{Z}$ and let t be a chosen generator. One says that K is a fibered knot if one is given a representative $p : X \longrightarrow S^1$ for t, which is a locally trivial (differentiable) fibration.

Remark : It is often nice to add the further restriction that $p|bX \longrightarrow S^1$ (which is, by hypothesis, a fibration) should be the projection onto the fiber associated with a trivialization of the sphere normal bundle to K in S^{n+2}. A useful remark due to S. Cappell shows that whenever $n \neq 2,3$, one can always change p such that this further requirement is satisfied. See Cappell (1969). In the sequel, we shall usually make this assumption.

The fiber of p is a codimension one submanifold W of S^{n+2}. It is connected because p represents a generator of $H^1(X;\mathbb{Z})$. (To see that, consider the end of the homotopy exact sequence of the fibration : $\pi_1(X) \longrightarrow \pi_1(S^1) \longrightarrow \pi_0(F) \longrightarrow 0$.)

If we add to W a collar inside the normal bundle to K in S^{n+2}, we get a Seifert surface V for K.

Choose now a point $1 \in S^1$ and remove a small open interval I centered in 1. Call J the big closed interval that remains. $p^{-1}(I)$ is a trivialised open neighborhood of W in X. So, $p^{-1}(J)$ is what we called Y in Chap. III § 1 . But, as J is contractible, $p^{-1}| \ Y \longrightarrow J$ is a trivial fibration. So, Y is homeomorphic to $W \times J$.

Looking at things a bit differently, we see that we can think of X as being obtained from $W \times [0,1]$ by $W \times \{0\}$ and $W \times \{1\}$ identified together via a homeomorphism $h : W \to W$. More precisely, X is the quotient of $W \times [0,1]$ by the equivalence relation $(x,0) \sim (h(x),1)$.

h is called "the" monodromy of the fibration. p being given, h is well defined up to isotopy. If we insist that p satisfies the restriction condition on bX, we shall get a monodromy map which is the identity on bW.

§ 2. The infinite cyclic covering of a fibered knot.

Let us consider the product $W \times \mathbb{R}$ and the equivalence relation $(x,a) \sim (h^j(x), a+j)$ for any $j \in \mathbb{Z}$. It is immediate to verify that the quotient space is homeomorphic to X. Moreover, the quotient map $W \times \mathbb{R} \longrightarrow X$ is a regular covering map, whose Galois group is C. So this is the infinite cyclic covering of X. We deduce from that :

1) $\widetilde{X}$ has the homotopy type of W, which is a compact C.W. complex.

2) The generator t of the Galois group C acts by the map $(x,a) \longmapsto (h(x), a+1)$. So t acts on $H_*(\widetilde{X})$ as h acts on $H_*(W)$.

As before, let us denote by $F_k(\widetilde{X})$ the torsion-free quotient of $H_k(\widetilde{X};\mathbb{Z})$. By 1), $F_k(\widetilde{X})$ is a finitely generated free abelian group ; and it is also a $\mathbb{Z}C$-module. Under these circumstances,a theorem of algebra says that a generator λ of the first elementary ideal of the $\mathbb{Z}C$-module $F_k(\widetilde{X})$ is just the characteristic polynomial of t . Moreover, it is not hard to see that λ is just the Alexander polynomial Δ_1 of $H_k(\widetilde{X};\mathbb{Z})$. (The lazy reader can look at Weber's paper in this book). Recalling that t acts like h_k we get the folklore theorem :

When a knot fibers, the Alexander polynomial of $H_k(\widetilde{X};\mathbb{Z})$is just the characteristic polynomial of the monodromy h_k acting on $F_k(W)$,

As it is a characteristic polynomial, its leading coefficient is +1 ; as h_k is an isomorphism on the finitely generated free abelian group $F_k(\widetilde{X})$ its last coefficient is ± 1 .

Remark : A simplified version of the above argument gives the following : Let $\mathbb{F}$ be a field. Then the order of the $\mathbb{F}C$-module $H_k(\widetilde{X};\mathbb{F})$ is just the characteristic polynomial of the automorphism

$h_k : H_k(W;\mathbb{F}) \longrightarrow H_k(W;\mathbb{F})$. Cf Milnor (1968a)and (1968b).

Using some more algebra, it is not hard to see that the minimal polynomial of the action of h_k on $F_k(W)$ is Δ_1/Δ_2 , Δ_i being the g.c.d. of the ith elementary ideal of $H_k(\tilde{X}; \mathbf{Z})$.

See R. Crowell : "The annihilator of a knot module" Proceedings AMS 15(1964) p. 696-700. This fact is much used by people working on singularities. For instances see N.A'Campo (1972a).

Let us close this paragraph by mentioning that fibered knots give a nice interpretation of the pairing of torsion submodules mentioned in Chap. II §3 : it is the linking pairing induced on the fiber by Poincaré duality (See J. Levine (1974) § 7) .

§ 3 : When does a knot fiber ?

We saw in this chapter § 1 that a knot fibers if and only if one can find a Seifert surface V such that Y is homeomorphic to $W \times [0,1]$. One can choose a homeomorphism which is the "identity" from W_+ to $W \times \{0\}$. The homeomorphism we get from W_- to $W \times \{1\}$ is just h .

Moreover i_+ and i_- are homotopy equivalences. So, $(i_+)_k$ and $(i_-)_k : H_k(W) \longrightarrow H_k(Y)$ are isomorphisms and :

$$h_k = (i_+)_k^{-1} \cdot (i_-)_k \quad \text{for all } k .$$

If the fibered knot is (2m-1)-dimensional, the Seifert pairing $A : F_m(W) \times F_m(W) \longrightarrow \mathbb{Z}$ associated with the fiber W is unimodular, because $(i_+)_m$ is an isomorphism.

Suppose now that, for a given (2m-1)-knot, we can find an (m-1)-connected Seifert surface W such that its Seifert pairing is unimodular. Then, if $m \geq 3$, by the h-cobordism theorem Y is homeomorphic to the product $W \times [0,1]$ and so the knot fibers. Moreover, using notations introduced in chap. III § 1 and § 3, the matrix for h_m is given by $(-1)^{m+1} A^{-1} \cdot A^T$.

For classical knots ($m = 1$), the unimodularity of a Seifert matrix is necessary for a knot to fiber, but it is not sufficient. See R. Crowell and D. Trotter (1962). The correct condition, due to L. Neuwirth and J. Stallings is that one should find a Seifert surface such that i_+ and i_- induce isomorphisms on the fundamental group.

It is harder to get useful fibration theorems for non-simple knots.

However, we saw in § 2 that a necessary condition for a knot to fiber is that the extremal coefficients of the Alexander polynomial for $H_k(\tilde{X};Z)$ should be $\pm$ 1 for all $k \geq 1$. A theorem due to D.W. Sumners says that the converse is true if $\pi_1(X) = \mathbb{Z}$ and $n \geq 4$. See Sumners (1971).

If one spins a fibered knot, one gets again a fibered knot. This fact has been used by J.J. Andrews and D.W. Sumners (1969).

§ 4. Twist-spinning.

An important and striking way to construct a fibered knot is E.C. Zeeman's twist-spinning.

We give a sketched description of the twist-spinning construction and for more details, we refer the reader to Zeeman's original paper (1963), where the geometry of the construction is beautifully described.

Look at the unit closed ball E^{n+2} as being the product $E^n \times E^2$. In E^2 use polar coordinates, (ρ,Φ) being mapped onto $\rho e^{2i\pi\Phi}$, $0 \leqslant \rho \leqslant 1, 0 \leqslant \Phi \leqslant 1$. So, a point in E^{n+2} will be described by a triple (x,ρ,Φ). Also, S^1 is the unit circle in E^2, with angular coordinate θ , $0 \leqslant \theta \leqslant 1$.

Suppose now that we have a subspace $A \subset E^{n+2}$. Let $r \in \mathbf{Z}$ be given. The full r twist of A is the subspace $A_r \subset E^{n+2} \times S^1$ consisting of the quadruples : $(x,\rho,\Phi + r\theta,\theta)$ for all $(x,\rho,\Phi) \in A$, $\theta \in [0,1]$. It is obvious that A_r is abstractly homeomorphic to $A \times S^1$.

Now, let an n-knot $K \subset S^{n+2}$ be given. Choose a small open (n+2)-disc neighborhood of a point belonging to K such that :

1) The intersection of the disc with K is an open n-disc.
2) The small disc pair thus obtained is standard.

Let us take the complementary pair (D^{n+2}, B). Identify D^{n+2} with E^{n+2}. $r \in \mathbf{Z}$ being given, look at the pair $(D^{n+2} \times S^1, B_r)$. On the boundary it is the standard $(S^{n+1} \times S^1, S^{n-1} \times S^1)$, because via the identification bB goes to $bE^n \times \{0\}$. Glue along the boundary the standard $(S^{n+1} \times D^2, S^{n-1} \times D^2)$ and you get an (n+1)-knot; because, abstractly for any $k \geqslant 0$ $(D^k \times S^1)\amalg(S^{k-1} \times D^2)$ glued along $S^{k-1} \times S^1$ yields S^{k+1}.

This is the r-twist spinning of the original knot.

If we are careful that the subball $B \subset D^{n+2}$ is standard near the boundary, the twist-spun knot will be differentiable, if we started with a differentiable knot.

It is not hard to see that, for $r = 0$, the construction is essentially Artin's spinning. Changing θ into $-\theta$ changes the r-twist spun into the $(-r)$ one, so the construction really depends on $|r|$. The properties of the twist-spinning operation are given by :

Zeeman's theorem : Suppose $r \neq 0$. Then :

1) The exterior of the r-twist spun knot fibers on S^1, in the sense of § 1.

2) The fiber W is the r-th cyclic branched covering of the original knot, minus an open (n+2)-disc.

3) Let f be a correctly chosen generator of the Galois action on the unbranched r-th cyclic covering of the exterior of the original knot. f extends to an automorphism $\overline{f}$ of the branched cyclic covering (the knot being fixed) and $\overline{f}$ restricts to an automorphism h of the punctered branched cyclic covering W. h is of order r and can be taken as the monodromy of the fibration. Beware : h is not quite the identity on bW.

4) There is an action of S^1 on S^{n+3} leaving the twist-spun knot invariant, and acting freely outside the knot. But the action on the knot is not the identity.

Comments :

a) Because of point 4), one is very close to counter-examples to the Smith conjecture, for multidimensional knots. Soon after Zeeman's paper, C.H. Giffen (1964) was able to produce such counter-examples, by using as a start the twist-spinning operation. Several other counter-examples are now known (all for non-classical knots !) .

b) By § 2, the infinite cyclic covering of the r-twist spun knot has the homotopy type of the punctered branched r-cyclic covering of the original knot, and the monodromy is "known". So, to compute the invariants of the new knot, one can use classical procedures about branched cyclic coverings.

c) A generalization of Artin's result about π_1 of a spun knot shows that π_1 of the r-twist spun knot is obtained from π_1 of the original knot by adding the relations saying that the r-th power of the meridian commutes with everybody. J. Levine has a very useful way to look at the twist-spinning contruction which yields this result very nicely. (Unpublished).

d) As the 1-branched cyclic covering of an n-knot is the (n+1)-sphere, 1-twist spun knots bound a disc and are thus trivial.

§ 5. Isolated singularities of complex hypersurfaces.

Let $f : \mathbb{C}^{m+1} \longrightarrow \mathbb{C}$ be a $\mathbb{C}$-polynomial map, such that $f(0) = 0$ and that $0 \in \mathbb{C}^{m+1}$ is an isolated singularity of f. (This means that the $\mathbb{C}$-gradient of f does not vanish in a neighborhood of 0 except at 0). J. Milnor (1968b) shows :

1) The intersection K of the hypersurface $f^{-1}(0) = H$ with sufficiently small spheres S_ε^{2m+1} in $\mathbb{C}^{m+1}$, centered in 0, is transversal. Thus K is a(real) codimension two submanifold of S_ε^{2m+1}, but not necessarily a sphere.

2) The exterior of K in S^{2m+1} fibers in the strong sense, i.e. the restriction of the fibration to bX is the projection onto S^1 associated to a trivialization of the normal bundle of K in S^{2m+1}.

3) The fiber W has the homotopy type of a wedge of m-dimensional spheres.

If we look at the homology exact sequence of the pair V mod.K, we see that K is not too far from being a homology sphere. Its only (possibly) non-vanishing homology groups are in dimensions (m-1) and m .Their vanishing depends on the intersection pairing on $H_m(V) = H_m(W)$. Moreover one can prove that if $m \geq 3$, K is simply-connected .

It is clear that there is a Seifert pairing for W, and that, if we agree to call "knots" submanifolds such as K, we have got an odd dimensional, fibered, simple knot. One can check that Levine's S-equivalence theory works in that case also. So, from a topological point of view, the situation is rather well understood. See A.F. Durfee (1973) for detail.

Remarks :

a) It is known (see Milnor's book) that locally around 0, the pair (C^{m+1},H) is homeomorphic to the cone on the pair (S^{2m+1},K). Thus, topologically, the singularity is determined by the knot.

b) If $f : U \longrightarrow \mathbb{C}$ is a holomorphic map with $f(0) = 0$, (U open neighborhood of 0 in $\mathbb{C}^{m+1}$) and with 0 as isolated singularity, then the germ of f at 0 is analytically equivalent to a polynomial. For a very detailed discussion of this kind of results see : "Remarks on finitely determined analytic germs" by J. Bochnak and S.Lojasiewicz in Springer Lecture Notes vol. 192 (1970) p. 262-270. So one can apply the theory also to holomorphic germs.

c) Define two holomorphic germs $f_i : U_i \longrightarrow C$, $i = 1,2$, U_i open neighborhood of 0 in $\mathbb{C}^{m+1}$, $f_i(0) = 0$, to be topologically equivalent if there exist a germ Φ of homeomorphism at $0 \in \mathbb{C}^{m+1}$, $\Phi(0)= 0$ and a germ φ of homeomorphism at $0 \in \mathbb{C}$, $\varphi(0) = 0$ such that :

$$f_2 = \varphi \circ f_1 \circ \Phi \quad \text{in a suitable neighborhood of } 0 \in C^{m+1}.$$

A recent theorem of H.C. King (1977) says that, if $m \neq 2$, two holomorphic germs with isolated singularities at 0 are topologically equivalent if and only if the knots they determine are isotopic. So, roughly speaking, the knot determines the topological type of the germ f, a result much stronger than the classical one stated in a) above.

The main question in the topological study of isolated singularities of complex hypersurfaces is to relate the topological invariants coming from knot theory and the invariants coming from algebraic geometry. (Here, when we say "topological" we mean as well "differential" as opposed to "algebraic" or "analytic").

For instance, the differential structure on K is an interesting invariant. More precisely :

1) One would like to compute the knot invariants from the algebraic data. Historically the whole story began (after O. Zariski's work in the thirties) when F. Pham (1965) and subsequently E. Brieskorn (1966) studied the sigularities :

$$f(z_o, z_1, \ldots, z_k) = (z_o)^{i_o} + (z_1)^{i_1} + \ldots + (z_k)^{i_k} \quad .$$

In that case, computations can be done. For other results, see P. Orlik and J. Milnor (1969).

2) One would also like to know when a given knot is obtained from a singularity. This can be first attacked by trying to determine which restrictions are imposed on the knot invariants when it is "algebraic", besides the fact that it is a fibered knot. Striking examples of such restrictions are :

a) The monodromy theorem, which says that the roots of the Alexander polynomial of $H_m(\tilde{X};\mathbb{Z})$ (which is the characteristic polynomial of h_m) are all roots of unity. See E. Brieskorn(1969).

b) There exists a basis for $H_m(W;\mathbb{Z})$such that the Seifert matrix is triangular. See A.F. Durfee (1973).

c) The trace of h_m is equal to $(-1)^{m+1}$. See N.A'Campo (1972b) and, more generally, N. A'Campo(1974).

All these are very deep results about singularities.

Since the beginning of the theory, a lot of work has been spent to get nice geometrical descriptions of some singularities. For recent results, look at L.H.Kauffman (1973) and also at L.H. Kauffman and W.D. Neumann(1976).

For a more detailed exposition and more references about the whole subject, the reader should see J. Milnor's book (1968b), A.H. Durfee (1975), M. Demazure (1974).

Historical remark : The theory of isolated singularities began in the late twenties by the study of singularities of complex plane curves, approximately at the same time as knot theory really started (exception being made for M. Dehn's papers). In fact, progresses were made in knot theory to understand O. Zariski's results about curves and conversely, algebraic geometers found beautiful applications of J.W. Alexander and K. Reidemeister's work. It is amusing to note that a remark (due to W. Wirtinger) about the singularity $z_1^2 + z_2^3 = 0$ being locally the cone on the trefoil knot appears already in E. Artin's paper (1925). This permits us to close this paper at the point where we started it.

B I B L I O G R A P H Y

This bibliography contains the articles on knot theory quoted in this paper. The articles indirectly pertaining to the subject, or of a more technical nature have been mentioned in the text and will not be listed here again.

The reader should be aware that this bibliography is incomplete.

The articles are dated either according to the year in which the work was done, or in case of insufficient information, according to the year of submission to the journal. The publication year appears in the reference.

1925 :

E. ARTIN : "Zur Isotopie zweidimensionaler Flächen in R^4". Abh.Math. Seminar Univ. Hamburg 4 (1925) p. 174-177.

1928 :

E. R. VAN KAMPEN : "Zur Isotopie zweidimensionaler Flächen in R^4". Abh. Math. Seminar Univ. Hamburg 6 (1928) p. 216.

1934 :

H. SEIFERT : "Ueber das Geschlecht von Knoten". Math. Annalen 110 (1934) p. 571-592.

1954 :

R.C. BLANCHFIELD : "Intersection theory of manifolds with operators, with applications to knot theory". Annals of Math. 65 (1957) p. 340-356.

R.H. FOX and J. MILNOR : "Singularities of 2-spheres in 4-spaces and equivalence of knots". Bulletin AMS 63 (1957) p. 406 and Osaka J.Math. 3 (1966) p. 257-267.

1959 :

J.J. ANDREWS and M.L. CURTIS : "Knotted 2-spheres in 4-spaces". Annals of Math. 70 (1959) p. 565-571.

D. EPSTEIN : "Linking spheres". Proc. Camb. Phil. Soc (1960) p.215-219.

L. NEUWIRTH : "The algebraic determination of the genus of a knot" Amer. Jour. Math. 82 (1960) p. 791-798.

E.C. ZEEMAN : "Linking spheres". Abh. Math. Seminar Univ. Hamburg 24 (1960) p. 149-153.

1960 :

S. KINOSHITA : "On the Alexander polynomials of 2-spheres in 4-spaces". Annals of Math. 74 (1961) p. 518-531.

H. TROTTER : "Homology of group systems with applications to knot theory". Annals of Math. 76 (1962) p. 464-498.

1961 :

a) R.H. FOX : "A quick trip through knot theory". Topology of 3-manifolds (M.K. Fort Jr Editor). Prentice Hall (1962) p. 120-167.

b) R.H. FOX : "Some problems in knot theory". Topology of 3-manifolds (M.K. Fort Jr Editor). Prentice Hall (1962) p. 168-176.

H. GLUCK : " The embeddings of the two-sphere in the four-sphere". Transactions AMS 104 (1962) p. 308-333.

1962 :

R.H. CROWELL and H. TROTTER : "A class of pretzel knots". Duke Math. Jour. 30 (1963) p. 373-377.

J. STALLINGS : "On topologically unknotted spheres". Annals of Math. 77 (1963) p. 490-503.

1963 :

M. KERVAIRE : "On higher dimensional knots". Differential and combinatorial topology. (S. Cairn edit). Princeton Univ. Press (1965) p. 105-120.

K. MURASUGI : "On a certain invariant of link types". Transactions AMS 117 (1965) p. 387-422.

H. TROTTER : "Non invertible knots exist". Topology 2 (1964), p.275-280.

E.C. ZEEMAN : "Twisting spun knots". Transactions AMS 115 (1965) p. 471-495.

1964 :

C.H. GIFFEN : "The generalized Smith conjecture". Amer.J.Math.88 (1966) p. 187-198.

M.W. HIRSCH and L. NEUWIRTH : "On piecewise regular n-knots". Annals of Math. 80 (1964) p. 594-612.

M. KERVAIRE : "Les noeuds de dimension supérieure". Bull. Soc. Math. de France 93 (1965) p. 225-271.

J. LEVINE : "Unknotting spheres in codimension two". Topology 4 (1965) p. 9-16.

C.T.C. WALL : "Unknotting tori in codimension one and spheres in codimension two". Proc. Camb. Phil. Soc. 61 (1965) p. 659-664.

<u>1965</u> :

F. PHAM : "Formules de Picard-Lefschetz généralisées et ramification des intégrales". Bull. Soc. Math. de France 93 (1965) p. 333-367.

<u>1966</u> :

E. BRIESKORN : "Beispiele zur Differentialtopologie von Singularitäten". Invent. Math. 2 (1966) p. 1-14.

W. BROWDER : "Diffeomorphisms of one-connected manifolds". Transactions AMS 128 (1967) p. 155-163.

J. LEVINE : "Polynomials invariants of knots of codimension two". Annals of Math. 84 (1966) p. 537-554.

<u>1967</u> :

J.L. SHANESON : "Embeddings with codimension two of spheres in spheres and h-cobordisms of $S^1 \times S^3$". Bulletin AMS 74 (1968) p. 467-471.

<u>1968</u>:

R.K. LASHOF and J.L. SHANESON : "Classification of knots in codimension two". Bulletin AMS 75 (1969) p. 171-175.

J. LEVINE : "Knot cobordism in codimension two". Comment. Math. Helv. 44 (1969) p. 229-244.

S.J. LOMONACO Jr : "The second homotopy group of a spun knot". Topology 8 (1969) p. 95-98 .

a) J. MILNOR : "Infinite cyclic coverings". Conference on the Topology of Manifolds. Prindle, Weber and Schmidt. (1968).

b) J. MILNOR : "Singular points of complex hypersurfaces". Annals of Math. Studies vol. 61 (1968).

1969 :

J.J. ANDREWS and D.W. SUMNERS : "On higher-dimensional fibered knots". Transactions AMS 153 (1971) p. 415-426.

E. BRIESKORN : "Die Monodromie der isolierten Singularitäten von Hyperflächen". Manuscripta Math. 2 (1970) p.103-161.

S. CAPPELL : "Superspinning and knot complements". Topology of manifolds. Georgia (1969). (J.C. Cantrell and C.H. Edwards edit.). Markham Publ. Comp. p. 358-383.

J. LEVINE : "An algebraic classification of some knots of codimension two". Comment. Math. Helv. 45 (1970) p. 185-198.

J. MILNOR and P. ORLIK : "Isolated singularities defines by weighted homogeneous polynomials". Topology 9 (1970) p. 385-393.

1970 :

M. KERVAIRE : "Knot cobordism in codimension two". Springer Lecture Notes vol. 197 (1970) p. 83-105.

H. TROTTER : "On the algebraic classification of Seifert matrices". Proceedings of the Georgia Topology Conference, Athens (1970) p. 92-103.

D.W. SUMNERS : "Polynomials invariants and the integral homology of coverings of knots and links". Invent. Math. 15 (1972) p. 78-90.

1972 :

G. BREDON : Regular O(n)-manifolds, suspensions of knots and knot periodicity, 79 (1973), p. 87-91.

N.A'CAMPO : "Sur la monodromie des singularités isolées d'hypersurfaces complexes". Invent. Math. 20 (1973) p. 147-169.

N.A'CAMPO : "Le nombre de Lefschetz d'une monodromie". Indagationes Math. 35 (1973) p. 113-118.

S. CAPPELL and J. SHANESON: "Submanifolds, group actions and knots I". Bulletin AMS 78 (1972), p. 1045-1048.

E. DYER and A.T. VASQUEZ :"The sphericity of higher dimensional knots". Can. J. Math. 25 (1973) p. 1132-1136.

C. KEARTON : "Non-invertible knots in codimension two". Proceedings AMS 40 (1973) p. 274-276.

H. TROTTER : "On S-equivalence of Seifert matrices ". Invent. Math. 20 (1973) p. 173-207.

1973 :

A.H. DURFEE : "Fibered knots and algebraic singularities". Topology 13 (1974) p. 47-59.

L. H. KAUFFMAN : "Branched cyclic coverings, open books and knot periodicity". Topology 13 (1974) p. 143-160.

1974 :

N. A'CAMPO : "La fonction zeta d'une monodromie". Comment. Math. Helv. 50 (1975) p. 233-248.

M. DEMAZURE : "Classification des germes à points critiques isolés et à nombre de modules 0 ou 1". Séminaire Bourbaki, Février 1974, exposé 443.

M. FARBER : "Linking coefficients and two-dimensional knots". Soviet Math. Dokl. 16 (1975) p. 647-650.

C. KEARTON : "Blanchfield duality and simple knots". Transactions AMS 202 (1975) p. 141-160.

J. LEVINE : "Knot modules I". Transactions AMS 229 (1977) p. 1-50.

1975:

S. CAPPELL and J.L. SHANESON : "There exist inequivalent knots with the same complement". Annals of Math. 103 (1976) p. 349-353.

A.H. DURFEE : "Knot invariants of singularities". Proc. of Symp. in Pure Math. AMS vol. 29 (1975) p. 441-448.

B. ECKMANN : "Aspherical manifolds and higher dimensional knots". Comm. Math. Helv. 51 (1976) p. 93-98.

C. McA. GORDON : "Knots in the 4-sphere". Comment. Math. Helv. 39 (1977) p. 585-596.

C. KEARTON : "An algebraic classification of some even-dimensional knots". Topology 15 (1976) p. 363-373.

1976 :

L.H. KAUFFMAN and D.W. NEUMANN : "Products of knots, branched fibrations and sums of singularities". Preprint (1976) 92 p.

N. STOLTZFUS : "Unraveling the integral knot concordance group" . Memoirs AMS vol. 192 (1977), 91 p.

1977 :

W.C. KING : "Topological type of isolated singularities". Preprint (1977) 22 p.

S. KOJIMA : "A classification of some even dimensional fibered knots". Preprint (1977) 23 p.

A. RANICKI : "The algebraic theory of surgery". Preprint (1977). 322 pages.

A Linking Invariant of Classical Link Concordance

by Deborah L. Goldsmith

Introduction

This paper describes a new invariant of link concordance for classical links in a 3-manifold. Let $L = K_1 \cup \ldots \cup K_n \subset M^3$ be an oriented link with components K_i, $1 \leq i \leq n$, and $n > 1$. Let $\pi : \tilde{M} \to M$ be a branched (or unbranched) covering space, whose branch set is a sublink of L. The invariant is a matrix $\Lambda_{\tilde{M}}(L)$ whose entries record the linking numbers $\ell_{\tilde{M}}(\tilde{K}_i, \tilde{K}_j)$ in $\tilde{M}$ of each pair of distinct components $\tilde{K}_i \subset \pi^{-1}(K_i)$, $\tilde{K}_j \subset \pi^{-1}(K_j)$, lying over L.

Here are the main theorems:

Theorem 4.16. Let $L \cup K_1 \cup \ldots \cup K_n \subset M$, $L' \cup K_1' \cup \ldots \cup K_n' \subset M'$ be $\mathbb{Z}$-concordant links in the $\mathbb{Z}$-homology 3-spheres M, M', respectively. Let $L \subset M$, $L' \subset M'$ be sublinks, and suppose that each K_i (K_i', respectively) is homologically split from L (L', respectively), $i = 1, \ldots, n$. Let $\tilde{M} \to M - L$, $\tilde{M}' \to M' - L'$ be the universal abelian covering spaces.

Then $\Lambda_{\tilde{M}}(K_1, \ldots, K_n) \equiv \Lambda_{\tilde{M}'}(K_1', \ldots, K_n')$.

Theorem 4.19. Let $L = K_1 \cup \ldots \cup K_n \subset M$, $L' = K_1' \cup \ldots \cup K_n' \subset M'$ be $\mathbb{Z}_p$-concordant links. Let $\tilde{M} \to M$ be a regular, iterated p-cyclic branched (unbranched) covering space, branched along a sublink of L. Suppose $\tilde{K}_i \subset \tilde{M}$, $i = 1, \ldots, n$, represent torsion cycles in $H_1(\tilde{M})$.

Then there is a unique related covering space $\tilde{M}' \to M'$, which depends on the concordance, and $\Lambda_{\tilde{M}}(L) \equiv \Lambda_{\tilde{M}'}(L')$.

Theorem 2.24. Let $L \cup K_1 \cup \ldots \cup K_n \subset M$ be a link in the $\mathbb{Z}_p$-homology 3-sphere M, where p is a prime, $p \neq 1$, and let $L \subset M$ be a sublink. Suppose $\tilde{M} \to M - L$ is an iterated p-cyclic, irregular covering space.

If $L \cup K_1 \cup \ldots \cup K_n \subset M$ is $\mathbb{Z}_p$-concordant to a link $L' \cup K_1' \cup \ldots \cup K_n' \subset M'$, such that $K_1' \cup \ldots \cup K_n' \subset M' - L'$ is a split link, then $\Lambda_{\tilde{M}}(K_1, \ldots, K_n) = 0$.

To illustrate, Theorem 4.16 could be applied as follows: the Whitehead link

K_1 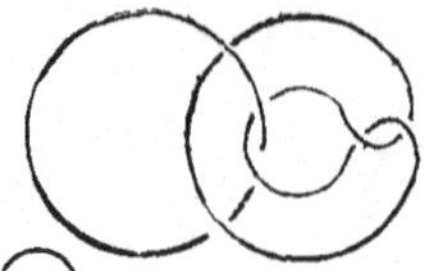 K_2 is not $\mathbb{Z}$-concordant to the trivial link  in S^3, since in the infinite cyclic covering space $\tilde{M} \to S^3 - K_1$, the lifts $\ldots \sigma^{-1}\tilde{K}_2$, $\tilde{K}_2$, $\sigma\tilde{K}_2$, $\ldots$ (where σ generates the cyclic group of covering translations of $\tilde{M}$) are linked:

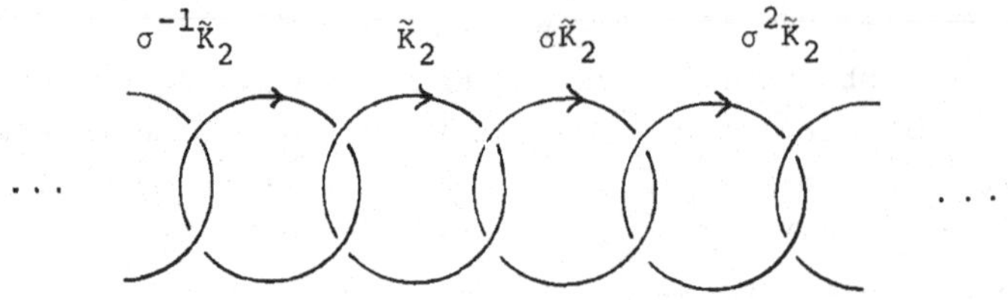

Concordance of links is usually defined thus:

<u>Definition</u>. Oriented links $L = K_1 \cup \ldots \cup K_n, L' = K_1' \cup \ldots \cup K_n' \subset M^3$ in an oriented 3-manifold M^3 are said to be <u>concordant</u> when there exist disjoint, locally flat, properly embedded annuli $F = F_1 \cup \ldots \cup F_n \subset M^3 \times I$ such that, for all $1 \leq i \leq n$, $\partial F_i = F_i \cap (\partial[M \times I]) = (L' \times 1) - (L \times 0)$. In this paper, I consider a weaker equivalence relation than concordance, called R-concordance:

<u>Definition</u>. Let R be a ring, and let $L = K_1 \cup \ldots \cup K_n \subset M$, $L' = K_1' \cup \ldots \cup K_n' \subset M'$ be oriented links in the compact, oriented 3-manifolds M, M', respectively. Then <u>$L \subset M$ is said to be R-concordant to $L' \subset M'$</u>, if there is a compact, oriented 4-manifold N^4 whose boundary is

$$\partial N^4 = (M' - M) \cup_{\partial M \times \{0,\ 1\}} (-[\partial M] \times I) ,$$

and disjoint, locally flat embedded annuli $F = F_1 \cup \ldots \cup F_n \subset N$ with boundary $\partial F_i = K_i' \cup (-K_i) \subset M' \cup (-M)$, such that for all $j \geq 0$, the maps

$$i_* : H_j(M - L;\ R) \to H_j(N - F;\ R)$$
$$i_*' : H_j(M' - L';\ R) \to H_j(N - F;\ R)$$

induced by inclusion are isomorphisms.

The rings R which are used in this paper are the integers $\mathbb{Z}$, the field $\mathbb{Z}_p = \mathbb{Z}/p\mathbb{Z}$, where p is a prime, and the rational numbers $\mathbb{Q}$. The algebraic equivalence relation of R-concordance is in general weaker than that of concordance; for if a pair of links in a $\mathbb{Z}$-homology

3-sphere M^3 are concordant, then they are R-concordant for every ring R .

The paper is divided into six sections:

§1. Covering Space Preliminaries
§2. Homology of Covering Spaces
§3. Related Covering Spaces
§4. A Linking Invariant of Concordance
§5. Computations and Examples
§6. Some Concluding Remarks

In §1, the different types of covering spaces are defined. In §2, the homological theorems are proved, which are used in §4 to show that $\Lambda_{\tilde{M}}(L)$ is an invariant of R-concordance. In §3, the idea of related (branched) covering spaces $\tilde{M} \to M$ and $\tilde{M}' \to M'$ is developed, in order to later compare the invariants $\Lambda_{\tilde{M}}(L)$, $\Lambda_{\tilde{M}'}(L')$ for R-concordant links $L \subset M$, $L' \subset M'$. In §4 the invariant $\Lambda_{\tilde{M}}(L)$ is defined, and an algorithm for calculating it is described in case $\tilde{M} \to M$ is an (unbranched) covering space. The last two sections are self-explanatory.

All spaces and maps are assumed to be in the P.L. category. Manifolds are assumed to be oriented, and submanifolds to be locally flat. All spaces and covering spaces are assumed to have basepoints (which project to basepoints under covering projections), to insure unique lifting of maps. The terminology developed to describe covering spaces will also be used to describe the associated branched covering spaces, without further explanation. (A branched covering space is a covering space outside of a set of points called the branch set. This covering space is the associated covering space to the branched covering space; the branched covering space is called the (unique) completion of its associated covering space (see [Fl]).) For example, a branched covering space is iterated cyclic if its associated, unbranched covering space is iterated cyclic; two branched covering spaces are related, if they, and their associated, unbranched covering spaces are related.

I would like to thank John Milnor, Eric Pederson and Lawrence Taylor for valuable conversations.

§1. Covering Space Preliminaries

All covering spaces are assumed to be connected, and to have a basepoint. Each connected, basepointed covering space $\pi : (\tilde{X}; \tilde{e}) \to (X; e)$ corresponds to a subgroup of $\pi_1(X; e)$ (see [Gr]); namely, it corresponds to the stabilizer S of $\tilde{e}$ under the action of $\pi_1(X; e)$ on the fiber $\pi^{-1}(e)$.

This action may be described as follows: let $g \in \pi_1(X; e)$ be represented by the e-based loop γ in X, let $p \in \pi^{-1}(e)$ be a point in the fiber over e, and let $\tilde{\gamma}_p$ denote the lift of γ from p. Define $p \cdot g = \tilde{\gamma}_p(1)$. The action of $\pi_1(X; e)$ on the fiber is right multiplication.

Here is an algebraic description of this action: the set $\pi^{-1}(e)$ corresponds to the set of right cosets of $\pi_1(X; e)$ by S (the correspondence is $Sg \leftrightarrow \tilde{e} \cdot g$), and the action of $\pi_1(X; e)$ is right multiplication.

<u>Proposition 1.1.</u> Let $\pi : (\tilde{X}; \tilde{e}) \to (X; e)$ be a covering space, and let $C \subseteq X$ be a path-connected subspace of X. Suppose the image $i_*(\pi_1(C)) \triangleleft \pi_1(X)$ of $\pi_1(C)$ under the inclusion homomorphism is a normal subgroup, modulo left multiplication by the stabilizer S.

For each path δ from e to C, there is an action of the fundamental group $\pi_1(X; e)$ on the components of $\pi^{-1}(C)$.

<u>Proof</u>: Let $\tilde{C} \subseteq \tilde{X}$ be the unique path component of $\pi^{-1}(C)$ such that the lift $\tilde{\delta}_{\tilde{e}}$ of the path δ from the basepoint, joins $\tilde{e}$ to $\tilde{C}$. For each $g \in \pi_1(X; e)$, denote by $\tilde{C} \cdot g$ the path component of $\pi^{-1}(C)$ which is joined to $\tilde{e} \cdot g$ by the lifted path $\tilde{\delta}_{\tilde{e} \cdot g}$.

Note that the map $g \to \tilde{C} \cdot g$ from $\pi_1(X; e)$ to the set of components of $\pi^{-1}(C)$ is unchanged by left multiplication by S (the stabilizer of $\tilde{e}$) and by right multiplication by H (the image $H = i_*(\pi_1(C; e))$ of the fundamental group $\pi_1(C; e)$ based at e by means of the path δ, under the inclusion homomorphism $i_* : \pi_1(C; e) \to \pi_1(X; e)$). The path components of $\pi^{-1}(C)$ are in one-to-one correspondence with the set of double cosets:

$$S \cdot g \cdot H \longleftrightarrow \tilde{C} \cdot g, \; g \in \pi_1(X; e) .$$

Since $H \triangleleft \pi_1(X; e)$ is normal, modulo left multiplication by S, the group $\pi_1(X; e)$ acts on the double cosets $\{S \cdot g \cdot H;\ g \in \pi_1(X; e)\}$, and therefore on the set of path components of $\pi^{-1}(C)$, by right multiplication.//

The covering space $\pi : (\tilde{X}; \tilde{e}) \to (X; e)$ is <u>regular</u> if the stabilizer S of e is normal. Then the quotient group $\pi_1(X; e)/S$ is naturally isomorphic to the group of covering translations of $\tilde{X}$, by the correspondence

$$[g] \longleftrightarrow \sigma_g, \quad g \in \pi_1(X; e) ,$$

where $\sigma_g : \tilde{X} \to \tilde{X}$ is the unique covering translation such that $\sigma_g(e) = \tilde{e} \cdot g$. Interestingly enough, the action of the group of covering translations on the fiber $\pi^{-1}(e)$, is induced by <u>left</u> multiplication of the set of right cosets $\{S \cdot g : g \in \pi_1(X; e)\}$.

<u>Proposition 1.2</u>. Let $\pi : (\tilde{X}; \tilde{e}) \to (X; e)$ be a regular covering space, and let $C \subseteq X$ be a path connected subspace of X.

The group of covering translations of $\tilde{X}$ acts on the components of $\pi^{-1}(C)$.

<u>Proof</u>: Geometrically, Proposition 1.2 is obvious. Algebraically, the action of the quotient group $\pi_1(X; e)/S$ on the set of double cosets $\{S \cdot g \cdot H; g \in \pi_1(X; e)\}$ is induced by left multiplication by $\pi_1(X; e)$. This action is independent of the choice of a path δ from e to C, and is, in general, different from the action of $\pi_1(X; e)$ described in Proposition 1.1.//

In general, a regular covering space corresponds to the kernel, S, of a representation (surjective homomorphism) $\pi_1(X; e) \twoheadrightarrow \Pi$ onto a group Π. This representation induces an isomorphism of Π with the group of covering translations of $\tilde{X}$, since it induces an isomorphism

$$\pi_1(X; e)/S \xrightarrow{\cong} \Pi$$

between the quotient group $\pi_1(X; e)/S$ (which, by a previous discussion, is naturally isomorphic to the group of covering translations) and Π.

If Π is abelian (free abelian), then $\pi : (\tilde{X}; \tilde{e}) \to (X; e)$ is said to be an <u>abelian (free abelian) covering space</u>. The <u>universal abelian</u>, or <u>homology covering space</u>, is the one whose stabilizer S is the kernel of the Hurewicz homomorphism $\pi_1(X; e) \to H_1(X; \mathbb{Z})$ (i.e., S is the commutator subgroup $[\pi_1, \pi_1]$). Its group of covering translations is isomorphic to $H_1(X; \mathbb{Z})$. In general, an abelian covering space corresponds to the kernel S of a representation $\pi_1(X; e) \twoheadrightarrow A$ onto an abelian group A; since S must contain the commutator subgroup $[\pi_1, \pi_1] \subseteq S$, the universal abelian covering space is the largest abelian covering space (in the same sense that the universal covering space of X is the largest covering space of X).

A <u>(finite) cyclic covering space</u> $\pi : (\tilde{X}; \tilde{e}) \to (X; e)$ is one which corresponds to the kernel, S, of a representation

$$\pi_1(X; e) \twoheadrightarrow \mathbb{Z}_n \cong \mathbb{Z}/n\mathbb{Z}$$

onto a (finite) cyclic group. If $n > 0$, this will be called an <u>n-cyclic covering space</u>.

Thus far, the concepts introduced are standard.

<u>Definition 1.3</u>. An <u>iterated cyclic covering space</u> $\pi : \tilde{X} \to X$ is the top of a tower

$$\tilde{X} = \tilde{X}_n \to \tilde{X}_{n-1} \to \dots \to \tilde{X}_0 = X$$

of cyclic covering spaces $\tilde{X}_i \to \tilde{X}_{i-1}$, $i = 1, \dots, n$.

<u>Definition 1.4</u>. An iterated cyclic covering space $\pi : \tilde{X} \to X$ is an <u>iterated k-cyclic covering space</u> if each of the covering spaces $\tilde{X}_i \to \tilde{X}_{i-1}$, $i = 1, \dots, n$, in Definition 1.3 has index k.

<u>Definition 1.5</u>. A covering space $\pi : \tilde{X} \to X$ is the <u>limit</u> $\lim_{i\to\infty} \tilde{X}_i \to X$ <u>of covering spaces</u> $\tilde{X}_i \to X$ if there is a sequence of subgroups

$$\pi_1(X; e) = S_0 \supseteq S_1 \supseteq \dots$$

such that the covering space $\tilde{X}_i \to X$ corresponds to the subgroup S_i, and the covering space $\tilde{X} \to X$ corresponds to the intersection $\bigcap_{i=0}^{\infty} S_i$.

<u>Proposition 1.6</u>. Let S be a group which is the bottom of a decreasing sequence

$$S = S_n \triangleleft S_{n-1} \triangleleft \dots \triangleleft S_0$$

of normal subgroups $S_i \triangleleft S_0$ of a group S_0, such that the successive quotients $S_{i-1}/S_i \cong \mathbb{Z}^{n_i}$ are free abelian groups of finite rank n_i, $i = 1, \dots, n$.

Then S is also the intersection $S = \bigcap_{i=0}^{\infty} S_i$ of an infinite sequence

$$\dots \triangleleft G_2 \triangleleft G_1 \triangleleft G_0 = S_0$$

of decreasing normal subgroups of S_0, whose successive quotients $G_{i-1}/G_i \cong \mathbb{Z}_p$ are finite cyclic groups of order p (where p is an arbitrary prime, $p \neq 1$).

<u>Proof</u>: The cosets of S_0 by the subgroup S, are in one-to-one correspondence with $\bigoplus_{i=1}^{n} S_{i-1}/S_i \cong \bigoplus_{i=1}^{n} \mathbb{Z}^{n_i}$. Define $H_{(r_1,\dots,r_n)} \subseteq S_0$ to be the subset of elements in the cosets corresponding to

$\bigoplus_{i=1}^{n} r_i\mathbb{Z}^{n_i} \subseteq \bigoplus_{i=1}^{n} \mathbb{Z}^{n_i}$. Then $H_{(r_1,\ldots,r_n)}$ is a normal subgroup of S_0.

Order the set of n-tuples $(r_1, \ldots, r_n)$ by lexicographical ordering. There is a cofinal, increasing sequence $\vec{a}(i) = (a_1(i), \ldots, a_n(i))$ of n-tuples of integers, such that consecutive terms differ in exactly one place, by p . (Let

$$a_k(i) = \begin{cases} (j+1)p & \text{if } k \leq r \\ jp & \text{if } k > r \end{cases}$$

where $i = jn + r$ and $0 \leq r < n$. This sequence begins $(0, \ldots, 0)$, $(p, 0, \ldots, 0)$, $(p, p, 0, \ldots, 0)$, ..., $(p, p, \ldots, p)$, $(2p, p, \ldots, p)$, $(2p, 2p, p, \ldots, p)$, etc.). Then the infinite sequence

$$\cdots \lhd H_{\vec{a}(2)} \lhd H_{\vec{a}(1)} \lhd H_{\vec{a}(0)} = S_0$$

is a decreasing sequence of normal subgroups of S_0 , whose successive quotients $H_{\vec{a}(1)}/H_{\vec{a}(i-1)} \cong \mathbb{Z}_{p^{s_1}}$ are finite cyclic groups of prime power order p^{s_i} , $i = 1, 2, \ldots,$.

Now, in the short exact sequence

$$1 \to H_{\vec{a}(i)} \to H_{\vec{a}(i-1)} \to \mathbb{Z}_{p^{s_i}} \to 1 ,$$

let $G_r^i \lhd H_{\vec{a}(i-1)}$ be the inverse image of the subgroup $p^r\mathbb{Z}_{p^{s_i}} \subseteq \mathbb{Z}_{p^{s_i}}$.
It is not hard to see that since $H_{\vec{a}(i)} \lhd S_0$ and $H_{\vec{a}(i-1)} \lhd S_0$ are normal subgroups, then $G_r^i \lhd S_0$ is also. Thus, for each $i = 1, 2, \ldots,$ there is a finite sequence

$$H_{\vec{a}(i)} = G_{s_i}^i \lhd G_{s_{i-1}}^i \lhd \cdots \lhd G_0^i = H_{\vec{a}(i-1)}$$

of normal subgroups of S_0 , lying between $H_{\vec{a}(i)}$ and $H_{\vec{a}(i-1)}$, whose successive quotients $G_{r-1}^i/G_r^i \cong \mathbb{Z}_p$ are finite cyclic groups of order p . Proposition 1.6 is immediate.//

<u>Theorem 1.7</u>. Suppose the covering space $\tilde{X} \to X$ is the top of a finite tower

$$\tilde{X} = \tilde{X}_n \to \tilde{X}_{n-1} \to \cdots \to \tilde{X}_0 = X$$

of free abelian (finitely generated) covering spaces $\tilde{X}_i \to \tilde{X}_{i-1}$, such that each covering space $\tilde{X}_i \to X$, $i = 1, \ldots, n$, is regular.

Then $\tilde{X} \to X$ is also the limit $\lim_{i\to\infty} \tilde{Y}_i \to X$ of an infinite sequence

$$\cdots \to \tilde{Y}_2 \to \tilde{Y}_1 \to \tilde{Y}_0 = X$$

of p-cyclic covering spaces $\tilde{Y}_i \to \tilde{Y}_{i-1}$, such that each covering space $\tilde{Y}_i \to X$ is regular.

Proof: The Theorem follows from Proposition 1.6.//

Example 1.8. Let X be a space whose first homology group $H_1(X;\mathbb{Z}) \cong \mathbb{Z}^n$ is free abelian and finitely generated. (For example, let $L \subset M$, $L' \subset M'$ be $\mathbb{Z}$-concordant links in the $\mathbb{Z}$-homology 3-spheres M, M', respectively, and let $F \subset N^4$ be a $\mathbb{Z}$-concordance between them. Then take either $X = M - L$, or $X = N - F$.)

The universal abelian covering space $\tilde{X} \to X$ is the limit $\lim_{i\to\infty} \tilde{X}_i \to X$ of a tower

$$\cdots \to \tilde{X}_2 \to \tilde{X}_1 \to \tilde{X}_0 = X$$

of p-cyclic covering spaces $\tilde{X}_i \to \tilde{X}_{i-1}$ (where p is an arbitrary prime, $p \neq 1$).

Example 1.9. Let X be a space whose fundamental group $\pi_1(X) \cong F(x_1, \ldots, x_n)$ is a free group of rank n. (For example, let $L \subset S^3$ be the trivial link of n components, and set $X = X^3 - L$.) Let $G = \pi_1(X)$, and let $G = G_0 \rhd G_1 \rhd \cdots$ be the lower central series of G. For each i, define the covering space $\tilde{X}_i \to X$ to correspond to the subgroup $G_i \lhd G$; let $\tilde{X}_{\aleph_0} \to X$ correspond to the intersection $G_{\aleph_0} = \bigcap_{i=0}^{\infty} G_i$. (See [S].)

Then each covering space $\tilde{X}_i \to X$ is the limit $\lim_{j\to\infty} \tilde{Y}^i_j \to X$ of a sequence

$$\cdots \to \tilde{Y}^i_2 \to \tilde{Y}^i_1 \to \tilde{Y}^i_0 = X_i$$

of p-cyclic covering spaces $\tilde{Y}^i_j \to \tilde{Y}^i_{j-1}$, such that $\tilde{Y}^i_j \to X$ is a regular covering space, $j = 0, 1, \ldots,$. (This follows from the fact (proved in [F2]) that the quotients G_{i-1}/G_i are finitely generated, free abelian groups.)

§2. Homology of Covering Spaces

Suppose $\pi_Y : \tilde{Y} \to Y$ is a covering space, and $i : X \hookrightarrow Y$ is inclusion of a subspace X into Y. In the commutative diagram

$$\begin{array}{ccc} \tilde{X} & \xrightarrow{\tilde{i}} & \tilde{Y} \\ \downarrow{\scriptstyle \pi_X} & & \downarrow{\scriptstyle \pi_Y} \\ X & \xrightarrow{i} & Y \end{array}$$

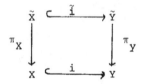

let $\tilde{X} = \pi_Y^{-1}(X)$ denote the inverse image of X under π_Y, let the map $\tilde{i} : \tilde{X} \hookrightarrow \tilde{Y}$ be the inclusion of $\tilde{X}$ into $\tilde{Y}$, and let the covering space $\pi_X : \tilde{X} \to X$ be the restriction $\pi_X = \pi_Y | \tilde{X}$.

Theorem 2.1. Let p^s be a positive power of a prime p. Let $\pi_Y : \tilde{Y} \to Y$ be a cyclic covering space of Y having index p^s, and suppose $i : X \hookrightarrow Y$ is inclusion of a subspace X into Y.

If the maps

$$i_* : H_j(X : \mathbb{Z}_p) \to H_j(Y; \mathbb{Z}_p)$$

induced by inclusion are isomorphisms for all $j \geq 0$, then the maps

$$\tilde{i}_* : H_j(\tilde{X}; \mathbb{Z}_p) \to H_j(\tilde{Y}; \mathbb{Z}_p)$$

are also.

Proof: Let $\Pi \cong \mathbb{Z}_{p^s}$ denote the cyclic group of covering translations of $\tilde{Y}$ and (by restriction) of $\tilde{X}$. Choose a generator $t \in \Pi$. Then the chain complexes

$$C_*(\tilde{X}; \mathbb{Z}_p) \; , \; C_*(\tilde{Y}; \mathbb{Z}_p) \quad \text{and} \quad C_*(\tilde{Y}, \tilde{X}; \mathbb{Z}_p) \; ,$$

and their corresponding homology groups $H_*(\tilde{X}; \mathbb{Z}_p)$, $H_*(\tilde{Y}; \mathbb{Z}_p)$ and $H_*(\tilde{Y}, \tilde{X}; \mathbb{Z}_p)$ are modules over the group algebra $\mathbb{Z}_p\Pi$.

The sequence

$$0 \to C_*(\tilde{Y}, \tilde{X} : \mathbb{Z}_p) \xrightarrow{1 + t + \ldots + t^{p^s-1}} C_*(\tilde{Y}, \tilde{X}; \mathbb{Z}_p) \xrightarrow{t - 1} C_*(\tilde{Y}, \tilde{X}; \mathbb{Z}_p)$$

$$\xrightarrow{\pi} C_*(Y, X; \mathbb{Z}_p) \to 0$$

of chain complexes is exact. Define $\bar{C}_*(\tilde{Y}, \tilde{X}; \mathbb{Z}_p) = C_*(\tilde{Y}, \tilde{X}; \mathbb{Z}_p)/\mathrm{Im}(1 + t + \ldots + t^{p^s-1})$. Then we obtain two short exact sequences:

$$0 \to \bar{C}_*(\tilde{Y}, \tilde{X}; \mathbb{Z}_p) \xrightarrow{(t-1)_*} C_*(\tilde{Y}, \tilde{X}; \mathbb{Z}_p) \xrightarrow{\pi} C_*(Y, X; \mathbb{Z}_p) \to 0$$

$$0 \to C_*(\tilde{Y}, \tilde{X}; \mathbb{Z}_p) \xrightarrow{1 + t + \ldots t^{p^s-1}} C_*(\tilde{Y}, \tilde{X}; \mathbb{Z}_p) \xrightarrow{\pi} \bar{C}_*(\tilde{Y}, \tilde{X}; \mathbb{Z}_p) \to 0 \, .$$

These give rise to long exact sequences

$$\ldots \xrightarrow{\pi_*} H_n(Y, X; \mathbb{Z}_p) \xrightarrow{\partial} H_{n-1}(\bar{C}) \xrightarrow{(t-1)_*} H_{n-1}(\tilde{Y}, \tilde{X}; \mathbb{Z}_p)$$

$$\xrightarrow{\pi_*} H_{n-1}(Y, X; \mathbb{Z}_p) \to \ldots$$

and

$$\cdots \xrightarrow{\pi_*} H_n(\bar{C}) \xrightarrow{\partial} H_{n-1}(\tilde{Y}, \tilde{X}; \mathbb{Z}_p) \xrightarrow{1 + t + \ldots + t^{p^s - 1}} H_{n-1}(\tilde{Y}, \tilde{X}; \mathbb{Z}_p)$$

$$\xrightarrow{\pi_*} H_{n-1}(\bar{C}) \to \cdots$$

of homology (compare Milnor, proof of Assertion 5, [M1]). By hypothesis, $H_*(Y, X; \mathbb{Z}_p) \cong 0$. Therefore the maps $H_*(\bar{C}) \xrightarrow{(t-1)_*} H_*(\tilde{Y}, \tilde{X}; \mathbb{Z}_p)$ are isomorphisms.

We now proceed by induction. It is always the case that $H_0(\tilde{Y}, \tilde{X}; \mathbb{Z}_p) \cong 0$. Suppose $H_{n-1}(\tilde{Y}, \tilde{X}; \mathbb{Z}_p) \cong 0$. Then $H_n(\tilde{Y}, \tilde{X}; \mathbb{Z}_p) \xrightarrow{\pi_*} H_n(\bar{C})$ is onto. Thus, the composition $(t-1)_* \circ \pi_* = (t-1)_* : H_n(\tilde{Y}, \tilde{X}; \mathbb{Z}_p) \to H_n(\tilde{Y}, \tilde{X}; \mathbb{Z}_p)$ is onto. Finally, $(t-1)^{p^s} = (t^{p^s} - 1)_* = 0$ is onto, and consequently, $H_n(\tilde{Y}, \tilde{X}; \mathbb{Z}_p) \cong 0$. The induction is now complete.//

Corollary 2.2. Let p^s be a positive power of a prime p. Let $\pi : \tilde{Y} \to Y$ be an iterated cyclic covering space of index p^s, and suppose $i : X \hookrightarrow Y$ is inclusion of the subspace X into Y.

If the maps

$$i_* : H_j(X; \mathbb{Z}_p) \to H_j(Y; \mathbb{Z}_p)$$

induced by inclusion are isomorphisms for all $j \geq 0$, then the maps

$$\tilde{i}_* : H_j(\tilde{X}; \mathbb{Z}_p) \to H_j(\tilde{Y}; \mathbb{Z}_p)$$

are also.

Proof: This Corollary may be proved by repeated applications of Theorem 2.1.//

Example 2.3. Let $L \subset M$, $L' \subset M'$ be $\mathbb{Z}_p$-concordant links, where p is a prime, and let $F \subset N^4$ be a $\mathbb{Z}_p$-concordance between them. Suppose $\pi : \tilde{N} \to N - F$ is an iterated cyclic covering space of the complement of the concordance, having prime power index p^s.

Let $X = M - L$, $\tilde{X} = \tilde{M}$, $Y = N - F$, $\tilde{Y} = \tilde{N}$ in Corollary 2.2. We then have that for all $j \geq 0$, the maps

$$\tilde{i}_* : H_j(\tilde{M}; \mathbb{Z}_p) \to H_j(\tilde{N}; \mathbb{Z}_p)$$

induced by inclusion, are isomorphisms.

Example 2.4. Define $\tilde{N}$ as in Example 2.3. Then the intersection form $H_2(\tilde{N}; \mathbb{Q}) \times H_2(\tilde{N}; \mathbb{Q}) \to \mathbb{Q}$ vanishes.

Proof: This intersection form is the composition

$$H_2(\tilde{N}; \mathbb{Q}) \times H_2(\tilde{N}; \mathbb{Q}) \xrightarrow{j \times id} H_2(\tilde{N}, \partial\tilde{N}; \mathbb{Q}) \times H_2(\tilde{N}; \mathbb{Q}) \to \mathbb{Q} ,$$

where $j : H_2(\tilde{N}; Q) \to H_2(\tilde{N}, \partial\tilde{N}; Q)$ is induced by inclusion, and the last map is Poincaré-Lefschetz Duality. It follows from Example 2.3 that $H_2(\tilde{N}, \partial\tilde{N}; \mathbb{Z}_p)$ (and hence $H_2(\tilde{N}, \partial\tilde{N}; Q))$ is 0 .//

Theorem 2.5. Let X^{4k} be a 4k-dimensional, oriented manifold. Suppose $\tilde{X} \to X$ is the limit $\lim_{n\to\infty} \tilde{X}_n \to X$ of an infinite tower

$$\cdots \to \tilde{X}_2 \to \tilde{X}_1 \to \tilde{X}_0 = X$$

of covering spaces such that for all n, the intersection form $H_{2k}(\tilde{X}_n; \mathbb{Q}) \times H_{2k}(\tilde{X}_n; \mathbb{Q}) \to \mathbb{Q}$ on the middle dimensional homology group vanishes.

Then the intersection form $H_{2k}(\tilde{X}; \mathbb{Q}) \times H_{2k}(\tilde{X}; \mathbb{Q}) \to \mathbb{Q}$ of the limit vanishes.

Proof: We will assume that all manifolds are triangulated, and all projection maps are simplicial. Let $\alpha, \beta \in H_{2k}(\tilde{X}; \mathbb{Q})$ be homology classes. We may further assume, without loss of generality, that α, β are represented by integral cycles A, B, respectively, which are oriented, 2k-dimensional, finite simplicial complexes, such that $A \cup B$ and the projections $\pi : A \cup B \to X$ (and hence, $p_n : A \cup B \to \tilde{X}_n$, $n = 0, 1, \ldots$) are in general position. Denote $A_n = p_n(A)$, $B_n = p_n(B)$.

This has the following consequences:

1. $A_n \cap B_n \subset \tilde{X}_n$ is a finite set of points, $n = 0, 1, \ldots$, and
2. the projections $p_n : A \cap B \to A_n \cap B_n$, $n = 0, 1, \ldots$, are $1 - 1$.

Let $|S|$ denote the cardinality of a set S. The monotonically non-increasing sequence

$$|\pi(A) \cap \pi(B)| = |A_0 \cap B_0| \geq |A_1 \cap B_1| \geq |A_2 \cap B_2| \geq \cdots$$

of non-negative integers, must be eventually constant. Hence, either $A \cap B = \phi$, or for $n \gg 0$, $|A_n \cap B_n| \cong |A \cap B|$. (Suppose not. Then there are points $q \in A$, $r \in B$, $q, r \notin A \cap B$, such that $p_n(q) = p_n(r) \in A_n \cap B_n$. Choose a point $s \in A \cap B$, and paths $\delta_p \subset A$ from s to q, and $\delta_r \subset B$ from s to r. Now $p_n(\delta_q^{-1}\delta_r) \in \tilde{X}_n$ is a loop for every n, but $\delta_q^{-1}\delta_r \subset \tilde{X}$ is not. This is impossible because $\tilde{X}$ is the limit $\lim_{n\to\infty} X_n \to X$.) Finally, $p_n : A \cap B \to A_n \cap B_n$ is orientation preserving and $1 - 1$ for $n \gg 0$, and $|A \cap B| = |A_n \cap B_n|$ or $A \cap B = \phi$. Since the total algebraic intersections of $A_n \cap B_n$ is zero, we can conclude that $\alpha \times \beta = 0$ .//

Example 2.6. Let $L \subset M$, $L' \subset M'$ be $\mathbb{Z}$-concordant links in the

$\mathbb{Z}$-homology 3-spheres M, M', respectively, and let $F \subset N^4$ be a $\mathbb{Z}$-concordance between them.

If $\pi : \tilde{N} \to N - F$ is the universal abelian covering space of the complement of the concordance, then the intersection form $H_2(\tilde{N}; \mathbb{Q}) \times H_2(\tilde{N}; \mathbb{Q}) \to \mathbb{Q}$ vanishes.

Proof: It follows from Example 1.8 that $\pi : \tilde{N} \to N - F$ is the limit $\lim_{i\to\infty} \tilde{N}_i \to N - F$ of a tower

$$\ldots \to \tilde{N}_2 \to \tilde{N}_1 \to \tilde{N}_0 = N - F$$

of p-cyclic covering spaces $\tilde{N}_i \to \tilde{N}_{i-1}$ (where p is an arbitrary prime, $p \neq 1$). By Example 2.4, the intersection forms $H_2(\tilde{N}_i; \mathbb{Q}) \times H_2(\tilde{N}_i; \mathbb{Q}) \to \mathbb{Q}$, $i = 0, 1, \ldots$, vanish; therefore, by Theorem 2.5, the intersection form $H_2(\tilde{N}; \mathbb{Q}) \times H_2(\tilde{N}; \mathbb{Q}) \to \mathbb{Q}$ vanishes.//

Example 2.7. Let $C^n \subset S^3$ be the trivial link of n components. Let $F \subset N^4$ be a $\mathbb{Z}$-concordance to a link $L = K_1 \cup \ldots \cup K_n \subset M^3$, where p is a prime $p \neq 1$. The covering spaces $\tilde{M}_n \to S^3 - C^n$ corresponding to the groups G_n, $n = 0, 1, 2, \ldots, \aleph_0$, in the lower central series for the free group $G_0 = \pi_1(S^3 - C^n)$ (see Example 1.9) extend uniquely to covering spaces $\tilde{N}_n \to N - F$.

The intersection forms

$$H_2(\tilde{N}_n; \mathbb{Q}) \times H_2(\tilde{N}_n; \mathbb{Q}) \to \mathbb{Q},$$

$n = 0, 1, 2, \ldots, \aleph_0$, vanish.

Proof: By Example 1.9, each $\tilde{M}_n \to S^3 - C^n$ is the limit of a sequence of p-cyclic, covering spaces. It is proved in the next section, Example 3.13, that $\tilde{M}_n$ extends uniquely to an iterated, p-cyclic covering $\tilde{N}_n \to N - F$. Let $\tilde{N}_{\aleph_0} = \lim_{n\to\infty} \tilde{N}_n$ $N - F$. The conclusion follows from Example 2.4 and Theorem 2.5.

§3. Related Covering Spaces

In this section we will investigate the existence and uniqueness of related covering spaces associated to R-concordant links.

Let $i : X \hookrightarrow Y$ and $i' : X \hookrightarrow Y$ be subspaces. A covering space $\tilde{X} \to X$ is compatible with Y if it has a unique extension to a covering space $\pi : \tilde{Y} \to Y$. In this case, the restriction of π to $\pi^{-1}(X') = \tilde{X}'$ is a covering space $\tilde{X}' \to X'$ which is related by Y to $\tilde{X} \to X$.

The Algebra of Related Covering Spaces.

Definition 3.1. Let $f : G \to H$ be a homomorphism of groups, and let $\pi : G \to A$ be a representation of G onto a group A. We say that π is compatible with f if there is a unique representation $\Pi : H \to A$ of H onto A, such that $\pi = \Pi \circ f$.

Definition 3.2. Let $f : G \to H$, $F' : G' \to H$ be homomorphisms of groups, and let $\pi : G \to A$, $\pi' : G' \to A$ be representations onto a group A. We say that π, π' are related by H, if there is a unique representation $\Pi : H \to A$ of H onto A, such that $\pi = \Pi \circ f$ and $\pi' = \Pi \circ f'$.

Proposition 3.3. Let $f : G \to H$ be a homomorphism, and let $\pi : G \to \mathbb{Z}_n$ be a representation of G onto $\mathbb{Z}_n$.

If f induces an isomorphism $f_* : H_1(G; \mathbb{Z}_n) \to H_1(H; \mathbb{Z}_n)$ of first homology groups, then π is compatible with f.

Proof: The representation $\pi : G \to \mathbb{Z}_n$ factors through $H_1(G; \mathbb{Z}_n)$:

$$G \xrightarrow{i_*} H_1(G; \mathbb{Z}_n) \xrightarrow{p} \mathbb{Z}_n, \qquad \pi = p \circ i_*$$

It is clear from the commutative diagram

$$\begin{array}{ccccc} G & \xrightarrow{i_*} & H_1(G; \mathbb{Z}_n) & \xrightarrow{p} & \\ \downarrow f & & \downarrow f_* & & \mathbb{Z}_n \\ H & \xrightarrow{i_*} & H_1(H; \mathbb{Z}_n) & \nearrow & \end{array}$$

that the representation of H onto $\mathbb{Z}_n$ is $p \circ f_*^{-1} \circ i_*$.//

Corollary 3.4. Let $f : G \to H$ and $\pi: G \to \mathbb{Z}_n$ satisfy the hypotheses of Proposition 3.3.

If $f' : G' \to H$ is a second group homomorphism, $f'_* : H_1(G'; \mathbb{Z}_n) \to H_1(H; \mathbb{Z}_n)$ its induced map on homology, then there is a unique representation $\pi' : G' \to \mathbb{Z}_n$ related to π by H.

Proof: Define $\phi : H_1(G'; \mathbb{Z}_n) \to H_1(G; \mathbb{Z}_n)$ by $\phi : f_*^{-1} \circ f'_*$, and let $i'_* : G' \to H_1(G'; \mathbb{Z}_n)$ be the usual map. The representation is $\pi' = p \circ i'_*$ (where $p : H_1(G; \mathbb{Z}_n) \to \mathbb{Z}_n$ is defined in Proposition 3.3).//

Proposition 3.5. Let $f : G \to H$ be a homomorphism, and let $\pi : G \to A$ be a representation of G onto an abelian group.

If f induces an isomorphism $f_* : H_1(G; \mathbb{Z}) \to H_1(H; \mathbb{Z})$ of first homology groups, then π is compatible with f.

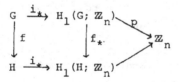

Proof: This is exactly the same as the proof of Proposition 3.3, with $\mathbb{Z}$ substituted everywhere for $\mathbb{Z}_n$ .//

Corollary 3.6. Let $f : G \to H$ and $\pi : G \to A$ satisfy the hypotheses of Proposition 3.5.

If $f' : G' \to H$ is a second homomorphism, $f'_* : H_1(G'; \mathbb{Z}) \to H_1(H; \mathbb{Z})$ its induced map on homology, then there exists a unique representation $\pi' : G' \to A$ related to π by H .

Proof: Same as the proof of Corollary 3.4, substituting $\mathbb{Z}$ for $\mathbb{Z}_n$ everywhere.//

Remark 3.7. Note that if $i : X \hookrightarrow Y$ and $i' : X' \hookrightarrow Y$ are subspaces, if $i_* : \pi_1(X) \to \pi_1(Y)$, $i'_* : \pi_1(X') \to \pi_1(Y)$ are the homomorphisms of fundamental groups induced by inclusion (by choosing paths from the basepoints of X , X' , respectively, to the basepoint of Y), and if $\tilde{X} \to X$, $\tilde{X}' \to X'$ are covering spaces corresponding to the representations $\pi : \pi_1(X) \to A$, $\pi' : \pi_1(X') \to A$, respectively, then the following pairs of statements are equivalent:

1. $\tilde{X} \to X$ is compatible with Y , or
1'. $\pi : \pi_1(X) \to A$ is compatible with $i_* : \pi_1(X) \to \pi_1(Y)$.
2. $\tilde{X} \to X$ and $\tilde{X}' \to X'$ are related by Y , or
2'. $\pi : \pi_1(X) \to A$ and $\pi' : \pi_1(X') \to A$ are related by $\pi_1(Y)$.

In view of this remark, and Propositions 3.3, 3.5 we have the following geometric theorems:

Theorem 3.8. Let $i : X \hookrightarrow Y$ be a subspace and let $\tilde{X} \to X$ be an n-cyclic covering space.

If the inclusion-induced map $i_* : H_1(X; \mathbb{Z}_n) \to H_1(Y; \mathbb{Z}_n)$ of homology groups is an isomorphism, then $\tilde{X} \to X$ is compatible with Y .

Corollary 3.9. Let $i : X \hookrightarrow Y$ be a subspace, and let $\tilde{X} \to X$ be an iterated, p-cyclic covering space, where p is a prime.

If the inclusion-induced maps $i_* : H_j(X; \mathbb{Z}_p) \to H_j(Y; \mathbb{Z}_p)$ on homology groups, $j \geq 0$, are isomorphisms, then $\tilde{X} \to X$ is compatible with Y .

Proof: The proof is by induction on the spaces $\tilde{X}_i$ in the tower

$$\tilde{X} = \tilde{X}_n \to \dots \to \tilde{X}_1 \to \tilde{X}_0 = X$$

of p-cyclic covering spaces $\tilde{X}_i \to \tilde{X}_{i-1}$. Suppose $\tilde{X}_i \to \tilde{X}_{i-1}$ extends uniquely to a covering space $\tilde{Y}_i \to \tilde{Y}_{i-1}$ such that the inclusion induced map $H_1(\tilde{X}_i; \mathbb{Z}_p) \to H_1(\tilde{Y}_i; \mathbb{Z}_p)$ on homology groups is an isomorphism. It then follows from Theorem 2.1, 3.8 that $\tilde{X}_{i+1} \to \tilde{X}_i$ extends uniquely to a covering space $\tilde{Y}_{i+1} \to \tilde{Y}_i$ with the same property.//

Corollary 3.10. Let $i : X \hookrightarrow Y$ and $\tilde{X} \to X$ satisfy the hypotheses of Corollary 3.9.

If $i' : X' \hookrightarrow Y$ is a second subspace, then there is a unique (iterated, p-cyclic) covering space $\tilde{X}' \to X'$ related to $\tilde{X} \to X$ by the concordance.

Proof: Since $\tilde{X} \to X$ is compatible with Y, by Corollary 3.9, let $\pi : \tilde{Y} \to Y$ be the unique extension to a covering space of Y. Then $\tilde{X}' = \pi^{-1}(X')$ .//

Theorem 3.11. Let $i : X \hookrightarrow Y$ be a subspace and let $\tilde{X} \to X$ be an abelian covering space.

If the inclusion-induced map $i_* : H_1(X; \mathbb{Z}) \to H_1(Y; \mathbb{Z})$ on homology groups is an isomorphism, then $\tilde{X} \to X$ is compatible with Y.

Corollary 3.12. Let $i : X \hookrightarrow Y$ and $\tilde{X} \to X$ satisfy the hypotheses of Theorem 3.11.

If $i' : X' \hookrightarrow Y$ is another subspace, then there is a unique (abelian) covering space $\tilde{X}' \to X'$ related to $\tilde{X} \to X$ by Y.

Proof: Same as the proof of Corollary 3.10.//

Example 3.13. Let $L \subset M$, $L' \subset M'$ be $\mathbb{Z}_p$-concordant links, where p is a prime, and let $F \subset N^4$ be a $\mathbb{Z}$-concordance between them. Suppose $\tilde{M} \to M - L$ is an iterated, p-cyclic covering space.

There is a unique (iterated, p-cyclic) covering space $\tilde{M}' \to M' - L'$ which is related by the concordance, to $\tilde{M} \to M - L$.

Proof: In Corollary 3.10, set $X = M - L$, $X' = M' - L'$, $\tilde{X} = \tilde{M}$, and $Y = N - F$ .//

Example 3.14. Let $L \subset M$, $L' \subset M$ be $\mathbb{Z}$ $(\mathbb{Z}_p)$-concordant links in the $\mathbb{Z}$ $(\mathbb{Z}_p)$-homology 3-spheres M , M' , respectively. Let $\tilde{M} \to M - L$ be an abelian (p-cyclic) covering space.

Then there is a unique abelian (p-cyclic) covering space $\tilde{M}' \to M' - L'$ which is related to $\tilde{M} \to M - L$ by every $\mathbb{Z}$ $(\mathbb{Z}_p)$-concordance between $L \subset M$ and $L' \subset M'$. (p is a prime, $p \neq 1$.)

Proof: First, let $F \subset N$ be a particular $\mathbb{Z}$ $(\mathbb{Z}_p)$-concordance between $L \subset M$ and $L' \subset M'$, and set $X = M - L$, $X' = M' - L'$, $\tilde{X} = \tilde{M}$, $\tilde{X}' = \tilde{M}'$, $Y = N - F$ in Corollary 3.12 (3.10). Next, observe that $\tilde{M}' \to M' - L'$ does not depend on the concordance:

Suppose that the covering space $\tilde{M} \to M - L$ is determined by the representation $\pi : \pi_1(M - L) \xrightarrow{i_*} H_1(M - L; \mathbb{Z}) \xrightarrow{g} A$. Then the related covering space $\tilde{M}' \to M' - L'$ is determined by the representation

$$\pi_1(M' - L') \xrightarrow{i_*} H_1(M' - L'; \mathbb{Z}) \xrightarrow{g'} A ,$$

where $g' = g \circ \phi$, and ϕ is defined in the proof of Corollary 3.4. Now, ϕ may be described independently of the concordance:

Let $L \subset M$ have components $L = K_1 \cup K_2 \cup \ldots \cup K_n$. Since M is a $\mathbb{Z}_p$-homology 3-sphere, Alexander duality implies that $H_1(M - L; \mathbb{Z}_p) \cong \underbrace{\mathbb{Z}_p \oplus \ldots \oplus \mathbb{Z}_p}_{n \text{ times}}$, generated by $0 \oplus \ldots \oplus \underset{i^{th} \text{ place}}{1} \oplus 0 \oplus \ldots \oplus 0 = [m_i]$, $i = 1, \ldots, n$, where m_i is a positive meridian in M about K_i. (A positive meridian m_i is a simple, closed curve on the boundary of a regular neighborhood N_i of K_i, which is contractible in N_i, and has linking number $\ell_M(m_i, K_i) = +1$ with K_i in M.) Let m_i', $i = 1, \ldots, n$, be positive meridians for the link $L' \subset M'$, with components $L' = K_1' \cup \ldots \cup K_n'$. The map ϕ is defined by $\phi([m_i']) = [m_i]$. //

§4. A Linking Invariant of Concordance

The invariant of the link $L \subset M$ which is described here is computed from particular branched, or unbranched, covering spaces $\tilde{M} \to M$, whose branch set is a sublink of L. The invariant is easiest to define when a full set of covering translations exist for $\tilde{M}$ (i.e., when $\tilde{M}$ is regular), and I will define it first for that special case. Since irregular covering spaces are sometimes useful, I define a similar object for them; however, the equivalence relation induced on these objects by concordance is impractical for computation, except in the one instance where $L \subset M$ is concordant to a split link (to be defined later).

Let $L = K_1 \cup \ldots \cup K_n \subset M$ be an oriented link in a 3-manifold M. Let $\pi : \tilde{M} \to M$ be a regular branched, or unbranched covering space, whose branch set is a sublink of L. Suppose $\pi : \tilde{M} \to M$ corresponds to the kernel of a representation $f : \pi_1(M) \to A$ onto a group A; then A acts on $\tilde{M}$ as its group of covering translations (see §1). Choose oriented components $\tilde{K}_i \subset \pi^{-1}(K_i)$, $i = 1, \ldots, n$, and let $\sigma\tilde{K}_i$, $\sigma \in A$, denote the image of $\tilde{K}_i$ under the action of $\sigma \in A$. Finally, assume that $\tilde{K}_i$, $i = 1, \ldots, n$, are closed curves, and that the homology classes $[\tilde{K}_i] \in H_1(\tilde{M}; \mathbb{Q})$ represented by them, vanish. The notation $\ell_{\tilde{M}}(C_1, C_2)$ denotes the linking number of the disjoint closed curves C_1, $C_2 \subset \tilde{M}$; we adopt the convention $\ell_{\tilde{M}}(C, C) = 0$.

The Invariant

Definition 4.1. Define $\Lambda_{\tilde{M}}(K_1, \ldots, K_n)$ to be the $n \times n$ matrix $[\lambda_{ij}]$ with entries $\lambda_{ij} \in \mathbb{Q}(A)$ in the group algebra $\mathbb{Q}(A)$, given by

$$\lambda_{ij} = \sum_{\sigma \in A} \ell_{\tilde{M}}(\tilde{K}_i, \sigma\tilde{K}_j) \cdot \sigma .$$

Properties of $\Lambda_{\tilde{M}}(K_1, \ldots, K_n)$

Let $I : A \to A$ be the involution which takes every element to its inverse, and let $I_* : \mathbb{Q}(A) \to \mathbb{Q}(A)$ be the involution induced on the group algebra by I. If $\Lambda = [\lambda_{ij}]$ is a matrix with entries $\lambda_{ij} \in \mathbb{Q}(A)$, let $\Lambda^* = [I_*(\lambda_{ij})]$, and let $\Lambda^t = [\lambda_{ji}]$.

Proposition 4.2. $\Lambda^t_{\tilde{M}}(K_1, \ldots, K_n) = \Lambda^*_{\tilde{M}}(K_1, \ldots, K_n)$.

Proof: $\lambda_{ji} = \sum_{\sigma \in A} \ell_{\tilde{M}}(\tilde{K}_j, \sigma\tilde{K}_i) \cdot \sigma = \sum_{\sigma \in A} \ell_{\tilde{M}}(\sigma\tilde{K}_i, \tilde{K}_j) \cdot \sigma =$

$$\sum_{\sigma \in A} \ell_{\tilde{M}}(\tilde{K}_i, \sigma^{-1}\tilde{K}_j) \cdot \sigma = \sum_{\sigma \in A} \ell_{\tilde{M}}(\tilde{K}_i, \sigma\tilde{K}_j) \cdot \sigma^{-1} = I_*(\lambda_{ij}) . //$$

Equivalence Relations on $\Lambda = [\lambda_{ij}]$

The matrix $\Lambda_{\tilde{M}}(K_1, \ldots, K_n)$ can be made independent of the choice $\tilde{K}_i$, $i = 1, \ldots, n$, by introducing the following equivalence relation:

Definition 4.3. Let Λ, Λ' be $n \times n$ matrices with entries in $\mathbb{Q}(A)$. Define Λ to be equivalent to Λ' (denoted $\Lambda \equiv \Lambda'$) if there is a diagonal matrix $B = \begin{bmatrix} \sigma_1 & & 0 \\ & \ddots & \\ 0 & & \sigma_n \end{bmatrix}$ with entries $\sigma_i \in A$, such that $\Lambda' = B^{-1}\Lambda B$.

Proposition 4.4. Let $\Lambda'_{\tilde{M}}(K_1, \ldots, K_n)$ be the matrix defined in Definition 4.1 for a new choice $\tilde{K}'_i \subset \pi^{-1}(K_i)$ of oriented component covering K_i, $i = 1, \ldots, n$. Then $\Lambda'_{\tilde{M}}(K_1, \ldots, K_n) \equiv \Lambda_{\tilde{M}}(K_1, \ldots, K_n)$.

Proof: Let $\tilde{K}'_i = \sigma_i \tilde{K}_i$, for some $\sigma_i \in A$. Then $\ell_{\tilde{M}}(\tilde{K}'_i, \sigma\tilde{K}'_j) =$

$\ell_{\tilde{M}}(\sigma_i\tilde{K}_i, \sigma\sigma_j\tilde{K}_j) = \ell_{\tilde{M}}(\tilde{K}_i, \sigma_i^{-1}\sigma\sigma_j\tilde{K}_j)$. Let $B = \begin{bmatrix} \sigma_1 & & 0 \\ & \ddots & \\ 0 & & \sigma_n \end{bmatrix}$. It is easily seen that $\Lambda'_{\tilde{M}}(K_1, \ldots, K_n) = B^{-1}\Lambda_{\tilde{M}}(K_1, \ldots, K_n)B$. //

Computation of $\Lambda_M(K_1, \ldots, K_n)$ for covering spaces $\tilde{M} \to M$.

Now assume that $\tilde{M} \to M$ is a covering space. Map compact, oriented surfaces $S_1, \ldots, S_n$ into M by maps $g_i : S_i \to M$ which

(1) are in general position with respect to each other on the interiors $\mathrm{int}(S_i)$, $i = 1, \ldots, n$

(2) are connected covering spaces $g_i : \partial S_i \to K_i$ on the boundary

(3) lift to maps $\tilde{g}_i : S_i \to \tilde{M}$ such that the restrictions $\tilde{g}_i : \partial S_i \to \tilde{K}_i$ are n_i-fold cyclic, orientation preserving covering spaces. (This can be done if $0 = [n_i\tilde{K}_i] \in H_1(\tilde{M}; \mathbb{Z})$.

Definition 4.5. A link $L = K_1 \cup \ldots \cup K_n \subset M$ in a 3-manifold M which is connected to the basepoint $e \in M$, consists of the link $L \subset M$, and a choice of paths δ_i, $i = 1, \ldots, n$, whose interiors are disjoint from the link and from each other, joining e to the components K_i, $i = 1, \ldots, n$, respectively.

Remark. Connecting a link to basepoint has the effect of choosing components $\tilde{K}_i \subset \pi^{-1}(K_i)$ in every branched and unbranched covering space $\pi : \tilde{M} \to M$, simultaneously. (Lift δ_i from the basepoint $\tilde{e} \in \tilde{M}$ of the covering space, and denote the lift by $\tilde{\delta}_i$. Then choose $\tilde{K}_i \subseteq \pi^{-1}(K_i)$ to be the unique component containing the point $\tilde{\delta}_i(1)$.)

Construction 4.6. Let the link $L = K_1 \cup \ldots \cup K_n \subset M$ be connected to basepoint by paths δ_i, $i = 1, \ldots, n$. For each i, choose a point $c_i \in \partial S_i$ such that $f_i(c_i) = \delta_i(1)$. For each pair of points $p \in \mathrm{int}(S_i)$, $q \in \partial S_j$ such that $g_i(p) = g_j(q)$, choose paths $\alpha_{i,p}$ in S_i from c_i to p, and $\beta_{j,q}$ in S_j from c_j to q. Denote by $\gamma_{i,p,q} \in \pi_1(M)$ the element represented by $\delta_i \alpha_{i,p} \beta_{j,q}^{-1} \delta_j^{-1}$.

Definition 4.7. For every point $p \in \mathrm{int}(S_i)$ such that $g_i(p) \in g_i(S_i) \cap K_j$, let sign(p) denote the orientation of p.

Definition 4.8. Define

$$w_{ij} = \frac{1}{n_i n_j} \sum_{\substack{p \in \mathrm{int}(S_i),\ q \in \partial S_j \\ g_i(p) = g_j(q)}} \mathrm{sign}(p) \cdot \gamma_{i,p,q} .$$

Definition 4.9. Let $f : A \to A'$ be a surjective homomorphism of groups. Then $f_* : \mathbb{Q}(A) \to \mathbb{Q}(A')$ is the additive homomorphism defined by

$$f_*(\gamma) = \begin{cases} f(\gamma) & \text{if } \gamma \neq 1 \\ 0 & \text{if } \gamma = 1 . \end{cases}$$

Theorem 4.10. Let the link $L = K_1 \cup \ldots \cup K_n \subset M$ be connected to the basepoint $e \in M$ by paths $\delta_1, \ldots, \delta_n$. Let $\tilde{M} \to M$ be a covering space determined by the representation $f : \pi_1(M) \to A$ onto a group A.

Then $\Lambda_{\tilde{M}}(K_1, \ldots, K_n) = [\lambda_{ij}]$ has entries $\lambda_{ij} = f_*(w_{ij})$.

Proof: $\ell_{\tilde{M}}(n_i\tilde{K}_i, n_j\sigma\tilde{K}_j) =$

$$\sum_{\{p \in \mathrm{int}\, S_i,\ q \in \partial S_j \,:\, f_*(\gamma_{i,p,q}) = \sigma\}} \mathrm{sign}(p) .//$$

Let $\tilde{M}' \to M$ be a covering space intermediate between M and $\tilde{M} \to M$. Then there is a group A', and a surjective homomorphism $p : A \to A'$ such that $\tilde{M}' \to M$ is determined by the representation $f' = p \circ f : \pi_1(M) \to A'$. //

Corollary 4.11. If $\tilde{M}' \to M$ is an intermediate covering space between M and $\tilde{M} \to M$, and if $\Lambda_{\tilde{M}}(K_1, \ldots, K_n) = [\lambda_{ij}]$, $\Lambda_{\tilde{M}}(K_1, \ldots, K_n) = [\lambda'_{ij}]$, then $\lambda'_{ij} = p_*(\lambda_{ij})$.

Proof: By Theorem 4.10, $\lambda_{ij} = f_*(w_{ij})$ and $\lambda'_{ij} = f'_*(w_{ij}) = p_* \circ f_*(w_{ij}) = p_*(\lambda_{ij})$. //

The Invariance of $[\Lambda_{\tilde{M}}(K_1, \ldots, K_n)]$ under R-concordance.

In Lemmas 4.12 and 4.13, let $L = K_1 \cup \ldots \cup K_n \subset M$, $L' = K'_1 \cup \ldots \cup K'_n \subset M'$ be R-concordant links, and let $F = F_1 \cup \ldots \cup F_n \subset N^4$ be an R-concordance between them. Let $\tilde{M} \to M$ and $\tilde{M}' \to \tilde{M}'$ be branched, or unbranched, covering spaces, which are related by the R-concordance. Thus, there is a unique extension of $\tilde{M} \to M$ and $\tilde{M}' \to M'$ to a covering $\pi : \tilde{N} \to N$. Let $\tilde{F}_i \subset \pi^{-1}(F_i)$,

$\tilde{K}_i \subset \pi^{-1}(K_i)$, $\tilde{K}_i' \subset \pi^{-1}(K_i')$ be oriented components such that $\partial\tilde{F}_i = \tilde{K}_i' - \tilde{K}_i$. Suppose $\tilde{M} \to M$, $\tilde{M}' \to \tilde{M}'$ correspond to the related representations $\pi_1(M) \to A$ and $\pi_1(M') \to A$, respectively. Finally, let $i_*: H_1(\tilde{M}; \mathbb{Q}) \to H_1(\tilde{N}; \mathbb{Q})$ and $i_*' : H_1(\tilde{M}'; \mathbb{Q}) \to H_1(\tilde{N}; \mathbb{Q})$ be induced by inclusion.

Lemma 4.12. Suppose $H_*(\tilde{N}, \tilde{M}'; \mathbb{Q}) = 0$. If $0 = [\tilde{K}_i] \in H_1(\tilde{M}; \mathbb{Q})$, then $0 = [\tilde{K}_i'] \in H_1(\tilde{M}'; \mathbb{Q})$, $i = 1, \ldots, n$.

Proof: Since $\partial\tilde{F}_i = \tilde{K}_i' - \tilde{K}_i$, we have that $i_*([\tilde{K}_i]) = i_*'([\tilde{K}_i']) \in H_1(\tilde{N}; \mathbb{Q})$. If $0 = [\tilde{K}_i] \in H_1(\tilde{M}; Q)$, then $0 = i_*[\tilde{K}_i] \in H_1(\tilde{N}; \mathbb{Q})$, and therefore $0 = i_*'[\tilde{K}_i'] \in H_1(\tilde{N}; \mathbb{Q})$. Now $0 = [\tilde{K}_i'] \in H_1(\tilde{M}'; \mathbb{Q})$, because $H_*(\tilde{N}, \tilde{M}'; \mathbb{Q}) = 0$. //

Lemma 4.13. Suppose the intersection form $H_2(\tilde{N}; \mathbb{Q}) \times H_2(\tilde{N}; \mathbb{Q}) \to \mathbb{Q}$ vanishes, and for each $i = 1, \ldots, n$, $0 = [\tilde{K}_i] \in H_1(\tilde{M}; Q)$ and $0 = [\tilde{K}_i'] \in H_1(\tilde{M}'; \mathbb{Q})$. Then for all $\sigma \in A$,

$$\ell_{\tilde{M}}(\tilde{K}_i, \sigma\tilde{K}_j) = \ell_{\tilde{M}'}(\tilde{K}_i', \sigma\tilde{K}_j') .$$

Proof: To compute $\ell_{\tilde{M}}(\tilde{K}_i, \sigma\tilde{K}_j)$ and $\ell_{\tilde{M}'}(\tilde{K}_i', \sigma\tilde{K}_j')$, choose 2-chains $D \in C_2(\tilde{M}; \mathbb{Q})$ and $D' \in C_2(\tilde{M}'; \mathbb{Q})$ such that $\partial D = \tilde{K}_j$ and $\partial D' = \tilde{K}_j'$. Then $[\tilde{F}_i] \in H_2(\tilde{N}, \partial\tilde{N}; \mathbb{Q})$ and $[D' - \sigma\tilde{F}_j - D] \in H_2(\tilde{N}; \mathbb{Q})$. In fact, $[\tilde{F}_i] \in \mathrm{Image}(i_*)$, where $i_* : H_2(\partial\tilde{N}; \mathbb{Q}) \to H_2(\tilde{N}, \partial\tilde{N}; \mathbb{Q})$ is induced by inclusion. (See Figure 1.) Intersecting these two cycles, we obtain

$$0 = [\tilde{F}_i] \times [D' - \sigma\tilde{F}_j - D] = [(\tilde{F}_i \cdot D') - (\tilde{F}_i \cdot \sigma\tilde{F}_j) - (\tilde{F}_i \cdot D)] =$$
$$[\tilde{F}_i \cdot D'] - [\tilde{F}_i \cdot D] = \ell_{\tilde{M}'}(\tilde{K}_i', \sigma\tilde{K}_j') - \ell_{\tilde{M}}(\tilde{K}_i, \sigma\tilde{K}_j) . //$$

Figure 1.

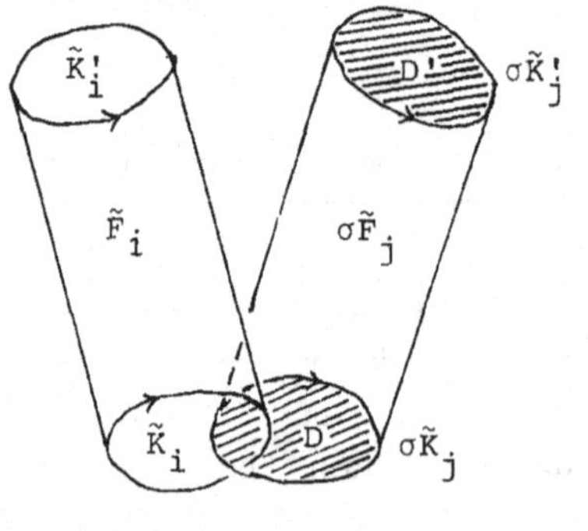

Corollary 4.14. Suppose $H_*(\tilde{N}, \tilde{M}'; \mathbb{Q}) = 0$. If $0 = [\tilde{K}_i] \in H_1(\tilde{M}; \mathbb{Q})$ for all $i = 1, \ldots, n$, then $0 = [\tilde{K}_i'] \in H_1(\tilde{M}'; \mathbb{Q})$ for all $i = 1, \ldots, n$, and $\Lambda_{\tilde{M}}(K_1, \ldots, K_n) \equiv \Lambda_{\tilde{M}'}(K_1', \ldots, K_n')$.

Proof: Since $H_*(\tilde{N}, \tilde{M}'; \mathbb{Q}) = 0$, it follows that the intersection form $H_2(\tilde{N}; \mathbb{Q}) \times H_2(\tilde{N}; \mathbb{Q}) \to \mathbb{Q}$ vanishes, and therefore the hypotheses of both Lemmas 4.12 and 4.13 are satisfied.//

Definition 4.15. Let $L \subset M$ be a link in a $\mathbb{Z}$-homology 3-sphere M, and let $K \subset M - L$ be a knot in its complement. The knot K is homologically split from L if each component $\tilde{K} \subseteq \pi^{-1}(K)$ lying over K in the universal abelian covering space $\pi : \tilde{M} \to M - L$, is null-homologous in $H_1(\tilde{M}; \mathbb{Z})$.

Remark. It is not hard to show that a necessary and sufficient condition for K to be homologically split from a knot K', is that K and K' span disjoint, oriented, compact surfaces in M.

Theorem 4.16. Let $L \cup K_1 \cup \ldots \cup K_n \subset M$, $L' \cup K_1' \cup \ldots \cup K_n' \subset M'$ be $\mathbb{Z}$-concordant links in the $\mathbb{Z}$-homology 3-spheres M, M', respectively. Let $L \subset M$, $L' \subset M'$ be sublinks, and suppose that each K_i (K_i', respectively) is homologically split from L (L', respectively), $i = 1, \ldots, n$. Let $\tilde{M} \to M - L$, $\tilde{M}' \to M' - L'$ be the universal abelian covering spaces.

Then $\Lambda_{\tilde{M}}(K_1, \ldots, K_n) \equiv \Lambda_{\tilde{M}'}(K_1', \ldots, K_n')$. (See Examples 5.1, 5.2.)

Proof: Let $F \subset N^4$ be a $\mathbb{Z}$-concordance between $L \subset M$ and $L' \subset M'$. It is a consequence of Example 3.14 that the covering spaces $\tilde{M} \to M - L$ and $\tilde{M}' \to M' - L'$ are related by N. Theorem 3.11 together with Proposition 3.5. imply that $\tilde{M} \to M - L$ and $\tilde{M}' \to M' - L'$ extend uniquely to the universal abelian covering space $\tilde{N} \to N - F$. From Example 2.6, it follows that the intersection form $H_2(\tilde{N}; \mathbb{Q}) \times H_2(\tilde{N}; \mathbb{Q}) \to \mathbb{Q}$ vanishes. Now apply Lemma 4.13.//

Theorem 4.17. Let $L = K_1 \cup \ldots \cup K_n \subset M$, $L' = K_1' \cup \ldots \cup K_n' \subset M'$ be $\mathbb{Z}$-concordant links in the $\mathbb{Z}$-homology 3-spheres M, M', respectively. Let $\pi : \tilde{M} \to M$ be an abelian branched (or unbranched) covering space, branched along a sublink of L. For each component $\tilde{K}_i \subseteq \pi^{-1}(K_i)$, $i = 1, \ldots, n$, suppose that $0 = [\tilde{K}_i] \in H_1(\tilde{M}; \mathbb{Q})$. Finally, let the index of $\tilde{M} \to M$ be a prime power p^s, $p \neq 1$.

Then there is a unique branched (or unbranched) abelian covering space $\pi' : \tilde{M}' \to M'$, branched along a sublink of L', which is related to $\pi : \tilde{M} \to M$. For each component $\tilde{K}_i' \subseteq (\pi')^{-1}(K_i')$,

$i = 1, \ldots, n$, it is the case that $0 = [\tilde{K}_i'] \in H_1(\tilde{M}'; \mathbb{Q})$; and

$$\Lambda_{\tilde{M}}(K_1, \ldots, K_n) \equiv \Lambda_{\tilde{M}'}(K_1', \ldots, K_n') .$$

Proof: The abelian branched (unbranched) covering space $\pi' : \tilde{M}' \to M'$ exists by Example 3.14. Let $\tilde{N} \to N$ be the unique extension of $\pi : \tilde{M} \to M$ and $\pi' : \tilde{M}' \to M'$. By Example 2.3 (since every abelian covering space is also iterated cyclic, and every $\mathbb{Z}$-concordance is also a $\mathbb{Z}_p$-concordance), the inclusion-induced maps $i_*' : H_j(\tilde{M}'; \mathbb{Z}_p) \to H_j(\tilde{N}; \mathbb{Z}_p)$ are isomorphisms for all $j \geq 0$. Thus $H_*(\tilde{N}, \tilde{M}'; \mathbb{Z}_p) = 0$, and consequently, the hypotheses of Corollary 4.14 is satisfied. The conclusion follows.//

Theorem 4.18. Let $L = K_1 \cup \ldots \cup K_n \subset M$, $L' = K_1' \cup \ldots \cup K_n' \subset M$ be $\mathbb{Z}_p$-concordant links in the $\mathbb{Z}_p$-homology 3-spheres M, M' , respectively, where p is a prime. Let $\tilde{M} \to M$ be a p-cylic, branched covering space, branched along a sublink of L . Suppose $0 = [\tilde{K}_i] \in H_1(\tilde{M}; \mathbb{Q})$, $i = 1, \ldots, n$.

Then there is a unique p-cyclic branched covering space $\tilde{M}' \to M'$, branched along a sublink of L' , which is related to $\tilde{M} \to M$. In it, $0 = [\tilde{K}_i'] \in H_1(\tilde{M}'; \mathbb{Q})$, $i = 1, \ldots, n$, and $\Lambda_{\tilde{M}}(K_1, \ldots, K_n) \equiv \Lambda_{\tilde{M}'}(K_1', \ldots, K_n')$. (See example 5.3.)

Proof: By Example 3.14, there is a unique regular, p-cyclic, branch covering space $\tilde{M}' \to M'$, which is related to $\tilde{M} \to M$. Let $\tilde{N} \to N$ be the unique extension of $\tilde{M}' \to M'$ and $\tilde{M} \to M$. By Example 2.3, we have that the inclusion induced maps $i_* : H_j(\tilde{M}'; \mathbb{Z}_p) \to H_j(N; \mathbb{Z}_p)$ are isomorphisms for all $j \geq 0$. The rest of the proof is the same as that of Theorem 4.11.//

Theorem 4.19. Let $L = K_1 \cup \ldots \cup K_n \subset M$, $L' = K_1' \cup \ldots \cup K_n' \subset M'$ be $\mathbb{Z}_p$-concordant links, and let $F \subset N^4$ be a particular $\mathbb{Z}_p$-concordance between them. Let $\tilde{M} \to M$ be a regular, iterated p-cyclic branched (unbranched) covering space, branched along a sublink of L . Suppose $0 = [\tilde{K}_i] \in H_1(\tilde{M}; \mathbb{Q})$, $i = 1, \ldots, n$.

Then there is a unique regular, iterated p-cyclic branched (unbranched) covering space $\tilde{M}' \to M'$, branched along a sublink of L' , which is related to $\tilde{M} \to M$ by N . In it, $0 = [\tilde{K}_i'] \in H_1(\tilde{M}'; \mathbb{Q})$, $i = 1, \ldots, n$, and $\Lambda_{\tilde{M}}(K_1, \ldots, K_n) \equiv \Lambda_{\tilde{M}'}(K_1', \ldots, K_n')$.

Proof: By Example 3.13, there is a unique, regular, iterated p-cyclic branched (unbranched) covering space $\tilde{M}' \to M'$, branched along a sublink of L' , which is related to $\tilde{M} \to M$ by N . Let $\tilde{N} \to N$ be the unique extension of $\tilde{M} \to M$ and $\tilde{M}' \to M'$. The rest of the proof is the same as that of Theorem 4.18.//

Defining $\Lambda_{\tilde{M}}(K_1, \ldots, K_n)$ for irregular covering spaces.

If the covering space $\tilde{M} \to M$ is irregular, we proceed exactly as for regular covering spaces, with the following changes:

We will assume that we can map surfaces $S_1, \ldots, S_n$ into M by maps $g_i : S_i \to M$, satisfying (1)-(3) in the Computation of $\Lambda_{\tilde{M}}(K_1, \ldots, K_n)$ for Regular Covering Spaces, and

(4) $\text{image}(g_i)_* \subseteq \pi_1(M)$ is a normal subgroup for $i = 1, \ldots, n$, where $(g_i)_* : \pi_1(S_i) \to \pi_1(M)$ is defining by joining K_i to the basepoint $e \in M$. (For example, if each closed curve K_i were homotopically trivial in M, then this would be possible.)

Definition 4.20. Define $\Lambda_{\tilde{M}}(K_1, \ldots, K_n)$ to be the $n \times n$ matrix $[\lambda_{ij}]$ with entries $\lambda_{ij} \in \mathbb{Q}(\pi_1(M))$, given by $\lambda_{ij} = w_{ij}$ (where w_{ij} is defined in Definition 4.8.).

To make $\Lambda_{\tilde{M}}(K_1, \ldots, K_n)$ independent of the way in which the link is connected to the basepoint $e \in M$, we introduce the following equivalence relation:

Definition 4.21. Let Λ, Λ' be $n \times n$ matrices with entries in $\mathbb{Q}(\pi_1(M))$. Define Λ to be equivalent to Λ' (denoted $\Lambda \sim \Lambda'$) if there is a diagonal matrix $B = \begin{bmatrix} \gamma_1 & & 0 \\ & \ddots & \\ 0 & & \gamma_n \end{bmatrix}$ with entries $\gamma_i \in \pi_1(M)$, such that $\Lambda' = B\Lambda B^{-1}$.

Proposition 4.22. Let $\Lambda^!_{\tilde{M}}(K_1, \ldots, K_n)$ be the matrix defined in Definition 4.20 for a different way of connecting the link $L \subset M$ to the basepoint $e \in M$. Then $\Lambda^!_{\tilde{M}}(K_1, \ldots, K_n) \sim \Lambda_{\tilde{M}}(K_1, \ldots, K_n)$.

Proof: Let $L = K_1 \cup \ldots \cup K_n \subset M$ be connected to the basepoint $e \in M$ by different paths $\delta_i^!$, $i = 1, \ldots, n$. For some elements $\gamma_i \in \pi_1(M, e)$, $i = 1, \ldots, n$, $[\delta_i^!] \simeq \gamma_i[\delta_i]$. Then,

$$\lambda_{ij}^! = w_{ij}^! = \frac{1}{n_i n_j} \sum_{\substack{p \in \text{int}(S_i),\ q \in \partial S_j \\ g_i(p) = g_j(q)}} \text{sign}(p) \cdot \gamma_{i,p,q}^!$$

where $\gamma_{i,p,q}^! = \gamma_i \cdot \gamma_{i,p,q} \cdot \gamma_j^{-1}$. (To see this, replace $[\delta_i]$ by

$\gamma_i[\delta_i]$ in the Construction.) Now let $B = \begin{bmatrix} \gamma_1 & & 0 \\ & \ddots & \\ 0 & & \gamma_n \end{bmatrix}$. It is easily seen that $\Lambda'_{\tilde{M}}(K_1, \ldots, K_n) = B\, \Lambda_{\tilde{M}}(K_1, \ldots, K_n) B^{-1}$ .//

It is significantly more complicated (and not very satisfactory for computations) to introduce a further equivalence relation which would make $[\Lambda_{\tilde{M}}(K_1, \ldots, K_n)]$ an invariant of R-concordance. One problem is that the matrices $\Lambda_{\tilde{M}}(K_1, \ldots, K_n)$ and $\Lambda_{\tilde{M}'}(K'_1, \ldots, K'_n)$ for R-concordant links $L = K_1 \cup \ldots \cup K_n \subset M$, $L' = K'_1 \cup \ldots \cup K'_n \subset M'$, do not have entries in the same group algebra. (It would be necessary to map $\pi_1(M)$, $\pi_1(M')$ to the set of right cosets of $\pi_1(M)$, $\pi_1(M')$, respectively, associated to the covering spaces $\tilde{M} \to M$, $\tilde{M}' \to M$, as in §1.)

I will not pursue this any further. Instead, let me state the following theorem, whose easy proof is left to the reader:

<u>Definition 4.23</u>. A link $L = K_1 \cup \ldots \cup K_n \subset M$ is a <u>split link</u>, if there are disjoint 3-cells $E_1, \ldots, E_n \subset M$ with $K_i \subset E_i$.

<u>Theorem 4.24</u>. Let $L \cup K_1 \cup \ldots \cup K_n \subset M$ be a link in the $\mathbb{Z}_p$-homology 3-sphere M , where p is a prime, $p \neq 1$, and let $L \subset M$ be a sublink. Suppose $\tilde{M} \to M - L$ is an iterated p-cyclic, irregular covering space.

If $L \cup K_1 \cup \ldots \cup K_n \subset M$ is $\mathbb{Z}_p$-concordant to a link $L' \cup K'_1 \cup \ldots \cup K'_n \subset M'$, such that $K'_1 \cup \ldots \cup K'_n \subset M' - L'$ is a split link, then $\Lambda_{\tilde{M}}(K_1, \ldots, K_n) = 0$.

§5. Computations and Examples

In this section, I compute the invariants discussed in Section 4 for particular links in S^3 .

Figure 2.

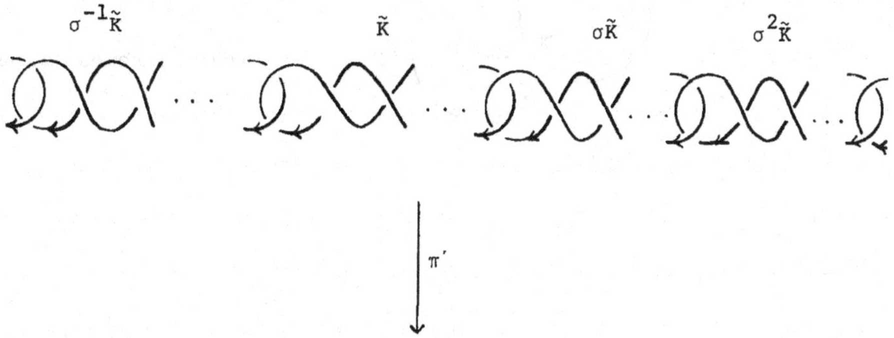

Figure 3.

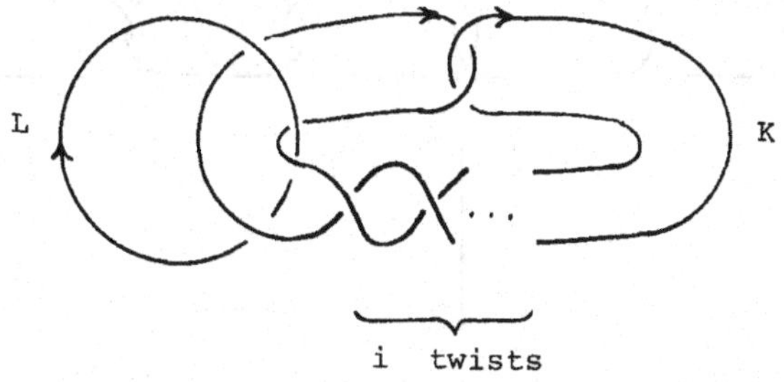

Example 5.1. Let $L \cup K \subset S^3$ be the link in Figure 3. Let $\pi : \tilde{M} \to S^3 - L$ be the infinite cyclic covering space, and let σ generate the cyclic group of covering translations of $\tilde{M}$. Note that $\tilde{M} \cong \mathbb{R}^3$. Choose an oriented component $\tilde{K} \subset \pi^{-1}(K)$; the link $\pi^{-1}(K)$ is illustrated in Figure 2.

$$\text{Then} \quad \Lambda_{\tilde{M}}(K) = [-\sigma^{-1} - \sigma] \ .$$

Application: The link $L \cup K \subset S^3$ in Figure 3, is not $\mathbb{Z}$-concordant to a trivial link of two components. (In fact, it is not $\mathbb{Z}$-concordant to a homologically split link of two components.)

Figure 4.

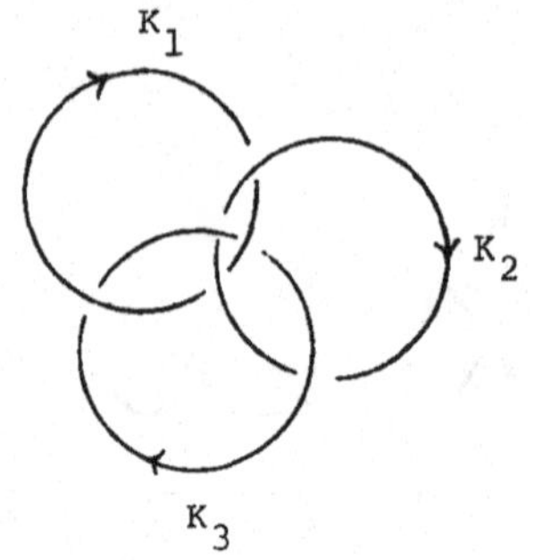

(the Borromean rings)

Figure 5.

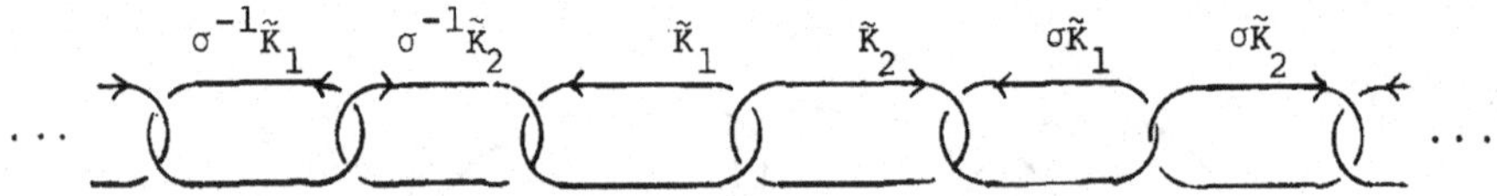

Figure 6.

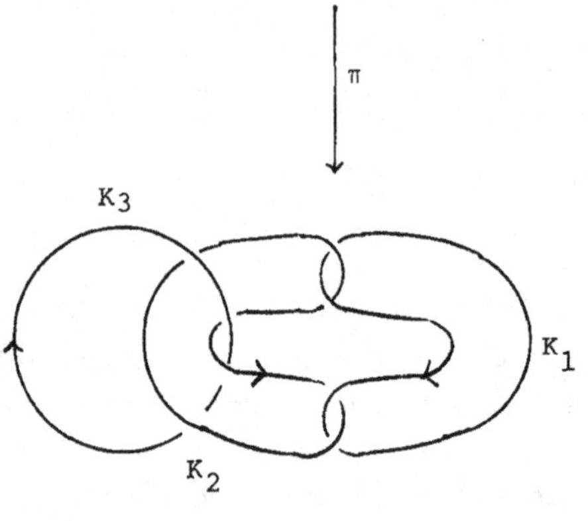

Example 5.2. Let $L \subset S^3$ be the Borromean Rings (Figure 4). It is not hard to see that this link is ambient isotopic to the link $K_1 \cup K_2 \cup K_3 \subset S^3$ in Figure 6. Let $\pi : \tilde{M} \to S^3 - K_3$ be the infinite cyclic covering space, and let σ generate the cyclic group of covering translations of $\tilde{M}$. Note that $\tilde{M} \simeq \mathbb{R}^3$. Choose oriented components $\tilde{K}_1 \subset \pi^{-1}(K_1)$ and $\tilde{K}_2 \subset \pi^{-1}(K_2)$; then the link $\pi^{-1}(K_1 \cup K_2)$ is illustrated in Figure 6.

$\Lambda_{\tilde{M}}(K_1 \cup K_2)$ is the matrix $\begin{bmatrix} 0 & 1-\sigma^{-1} \\ 1-\sigma & 0 \end{bmatrix}$.

<u>Application</u>: The Borromean Rings are not $\mathbb{Z}$-concordant to a trivial link (or a split link).

<u>Figure 7</u>.

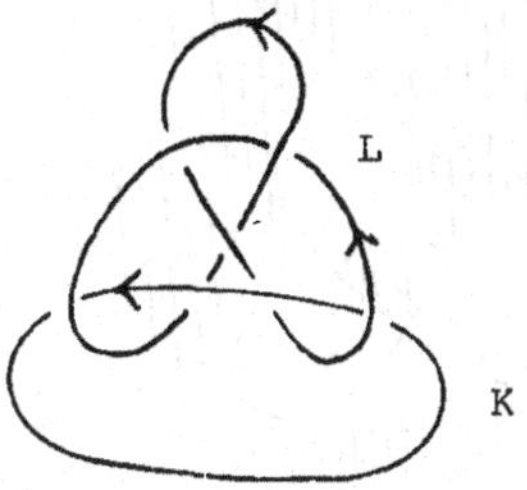

<u>Example 5.3</u>. The main point of this example is to compute the matrix $\Lambda_{\tilde{M}}(K)$ in a slightly more difficult case, when the branched covering space $\tilde{M}$ is not S^3 or $\mathbb{R}^3$.

Let $L \cup K \subset S^3$ be the link in Figure 7. Let $\hat{M}_n \to S^3$ be the n-fold cyclic branched covering space of S^3 branched along L , and let $\tilde{M}_n \to S^3 - L$ be the associated, unbranched covering space; let $\pi : \tilde{M} \to S^3 - L$ be the infinite cyclic covering space. Let t generate the infinite cyclic group of covering translations of $\tilde{M}$; let a generator for the finite cyclic group of covering translations of $\hat{M}_n$ and (by restriction) $\tilde{M}_n$ be denoted by t as well.

We will need the following information about the trefoil knot L :

<u>Property 1</u>. There is a compact, oriented surface $F \subset S^3$ with $\partial F = L$ (see Figure 8), and a homeomorphism $\tau : F \to F$ satisfying $\tau | \partial F = \mathrm{id}$, such that S^3 is the space $F \times I/(*)$ with identifications

$$(*) \quad \begin{cases} (x, 1) \sim (\tau(x), 0) & \text{for all } x \in F \\ (x, s) \sim (x, 0) & \text{for all } 0 \leq s \leq 1 \text{ , and } x \in \partial F \text{ .} \end{cases}$$

(We say that L is a fibered knot with fiber F , or that S^3 has a book structure with binding L , leaf F and monodromy $\tau : F \to F$.)

Figure 8.

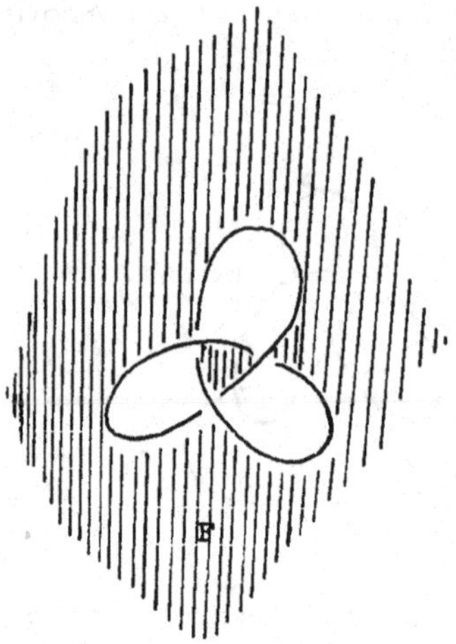

Figure 9.

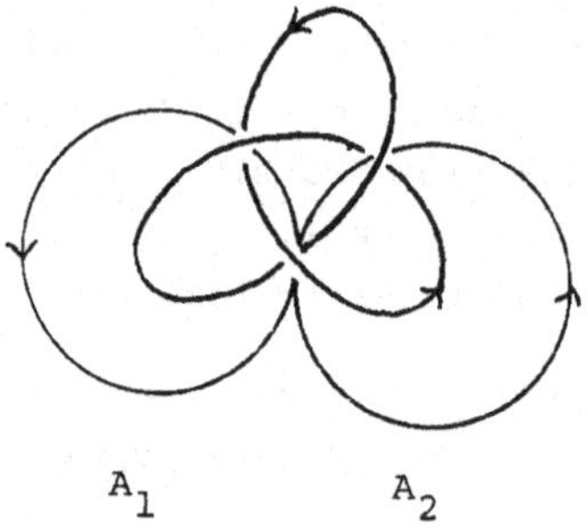

Figure 10.

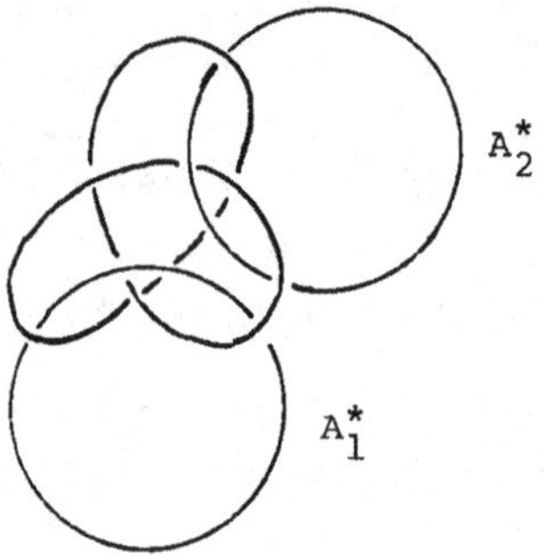

Property 2. If $A_1, A_2 \subset F$ are simple closed curves on F representing a basis for $H_1(F; \mathbb{Z})$ (see Figure 9), then the monodromy matrix $T = \tau_* : H_1(F; \mathbb{Z}) \to H_1(F; \mathbb{Z})$ in terms of this basis is

$$T = \begin{bmatrix} 0 & -1 \\ 1 & 1 \end{bmatrix} .$$

Note that $T^6 = I$.

Claim. $\tilde{M} \cong F \times \mathbb{R}$, and the covering translation $t : \tilde{M} \to \tilde{M}$ is defined by $t[x, s] = [\tau(x), s + 1]$.

Claim. $\tilde{M}_n$ is homeomorphic to the space $F \times I/(**)$ with identifications

$(**)\quad (x, 1) \sim (\tau^n(x), 0)$ for all $x \in F$;

$\hat{M}_n$ is homeomorphic to the space $F \times I/(**)$, $(**)'$ with identifications (**) and

$(**)'\quad (x, s) \sim (x, 0)$ for all $0 \leq s \leq 1$ and $x \in \partial F$.

Proof: Let $p_n : \tilde{M} \to M_n$ be the infinite cyclic covering space of $\tilde{M}_n$. Then the cyclic group of covering translations is generated by $t^n : \tilde{M} \to \tilde{M}$. //

Claim. The branched covering space $\tilde{M}_n$ is a $\mathbb{Q}$-homology 3-sphere if $n \neq 0$ (6), and $\hat{M}_n$ is a $\mathbb{Z}$-homology 3-sphere if $n = 1, 5$ (6).

The unbranched covering space $\tilde{M}_n$ is a $\mathbb{Q}$-homology circle if $n \neq 0$ (6), and $\tilde{M}_n$ is a $\mathbb{Z}$-homology circle if $n = 1, 5$ (6).

Proof: A relation matrix for $H_1(\hat{M}_n; \mathbb{Z})$ in terms of the generators $[A_1]$, $[A_2]$, is $T^n - I$. This matrix is $\mathbb{Q}$-invertible if $n \neq 0$ (6), and it is $\mathbb{Z}$-invertible if $n = 1, 5$ (6). //

Now, if $\gamma \subset F$ is a simple closed curve, let $\gamma^* \subset \tilde{M}_n - F$ denote γ pushed off of F in a positive normal direction, and let γ_* denote the curve γ pushed off of F in the opposite direction. Let the Seifert pairing $\theta_n : H_1(F, \mathbb{Z}) \times H_1(F; Z) \to \mathbb{Q}$ be defined by $\theta_n(x, y) = \ell_{\tilde{M}_n}(x, y^*)$.

The short exact sequence

$$0 \to C_*(\tilde{M}) \xrightarrow{t^n - 1} C_*(\tilde{M}) \xrightarrow{p_n} C_*(\tilde{M}_n) \to 0$$

of chain complexes, where $p_n : \tilde{M} \to \tilde{M}_n$ is the infinite cyclic covering space, gives rise to a long exact sequence

$$(***)\quad \cdots \xrightarrow{p_n} H_2(\tilde{M}_n) \xrightarrow{\partial} H_1(\tilde{M}) \xrightarrow{t^n - 1} H_1(\tilde{M}) \xrightarrow{p_n} H_1(\tilde{M}_n) \xrightarrow{\partial} \cdots$$

of homology, as in Assertion 5 of [M1]. Let $i : F \hookrightarrow \tilde{M}$ be the inclusion $F \to F \times 0 \subset F \times \mathbb{R}$, and let $i : F \hookrightarrow \tilde{M}_n$ be the composition of i with projection p_n . When $n \neq 0$ (6), define the isomorphism $f_n : H_1(F; \mathbb{Q}) \to H_1(F; \mathbb{Q})$ to make the diagram

$$\begin{array}{ccc} H_1(\tilde{M}) & \xrightarrow{t^n - 1} & H_1(\tilde{M}) \\ \uparrow i_* & & \uparrow i_* \\ H_1(F) & \xleftarrow{f_n} & H_1(F) \end{array}$$

commutative. (The map $t^n - 1$ is an isomorphism of $H_1(\tilde{M}; \mathbb{Q})$ to itself if $n \neq 0$ (6), since a matrix for $t^n - 1$ in terms of the basis $[A_1]$, $[A_2]$, is $T^{-n} - I$, which is $\mathbb{Q}$-invertible. Similarly, $t^n - 1$ is an isomorphism of $H_1(\tilde{M}; \mathbb{Z})$ to itself if $n = 1, 5$ (6), for then $T^{-n} - I$ is $\mathbb{Z}$-invertible.

Property 3. A matrix F_n for the linear transformation $f_n : H_1(F; \mathbb{Q}) \to H_1(F; \mathbb{Q})$, in terms of the basis $[A_1]$, $[A_2]$, is $F_n = (T^{-n} - I)^{-1}$; thus:

$$F_n = \begin{bmatrix} -1 & -1 \\ 1 & 0 \end{bmatrix} \quad \text{if} \quad n = 1 \ (6)$$

$$F_n = \frac{1}{3}\begin{bmatrix} -2 & -1 \\ 1 & 1 \end{bmatrix} \quad \text{if} \quad n = 2 \ (6)$$

$$F_n = -\frac{1}{2} I \quad \text{if} \quad n = 3 \ (6)$$

$$F_n = \frac{1}{3}\begin{bmatrix} -1 & 1 \\ -1 & -2 \end{bmatrix} \quad \text{if} \quad n = 4 \ (6)$$

$$F_n = \begin{bmatrix} 0 & 1 \\ -1 & -1 \end{bmatrix} \quad \text{if} \quad n = 5 \ (6)$$

Claim. If $n \neq 0$ (6), the Seifert pairing $\theta_n : H_1(F; \mathbb{Q}) \times H_1(F; \mathbb{Q}) \to \mathbb{Q}$ is given by

$$\theta_n(x, y) = \langle x, f_n(y) \rangle ,$$

where $\langle , \rangle$ is the intersection pairing on $H_1(F; \mathbb{Q})$.

Claim. If $n \neq 0$ (6), then the linking pairing $\ell_{\tilde{M}_n}(x, t^i y^*)$ is given by $\theta(x, T^{-i} y) = \theta(T^i x, y)$.

In terms of the basis $[A_1]$, $[A_2]$, a matrix for the bilinear pairing $\ell_{\tilde{M}_n}(x, t^i y^*)$ is

$$T^i J F_n^t$$

where F_n^t is the matrix F_n transposed, and where $J = \begin{bmatrix} 0 & 1 \\ -1 & 0 \end{bmatrix}$ is a matrix for the intersection pairing $\langle , \rangle$ on $H_1(F; \mathbb{Q})$, in terms of the basis $[A_1]$, $[A_2]$.

Property 4. A matrix for the bilinear pairing $\ell_{\tilde{M}_n}(x, t^i y^*)$ on $H_1(F; \mathbb{Q})$ in terms of the basis $[A_1]$, $[A_2]$, is

	$n = 1$ (6)	$n = 2$ (6)
$i = 0$ (6)	$\begin{bmatrix} -1 & 0 \\ 1 & -1 \end{bmatrix}$	$\frac{1}{3}\begin{bmatrix} -1 & -1 \\ 2 & -1 \end{bmatrix}$
$i = 1$ (6)	$\begin{bmatrix} -1 & 1 \\ 0 & -1 \end{bmatrix}$	$\frac{1}{3}\begin{bmatrix} -2 & 1 \\ 1 & -2 \end{bmatrix}$
$i = 2$ (6)	$\begin{bmatrix} 0 & 1 \\ -1 & 0 \end{bmatrix}$	$\frac{1}{3}\begin{bmatrix} -1 & 2 \\ -1 & -1 \end{bmatrix}$
$i = 3$ (6)	$\begin{bmatrix} 1 & 0 \\ -1 & 1 \end{bmatrix}$	$\frac{1}{3}\begin{bmatrix} 1 & 1 \\ -2 & 1 \end{bmatrix}$
$i = 4$ (6)	$\begin{bmatrix} 1 & -1 \\ 0 & 1 \end{bmatrix}$	$\frac{1}{3}\begin{bmatrix} 2 & -1 \\ -1 & 2 \end{bmatrix}$
$i = 5$ (6)	$\begin{bmatrix} 0 & -1 \\ 1 & 0 \end{bmatrix}$	$\frac{1}{3}\begin{bmatrix} 1 & -2 \\ 1 & 1 \end{bmatrix}$

	$n = 3\ (6)$	$n = 4\ (6)$
$i = 0\ (6)$	$\begin{bmatrix} 0 & -1 \\ 1 & 0 \end{bmatrix}$	$\frac{1}{3}\begin{bmatrix} 1 & -2 \\ 1 & 1 \end{bmatrix}$
$i = 1\ (6)$	$\begin{bmatrix} -1 & 0 \\ 1 & -1 \end{bmatrix}$	$\frac{1}{3}\begin{bmatrix} -1 & -1 \\ 2 & -1 \end{bmatrix}$
$i = 2\ (6)$	$\begin{bmatrix} -1 & 1 \\ 0 & -1 \end{bmatrix}$	$\frac{1}{3}\begin{bmatrix} -2 & 1 \\ 1 & -2 \end{bmatrix}$
$i = 3\ (6)$	$\begin{bmatrix} 0 & 1 \\ -1 & 0 \end{bmatrix}$	$\frac{1}{3}\begin{bmatrix} -1 & 2 \\ -1 & -1 \end{bmatrix}$
$i = 4\ (6)$	$\begin{bmatrix} 1 & 0 \\ -1 & 1 \end{bmatrix}$	$\frac{1}{3}\begin{bmatrix} 1 & 1 \\ -2 & 1 \end{bmatrix}$
$i = 5\ (6)$	$\begin{bmatrix} 1 & -1 \\ 0 & 1 \end{bmatrix}$	$\frac{1}{3}\begin{bmatrix} 2 & -1 \\ -1 & 2 \end{bmatrix}$

$n = 5\ (6)$

$i = 0\ (6)$	$\begin{bmatrix} 1 & -1 \\ 0 & 1 \end{bmatrix}$	$i = 3\ (6)$	$\begin{bmatrix} -1 & 1 \\ 0 & -1 \end{bmatrix}$
$i = 1\ (6)$	$\begin{bmatrix} 0 & -1 \\ 1 & 0 \end{bmatrix}$	$i = 4\ (6)$	$\begin{bmatrix} 0 & 1 \\ -1 & 0 \end{bmatrix}$
$i = 2\ (6)$	$\begin{bmatrix} -1 & 0 \\ 1 & -1 \end{bmatrix}$	$i = 5\ (6)$	$\begin{bmatrix} 1 & 0 \\ -1 & 1 \end{bmatrix}$

Finally, we can now compute $\Lambda_{\tilde{M}_n}(K) = \sum_{i=0}^{n-1} \ell_{\tilde{M}_n}(K, t^i K) \cdot t^i =$
$\sum_{i=1}^{n-1} \ell_{\tilde{M}_n}(K, t^i K^*) \cdot t^i$.

(The only property of K which is essential for this computation, is that $K = A_1^* \subset S^3 - L$. It is a simple matter to compute $\Lambda_{\tilde{M}_n}(K_1 \cup \ldots \cup K_m)$ for any link $K_1 \cup \ldots \cup K_m \subset S^3 - L$, such that $K_i = \gamma^*$ for some simple closed curve $\gamma \subset F$.)

<u>Computation</u>. $\Lambda_{\tilde{M}_n}(K)$ is

$n = 6k+1$: $(-t + t^3 + t^4 - t^6) \cdot (1 + t^6 + t^{12} + \ldots + t^{6(k-1)})$

$n = 6k+2$: $(-\frac{2}{3}t - \frac{1}{3}t^2 + \frac{1}{3}t^3 + \frac{2}{3}t^4 + \frac{1}{3}t^5 - \frac{1}{3}t^6) \cdot (1 + t^6 + t^{12} + \ldots$
$\ldots + t^{6(k-1)}) - \frac{2}{3}t^{6k+1}$

$n = 6k+3$: $(-\frac{1}{2}t - \frac{1}{2}t^2 + \frac{1}{2}t^4 + \frac{1}{2}t^5) \cdot (1 + t^6 + t^{12} + \ldots + t^{6(k-1)})$
$- \frac{1}{2}t^{6k+1} - \frac{1}{2}t^{6k+2}$

$n = 6k+4$: $(-\frac{1}{3}t - \frac{2}{3}t^2 - \frac{1}{3}t^3 + \frac{1}{3}t^4 + \frac{2}{3}t^5 + \frac{1}{3}t^6) \cdot (1 + t^6 + t^{12} + \ldots$
$\ldots + t^{6(k-1)}) - \frac{1}{3}t^{6k+1} - \frac{2}{3}t^{6k+2} - \frac{1}{3}t^{6k+3}$

$n = 6k+5$: $(-t^2 - t^3 + t^5 + t^6) \cdot (1 + t^6 + t^{12} + \ldots + t^{6(k-1)})$
$- t^{6k+2} - t^{6k+3}$.

<u>Application</u>. The link $L \cup K$ in Figure 7 is not $\mathbb{Z}$-concordant to the split link $L' \cup K' \subset S^3$ in Figure 11.

<u>Figure 11</u>.

§6. Some Concluding Remarks

This section contains the statements of two theorems, proofs omitted, as well as the posing of two problems, which naturally arise from this research, and whose solutions would be interesting.

Theorem 6.1. The invariant $[\Lambda_{\tilde{M}}(L)]$ of Definition 4.1 is an invariant of the isotopy class of $L \subset M$ (see [M3]).

Theorem 6.2. The invariant $[\Lambda_{\tilde{M}}(K_1, \ldots, K_n)]$ of Theorem 4.16 is an invariant of the homotopy class of the link $K_1 \cup \ldots \cup K_n \subset M - L$ (see [M2]).

The theme of this paper has been to derive invariants of various equivalence relations on links (notably concordance) from branched and unbranched covering spaces associated with these links.

Problem 6.3. Let $\equiv$ denote an equivalence relation on classical links $L \subset M^3$. Describe all branched and unbranched covering spaces $\tilde{M} \to M$ for which 'related' covering spaces $\tilde{M}' \to M'$ exist, and may be algorithmically constructed, for any equivalent link $L' \subset M'$, $L \equiv L'$.

Problem 6.4. Find invariants of the covering spaces described in Problem 6.3, or some subclass thereof, which are identical for all links in the same $\equiv$ class.

For example, if $\equiv$ denotes $\mathbb{Z}$-concordance, one may take abelian branched covering space whose branch set is a sublink of L, as a solution to Problem 6.3; for Problem 6.4, one may restrict attention only to those abelian branched covering spaces having prime power index p^s, and choose for the invariant, $[\Lambda_{\tilde{M}}(L)]$.

REFERENCES

[C-S] Cappell, S. and Shaneson, J., "Invariants of 3-Manifolds", Bull. Amer. Math. Soc., 81 (1975), 559-562.

[Cl] Casson, A.J., "Link Cobordism and Milnor's Invariant", Bull. London Math. Soc ., 7 (1975), 39-40.

[C-G] Casson, A.J. and Gordon, Cameron Mc.A., "Classical Knot Cobordism", Lectures given in Jan.-Feb. 1975 at Orsay, France.

[C-F-L] Chen, K.T., Fox, R.H., and Lyndon, R.C., "Free Differential Calculus, IV. The Quotient Groups of the Lower Central Series", Annals of Mathematics, 68 (1958), 81-95.

[D-K] Durfee, Alan and Kauffman, L.H., "Periodicity of Branched Cyclic Covers", Math. Ann., 218 (1975) 157-174.

[F] Fox, R.H., "Covering Spaces with Singularities", Algebraic Geometry and Topology (a symposium in Honor of S. Lefschetz), 243-257 (Princeton, 1957).

[Gi] Giffen, Charles, "Link Concordance Implies Link Homotopy", to appear.

[Go 1] Goldsmith, Deborah L., "Concordance Implies Homotopy for Classical Links in M^3", to appear.

[Go 2] __________, "A Signature Invariant of Classical Link Concordance", to appear.

[Gon] Gonzalez-Acuna, F., "Dehn's Construction on Knots", Bol. Soc. Mat. Mexicana, 15, No. 2 (1970) 58-79.

[Gr] Greenberg, M., "Lectures on Algebraic Topology", W.A. Benjamin, Inc., New York, 1967.

[H] Hirzebruch, F., "The Signature of Ramified Coverings", Global Analysis Papers in honor of K. Kadaira, University of Tokyo Press, Princeton University Press, 1969, 253-265.

[K] Kauffman, L.H., "Open Books, Branched Covers and Knot Periodicity", Topology, 13 (1974), 143-160.

[K-T] Kauffman, L.H. and Taylor, L., "Signature of Links", Trans. Amer. Math. Soc., 216 (1976), 351-365.

[L] Laufer, Henry B., "Some Numerical Link Invariants", Topology, 10 (1971), 119-131.

[Le] Levine, J., "An Algebraic Classification of some Knots of

Codimension Two", Comm. Math. Hel., 45 (1970), 185-198.

[M 1] Milnor, J., "Infinite Cyclic Coverings", Conference on the Topology of Manifolds at Michigan State University, 1967, edited by J.G. Hocking; Boston, Prindle, Weber & Schmidt (1968).

[M 2] ______, "Link Groups", Ann. of Math., 59 (1954), 177-195.

[M 3] ______, "Isotopy of Links", Algebraic Geometry and Topology (a symposium in honor of S. Lefschetz), 280-306 (Princeton, 1957).

[Mu] Murasugi, K., "On a Certain Numerical Invariant of Link Types", Trans. Amer. Math. Soc., 117 (1965), 387-422.

[R] Robertello, R.A., "An Invariant of Knot Cobordisms", Comm. Pure Appl. Math., (1965), 115.

[S] Stallings, John, "Homology and Central Series of Groups", Journal of Algebra, 2 (1965), 170-181.

[T] Tristram, A.G., "Some Cobordism Invariants for Links", Proc. Camb. Phil. Soc., 66 (1969), 251-264.

[Pr] Problems in Low Dimensional Manifold Theory, Stanford conference in algebraic topology (Aug. 1976), edited by R. Kirby.

University of Michigan
Ann Arbor, Michigan 48109

N O E U D S   A N T I S I M P L E S

Jean-Claude HAUSMANN

1. Introduction

Soit N un n-noeud, i.e. une sous-variété close de S^{n+2} homéomorphe à S^n.

Le concept classique de "noeud simple" peut s'exprimer par le fait que la variété complémentaire $X(N) = S^{n+2}-N$ s'obtient, à partir de $S^1 \times D^{n+1}$, par attachement d'anses dont l'indice avoisine le milieu de la dimension. Les noeuds simples ont été classifiés à isotopie près par J. Levine [L 3] .

Les noeuds étudiés ici, dits "antisimples", ont au contraire leur X(N) constructible sans anses au milieu de la dimension. On donne une classification des noeuds antisimples à isotopie près (et aussi à homéomorphisme de complémentaire près). Sous certaines hypothèses, le problème est ramené à la connaissance d'objets de nature purement homotopique. Comme application, on en déduit notamment un théorème de finitude pour les classes d'isotopie de noeuds dont le complémentaire a un type d'homologie tordue donné.

Les méthodes utilisées sont des cas particuliers d'une théorie générale des variétés sans anses au milieu de la dimension [H 2]. Cependant, le fait que l'on a affaire à des complémentaires de noeuds et quelques hypothèses apportent à la théorie générale des simplifications considérables et des particularités qui en font une théorie indépendante et intéressante en soi. Aussi cet article est-il essentiellement "self-contained".

Terminons cette introduction par la table des matières :

2. Définitions et notations

Nous travaillons dans la catégorie des variétés semi-linéaires (PL). La sphère standard de dimension n est notée comme d'habitude S^n. Les sous-variétés sont toujours localement plates.

(2.1.) Un n-noeud est une sous-variété orientée N de S^{n+2} qui est homéomorphe à S^n. Le complémentaire d'un voisinage ouvert de N dans S^{n+2} est noté X(N) ou X s'il n'y a pas de risque de confusion. Les orientations de N et de S^{n+2} déterminent un élément "méridien" $\alpha(N) \in \pi_1(x)$ et, si $n \geqslant 2$, une unique trivialisation du fibré normal à N.

(2.2) Deux n-noeuds N_0 et N_1 sont dits isotopes s'il existe un automorphisme orienté H de $S^{n+2} \times I$ préservant les niveaux tel que $H(x,o) = (x,o)$ et $H(N_0 \times 1) = N_1 \times 1$. On note par Ω_n l'ensemble des classes d'isotopie de n-noeuds.

(2.3) La relation d'équivalence entre n-noeuds suivante es· également considérée : N_0 est équivalent à N_1 s'il existe un homéomorphisme orienté $h : X(N_0) \longrightarrow X(N_1)$ tel que $\pi_1 h(\alpha(N_0)) = \alpha(N_1)$. On dénote par Ω_n^c l'ensemble des classes d'équivalence de n-noeuds pour cette relation. On a une surjection naturelle $\Omega_n \longrightarrow\!\!\!\rightarrow \Omega_n^c$ dont la pré-image d'un élément contient un ou deux éléments. [LS, Corollary 2.2.].

(2.4) Un n-noeud $N \subset S^{n+2}$ est dit <u>antisimple</u> si $X(N)$ a une décomposition en anses (partant d'une anse d'indice 0 dans int $X(N)$) sans anses d'indice :

- $k+1$ si $n = 2k$
- $k+1$ et $k+2$ si $n = 2k+1$.

Cette propriété est fermée pour les relations d'équivalence ci-dessus. Nous noterons par ${}^{as}\Omega_n \subset \Omega_n$ et ${}^{as}\Omega_n^c \subset \Omega_n^c$ les sous-ensembles formés des classes contenant un noeud antisimple.

(2.5) Soit K un polyèdre de dimension k tel que $H_*(K) = H_*(S^1)$ et que $\pi_1(K)$ soit la clôture normale d'un élément α . Pour tout $n \geq 2k$ le <u>n-noeud de Kervaire</u> <u>$N_n(K,\alpha)$</u> est construit de la manière suivante : on considère un voisinage régulier E_K^{n+3} d'un plongement de K dans R^{n+3} (un tel plongement est unique à isotopie près). La classe α est représentable par un plongement $\alpha : S^1 \longrightarrow \partial E_K^{n+3}$ et la parallélisation de ∂E_K^{n+3} induite par le plongement de E_K^{n+3} dans R^{n+3} détermine une trivialisation du fibré normal à α . On obtient donc un plongement $\bar{\alpha} : S^1 \times D^{n+1} \longrightarrow \partial E_K^{n+3}$ à l'aide duquel on effectue une chirurgie. La variété chirurgisée $(\partial E_K^{n+3} - S^1 \times \mathring{D}^{n+1}) \cup D^2 \times S^n$ est homéomorphe à

S^{n+2} dans laquelle $0 \times S^n \subset D^2 \times S^n$ donne le n-noeud $N_n(K, \alpha)$. Observons que $\pi_1(X(N_n(K, \alpha)) = \pi_1(K)$ et que $N_n(K, \alpha)$ est antisimple. La construction de $N_n(K, \alpha)$ est simplement une version canonique du procédé de [K1, démonstration du Théorème 1.1.] .

(2.6) Considérons des paires (K, α) comme dans (2.5). Deux telles paires (K_0, α_0) et (K_1, α_1) seront réputées équivalentes s'il existe une équivalence d'homotopie $f : K_0 \longrightarrow K_1$ telle que $\pi_1 f(\alpha_0) = \alpha_1$. On dénote par $\mathcal{K}_n$ l'ensemble des classes d'équivalence de paires (K, α) où K est de dimension $\leqslant [\frac{n}{2}]$.

(2.7) Toujours pour de telles paires (K, α), on choisit une fois pour toutes un voisinage régulier E_K^{2k+1} d'un plongement de K dans $\mathbb{R}^{2k+1}$ $(k = \dim K)$ et on définit par récurrence $E_K^m = E_K^{m-1} \times I$ pour $m > 2k+1$. Comme dans (2.5), α détermine un plongement $\overline{\alpha} : S^1 \times D^{m-1} \longrightarrow E_k^m$ et on note $\overline{\alpha}|S^1 \times 0 = \alpha$. Nous utiliserons les groupes suivants :

- $[E_K^m, E_K^m]_{rel\, \partial}$ = ensemble des classes d'homotopie (homotopies fixes sur ∂E_K^m) d'applications continues $f : E_K^m \longrightarrow E_K^m$ telles que $f|\, \partial E_K^m = \mathrm{id}$. Pour $m \geqslant 2k+2$, f est une équivalence d'homotopie puisque $E_K^m = E_K^{m-1} \times I$. $[E_K^m, E_K^m]_{rel\,\partial}$ est donc un groupe pour la composition des applications ; ce groupe est abélien car on peut disjoindre les supports.

- $\mathrm{Aut}(E_K^m)_{\mathrm{rel}\,\partial}$ = groupes des classes de concordance (concordances fixes sur ∂E_K^m) d'automorphismes orientés h de E_K^m tels que $h|\ \partial E_K^m = \mathrm{id}$.

- $\mathrm{Aut}(E_K^m, \bar{\alpha})$ = groupe des classes de concordance des automorphismes orientés h de E_K^m tels que $h \circ \bar{\alpha} = \bar{\alpha} : S^1 \times D^{m-1} \longrightarrow E_K^m$

- $\mathrm{Aut}(E_K^m, \alpha)$ = groupe des classes de concordance d'automorphismes h de E_K^m tels que $h \circ \alpha = \alpha : S^1 \times 0 \longrightarrow E_K^m$.

La correspondance $h \longrightarrow h \times \mathrm{id}_I$ donne des homomorphismes

$$\mathrm{Aut}(E_K^m, \bar{\alpha}) \longrightarrow \mathrm{Aut}(E_K^{m+1}, \bar{\alpha})$$

$$\mathrm{Aut}(E_K^m, \alpha) \longrightarrow \mathrm{Aut}(E_K^{m+1}, \alpha)$$

Par position générale, ces homomorphismes sont des isomorphismes. Les groupes $\mathrm{Aut}(E_K^m, \bar{\alpha})$ et $\mathrm{Aut}(E_K^m, \alpha)$ agissent à droite sur $[E_K^m, E_K^m]_{\mathrm{rel}\,\partial}$ et sur $\mathrm{Aut}(E_K^m)_{\mathrm{rel}\,\partial}$ par

$$f \cdot \beta = \beta^{-1} f \beta$$

et c'est une action par automorphisme de groupes .

3. Enoncés des théorèmes de classification :

Soit $N \subset S^{n+2}$ un noeud antisimple et $\alpha : S^1 \to X(N) = X$ l'élément méridien.

Choisissons une décomposition en anses de X sans anse au milieu de la dimension. La réunion des anses d'indice $\leqslant [\frac{n}{2}]$ donne un complexe fini K. Comme $\pi_1(K) = \pi_1(X)$, on peut considérer α comme élément de $\pi_1(K)$ et la paire (K, α) représente un élément de $\mathcal{K}_n$.

Lemme 3.1. La classe de $\mathcal{K}_n$ ainsi obtenue ne dépend que de la classe de N dans Ω_n^c. Les applications $\rho^c : {}^{as}\Omega_n^c \longrightarrow \mathcal{K}_n$ et $\rho : {}^{as}\Omega_n \longrightarrow \mathcal{K}_n$ déterminées de cette façon sont surjectives.

Démonstration : Soit (K^1, α^1) la paire obtenue à l'aide d'une décomposition en anse de $X^1 = X(N^1)$, où N^1 est un noeud antisimple représentent la même classe de Ω_n^c que N. Soit $h : X \longrightarrow X^1$ un homomorphisme orienté tel que $h \circ \alpha \sim \alpha^1$. Par position générale, on peut supposer que $h(K^*) \subseteq K^1$. Comme les inclusions $K \subset X$ et $K^1 \subset X^1$ sont $([\frac{n}{2}]+1)$-connexes, $h|K : K \longrightarrow K^1$ est une application $([\frac{n}{2}]+1)$-connexe entre complexes de dimensions $\leqslant [\frac{n}{2}]$, donc une équivalence d'homotopie. Les paires (K, α) et (K^1, α^1) représentent donc le même élément de $\mathcal{K}_n$.

La surjectivité de ρ et de ρ^c s'obtient en observant que $(K, \alpha) \in \mathcal{K}_n$ est l'image par ρ (ou ρ^c) du noeud de Kervaire $N_n(K, \alpha)$.

La classification des noeuds antisimples se réduit donc **aux deux problèmes suivants :**

1) la détermination de $\mathcal{K}_n$, problème homotopique qui ne sera pas discuté ici;

2) la détermination des préimages $\rho^{-1}(K, \alpha)$ et $(\rho^c)^{-1}(K, \alpha)$. Notons par ${}^{as}\Omega_n(K, \alpha)$ et ${}^{as}\Omega_n^c(K, \alpha)$ ces préimages. Leur calcul sous certaines hypothèses fait l'objet des théorèmes 3.2 et 3.3 ci-dessous.

Notons $\pi = \pi_1(K)$. $Wh(\pi)$ désigne le groupe de Whitehead de π et $L_n(\pi)$ le groupe d'obstruction à la chirurgie de Wall (cas orientable).

<u>Théorème 3.2</u>. Supposons que $Wh(\pi) = 0$. Alors, pour tout $n \geq 4$ il existe des bijections

$$ {}^{as}\Omega_n^c(K, \alpha) = Aut(E_K^{n+1})_{rel\,\partial} \big/ \text{action de } Aut(E_K^{n+1}, \bar{\alpha}) $$

$$ {}^{as}\Omega_n^c(K, \alpha) = Aut(E_K^{n+1})_{rel\,\partial} \big/ \text{action de } Aut(E_K^{n+1}, \alpha) \cdot $$

La classe de idE_K^{n+1} est celle du noeud de Kervaire $N_n(K, \alpha)$

Théorème 3.3. Supposons que $Wh(\pi) = 0$ et que l'homomorphisme $L_n(\pi) \longrightarrow L_n(\mathbb{Z})$ est un isomorphisme. Alors, pour tout $n \geq 4$, il existe des bijections

$$as_{\Omega_n}(K, \alpha) = [E_K^{n+1}, E_K^{n+1}]_{rel\partial} \Big/ \begin{array}{l}\text{action de}\\ \text{Aut}(E_K^{n+1}; \bar{\alpha})\end{array}$$

$$as_{\Omega_n^c}(K, \alpha) = [E_K^{n+1}, E_K^{n+1}]_{rel\partial} \Big/ \begin{array}{l}\text{action de}\\ \text{Aut}(E_K^{n+1}, \alpha)\end{array} .$$

La classe de idE_K^{n+1} est celle du noeud de Kervaire $N(K, \alpha)$

Remarques : La condition $Wh(\pi) = 0$ est toujours satisfaite pour les noeuds classiques d'après [Wn]. Elle semble indispensable à la théorie si l'on travaille avec les relations d'équivalence usuelles dans les noeuds, comme celles considérées ici.

Une relation du genre "h-cobordisme entre les complémentaires" ne nécessiterait pas cette hypothèse mais alors le fait que l'on a affaire à des noeuds ne paraît pas apporter de simplifications substantielles à la théorie générale de [H 2] .

Quant à la condition sur les groupes de chirurgie, elle est remplie pour les noeuds classiques fibrés.[C].

4. Les noeuds vus comme variétés closes

Soit $N \subset S^{n+2}$ un noeud. On a donc un plongement $\eta : S^n \longrightarrow S^{n+2}$ unique à isotopie près tel que η soit un homomorphisme orienté entre S^n et N. De plus, si $n \geq 2$, il existe une unique trivialisation du fibré normal à η compatible avec les orientations. Ceci permet de faire une chirurgie sur η et d'obtenir une variété close orientée V_N^{n+2}, réunion de X et de $S^1 \times D^{n+1}$ collés le long de $S^1 \times S^n$. On note par $\overline{\beta}$ le plongement de $S^1 \times D^{n+1} \longrightarrow V_N$ ainsi obtenu. Les points suivants sont faciles à vérifier :

1) $H_*(V_N) \simeq \begin{cases} Z & \text{si } * = 0,1,n+1,n+2 \\ 0 & \text{sinon} \end{cases}$

2) $\overline{\beta}$ engendre normalement $\pi_1(V_N) = \pi_1(X(N))$

3) Un noeud N est antisimple si et seulement si V_N admet une décomposition en anses sans anse d'indice k+1 si n = 2k et sans anse d'indice k+1 et k+2 si n = 2k+1 . On dira que V est sans anse au milieu de la dimension.

4) Ω_n est en bijection avec les classes d'équivalence de paires $(V_1^{n+2}, \overline{\beta})$ où V^{n+2} est une variété close orientée, $\beta : S^1 \times D^{n+1} \longrightarrow V$ un plongement, V et β satisfaisant 1) et 2) ci-dessus et où

$(V_1, \bar{\alpha}_1)$ est considéré comme équivalent à $(V_2, \bar{\alpha}_2)$ s'il existe un homéomorphisme orienté $h : V_1 \longrightarrow V_2$ tel que $h \circ \bar{\alpha}_1 = \bar{\alpha}_2$.

Si l'on veut obtenir Ω_n^c, il faut prendre la même relation d'équivalence mais en remplaçant la condition $h \circ \bar{\beta}_1 = \bar{\beta}_2$ par $h \circ \beta_1 = \beta_2$ où $\alpha_1 = \bar{\alpha}_1 | S^1 \times \{0\}$.

5) A une surface de Seifert Σ de N correspond une sous-variété close orientée de codimension 1 Σ_o de V, réunion de Σ et de $1 \times D^{n+1}$. Cette sous-variété représente le générateur de $H_{n+1}(V)$ dual de $\beta^* \in H^1(V)$.

6) Le noeud N est fibré si et seulement si V_N est un fibré sur S^1 de fibre Σ_o.

5. Démonstration des théorèmes 2.2 et 2.3

Soit (K, α) représentant un élément de $\mathcal{K}_n$, k=dimK. Considérons les paires (V, f) où :

- V est une variété close orientée de dimension m.
- $f : E_K^m \longrightarrow V$ est un plongement orienté (m-k)-connexe.

Deux telles paires (V_1, f_1) et (V_2, f_2) sont équivalentes s'il existe un homéomorphisme orienté $h : V_1 \longrightarrow V_2$ tel que $h \circ f_1 = f_2$. Notons par Γ_m^K l'ensemble des classes d'équivalence de telles paires. Les groupes $Aut(E_K^m, \bar{\alpha})$ et $Aut(E_K^m, \alpha)$ agissent sur Γ_m^K par $(V, f) . h = (V, f \circ h)$.

Proposition 5.1 Si $Wh(\pi) = 0$, on a, pour tout $n \geqslant 4$:

a) ${}^{as}\Omega_n(K, \alpha) = \Gamma_{n+2}^K / \text{action de } Aut(E_K^{n+2}, \bar{\alpha})$

b) ${}^{as}\Omega_n^c(K, \alpha) = \Gamma_{n+2}^K / \text{action de } Aut(E_K^{n+2}, \alpha)$

Démonstration : Soit $(V, f) \in \Gamma_{n+2}^K$. Le couple $(V, f \circ \bar{\alpha})$ représente un noeud au sens du § 4 qui est antisimple par la condition que f est (n+2-k)-connexe. Ceci définit une application $\Gamma_{n+2}^K \longrightarrow {}^{as}\Omega_n(K, \alpha)$.

Cette application est surjective car si $(V, \overline{\beta}) \in {}^{as}\Omega_n(K, \alpha)$, $\overline{\beta}$ s'étend en un plongement $f_0 : K \longrightarrow V$ qui est $(n+2-k)$-connexe. Par [W 1,proposition 5], les épaississements orientables stables de K sont en bijection avec $[K, BSPL] \simeq \pi_1(BSPL) = \{1\}$. Le plongement f_0 s'étend donc en un plongement $f : E^{n+2}_K \longrightarrow V$ tel que $f \circ \alpha = \beta$. On obtiendra $f \circ \overline{\alpha} = \overline{\beta}$ en composant éventuellement f par l'automorphisme de E^{n+2} qui change la trivialisation du fibré tangent (cf § 8) .

Soit (V_0, f_0) et (V, f) représentant deux classes de $\Gamma^K_n(K, \alpha)$ mais donnant la même classe dans ${}^{as}\Omega(K, \alpha)$. Il existe donc un homéomorphisme orienté $h : V_0 \longrightarrow V_0$ tel que $h \circ f_0 \circ \overline{\alpha} = f_1 \circ \overline{\alpha}$. Par position générale, on peut supposer que $h \circ f_0(E^{n+2}_K) \subset f_1(E^{n+2}_K)$. Comme $Wh(\pi) = 0$, la région $f_1(E^{n+2}_K) - \text{int}\, h \circ f_0(E^{n+2}_K)$ est un s-cobordisme. On peut donc par une isotopie ambiante se ramener à $f_1(E^{n+2}_K) = h \circ f_0(E^{n+2}_K)$, d'où $f_0 = f_1 \circ (f_1^{-1} \circ h \circ f_0)$ et $f_1^{-1} \circ h \circ f_0 \in Aut(E^{n+2}_K, \overline{\alpha})$.

Ceci démontre la partie a) de la proposition 5.1. La partie b) se démontre de même en utilisant en plus l'existence d'un élément de $Aut(E^n_K, \alpha)$ qui renverse la trivialisation du fibré tangent. (cf § 8).

<u>Proposition 5.2</u>. Si $Wh(\pi) = 0$, on a, pour tout $6 \leqslant m \geqslant 2\dim K+2$, une bijection

$$\gamma : \Gamma_m^K \xrightarrow{\ \simeq\ } (\mathrm{Aut}\ E_K^{m-1})_{rel}$$

et l'action d'un élément des groupes $\mathrm{Aut}(E_K^m, \overline{\alpha})$ (où $\mathrm{Aut}(E_K^m, \alpha)$) correspond à l'action de l'élément correspondant de $\mathrm{Aut}(E_K^{m-1}, \overset{(-)}{\alpha})$ sur $(\mathrm{Aut}\ E_K^{m-1})_{rel\ \partial}$ par conjugaison.

Remarque : le théorème 2.2 découle immédiatement des propositions 5.1 et 5.2.

Démonstration : La bijection est obtenue de la manière suivante : soit $(V, f) \in \Gamma_m^K$. On pose $A = V - \mathrm{int}(E_K^m)$. A est un h-cobordisme entre $f(E_K^{m-1} \times 0)$ et $f(E_K^{m-1} \times 1)$. Comme $Wh(\pi) = 0$, il existe un homéomorphisme orienté $g: E_K^{m-1} \times I \longrightarrow A$ tel que $g \mid E_K^{m-1} \times 0 \cup \partial E_K^{m-1} \times I = f$. A la paire (V, f) on peut ainsi faire correspondre $f^{-1} \circ g \mid E_K^{m-1} \times 1$ qui est un automorphisme de $E_K^{m-1} \times I$ et qui représente une classe de $\mathrm{Aut}(E_K^{m-1})_{rel\,\partial}$. On vérifie facilement que l'application est bien définie et est bijective.

Soit maintenant $\overline{h} \in \mathrm{Aut}(E_K^m, \overset{(-)}{\alpha})$ que l'on peut supposer de la forme $\overline{h} = h \times id$, $h \in \mathrm{Aut}(E_K^{m-1}, \overset{(-)}{\alpha})$. L'homéomorphisme $g\ \ \overline{h} : E_K^{m-1} \times I \longrightarrow A$ satisfait à $g \circ \overline{h}^{-1} \mid E_K^{m-1} \times 0 \cup \partial E^{m-1} \times I = f \circ \overline{h}$. A la paire $(V, f \circ \overline{h})$ on associera donc $(\overline{h}^{-1} \circ f^{-1}) \circ g \circ \overline{h} \mid E_K^{m-1} \times 1$: $h^{-1} \circ (f^{-1} \circ g) \circ h$ ce qui prouve la proposition .

Remarque : Il ressort de notre démonstration que, si $(V,f) \in \Gamma_m^K$, alors V est un "open book" [Wi] de page

E^K_{m-1} . Sa classe d'homomorphisme est alors déterminée par la classe de conjugaison de sa monomdromie qui est $\gamma(V,f)$.

Le théorème 3.3 découle de la proposition 5.3 ci-dessous, dans laquelle $\widetilde{L}_n(\pi)$ désigne le noyau de $L^s_n(\pi) \to L_n(Z)$.

Proposition 5.3. Il existe une suite exacte de groupe abélien

$$\longrightarrow [E^m_K , E^m_K]_{rel\partial} \xrightarrow{\rho} \widetilde{L}_{m+1}(\pi) \xrightarrow{\delta} \mathrm{Aut}(E^{m-1}_K)_{rel\ \partial} \longrightarrow [E^{m-1}_K, E^{m-1}_K]_{rel\partial}$$

pour tout m tel que $6 \leqslant m \geqslant 2\dim K+2$.

La démonstration de cette dernière proposition occupera le reste du paragraphe.

Pour une paire de Poincaré [W 2, chapitre 10] $(X, \partial X)$ de dimension m où ∂X est une variété PL, on désigne par $\mathcal{S}(X \text{ rel } \partial X)$ l'ensemble des triangulations homotopiques de $(X \text{ rel } \partial X)$, i.e. les classes de concordance d'équivalences d'homotopie simples de degré 1 $f: (M,\partial M) \longrightarrow (X,\partial X)$ où M est une variété compacte et $f|\partial M$ un homéomorphisme. Si X est une variété, $\mathcal{S}(X \text{ rel } \partial X)$ contient la classe de id_X et prend place dans la suite exacte de la chirurgie de Sullivan-Wall [W, chapitre 10] :

$$[\ \Sigma(X|\partial X) : G/_{PL}\ ] \longrightarrow L_{m+1}(\pi(X)) \xrightarrow{a} \mathcal{S}(X \text{ rel } \partial X) \longrightarrow$$
$$[X|\partial X, G/_{PL}\ ] \longrightarrow L_m(\pi_1(X))$$

où l'application a est donnée par l'action de $L_{m+1}(\pi_1(X))$

sur id_X.

Lemme 5.4. Pour $(X, \partial X) = (E_K^m, \partial E_K^m)$, l'application a restreinte à $\widetilde{L_{m+1}}(\pi)$ donne une bijection $\widetilde{L_{m+1}}(\pi) \xrightarrow{\simeq} \mathcal{S}(E_k^m \text{ rel } \partial E_k^m)$.

Démonstration Considérons $Y = S^1 \times D^{m-1}$ inclus dans $\text{int}E_K^m$ en utilisant n'importe quelle section s de $\pi \longrightarrow \mathbb{Z}$. Tout élément de $\mathcal{S}(Y \text{ rel } \partial Y)$ détermine un élément de $\mathcal{S}(E_K^m \text{ rel } \partial E_K^m)$ (par recollement de $E_K^m - \text{int}Y$). De l'inclusion $Y \longrightarrow \text{int}E^m$ on déduit une application $Y|\partial Y \longrightarrow E_K^m/\partial E_K^m$ qui induit des applications $[E_K^m/\partial E_K^m, G/_{PL}] \longrightarrow [Y/\partial Y, G/_{PL}]$ et

$[\Sigma(E_K^m/\partial E_K^m), G/_{PL}] \longrightarrow [\Sigma(Y/\partial Y, G/_{PL}]$ qui, par

théorie des obstructions, sont des bijections. Prenant la décomposition $L_{m+1}(\pi) \simeq L_{m+1}(\mathbb{Z}) \oplus \widetilde{L}_{m+1}(\pi)$ déterminée par la section s, on obtient un diagramme

$$\begin{array}{ccccccccc}
[\Sigma(E_K^m | \partial E_K^m), {}^G/_{PL}] & \rightarrow & L_{m+1}(\mathbb{Z}) \oplus L_{m+1}(\pi) & \longrightarrow & \mathcal{S}(E_K^m \text{rel} \partial E_K^m) & \longrightarrow & [E_K^m | \partial E_K^m, {}^G/_{PL}] & \longrightarrow & L_m(\pi) \\
\downarrow & & \uparrow & & \uparrow & & \downarrow & & \uparrow \\
[\Sigma(Y/\partial Y), {}^G/_{PL}] & \longrightarrow & L_{m+1}(\mathbb{Z}) & \longrightarrow & (Y \text{ rel} \partial Y) & \longrightarrow & (Y/\partial Y, {}^G/_{PL}] & \rightarrow & L_m(\mathbb{Z})
\end{array}$$

Le lemme 5.4 découle alors du fait que $\mathcal{S}(Y \text{ rel } \partial Y) = \{id_y\}$ ce qui peut s'obtenir de la même manière que $\mathcal{S}(\partial Y) = \{id_{\partial y}\}$ [L-S].

Démonstration de la proposition 5.3

On a la suite exacte qui est discutée en détail dans [H 2] :

$$[E^m_K, E^m_K]_{rel\,\partial} \longrightarrow \mathcal{S}(E^m \text{ rel } \partial E^m_K) \xrightarrow{\delta} \mathrm{Aut}(E^{m-1}_K)_{rel\,\partial} \longrightarrow [E^{m-1}_K, E^{m-1}_K]_{rel\,\partial}$$

Rappelons que la seule application non-évidente est l'application δ qui est définie de la manière suivante : Soit $f : M \longrightarrow E^m_K$ une équivalence d'homotopie telle que $f/\partial M : \partial M \longrightarrow \partial E^m_K$ soit un homéomorphisme . Comme M est alors un s-cobordisme de base $f^{-1}(E^{m-1}_K \times 0)$,

$f^{-1} \mid E^{m-1}_K \times 0 \cup \partial E^{m-1}_K \times f$ s'étend donc en un homéomorphisme $g : E^m_K \longrightarrow M$. On pose alors $\delta(M,f) = f^{-1} \circ g$.

On a une loi d'addition dans $\mathcal{S}(E^m_K \text{rel } \partial E^m)$ par juxtaposition et reparamétrisation de E^m_K. Avec cette structure de groupe abélien sur $\mathcal{S}(E^m_K \text{rel } \partial E^m_K)$, δ est un homomorphisme et la bijection $\widetilde{L_{m+1}}(\pi) \xrightarrow{\simeq} (E^m_K \text{rel } \partial E^m_K)$ du lemme 5.4 est un isomorphisme.

Ceci prouve la proposition 5.3.

Remarque : l'homomorphisme $\widetilde{L_{m+1}}(\pi) \longrightarrow \mathrm{Aut}(E^m_K)_{rel\,\partial}$ est l'homomorphisme ρ_n de [H 1, théorème 4.3.1].

6. Noeuds antisimples avec groupe Z

Soit $N \subset S^{n+2}$ un noeud de complémentaire X. Notons par $\tilde{X}$ le revêtement cyclique infini de X. Les groupes d'homologie $H_i(\tilde{X})$ sont des modules sur l'anneau $\Lambda = Z[t, t^{-1}]$

Proposition 6.1

a) Si N est antisimple, alors

$$H_k(\tilde{X}) = H_{k+1}(\tilde{X}) = 0 \text{ si } n = 2k$$
$$H_k(\tilde{X}) = H_{k+1}(\tilde{X}) = H_{k+2}(\tilde{X}) = 0 \text{ si } n = 2k+1$$

b) Si $\pi_1(X) = Z$, les conditions homologiques de a) sont équivalentes au fait que N est antisimple.

Démonstration : Notons par $A_k = H_k(\tilde{X})$, par $\mathrm{tor}A_k$ la Z -torsion de A_k et par $FA_k = A_k/_{\mathrm{tor}A}$. Ces différents groupes satisfont aux rapports de dualité de Poincaré suivants [L 2] :

$$F A_q = \mathrm{Ext}_\Lambda(\overline{F A_{n+1-q}}, \Lambda)$$
$$\mathrm{tor}A_q = \mathrm{Ext}^2_\Lambda(\overline{\mathrm{tor}A_{n-q}}, \Lambda)$$

où, si M est un Λ-module, $\overline{M}$ désigne le Λ-module sur le même groupe abélien que M mais où l'action de t est définie par la multiplication par t^{-1}.

Dans le cas $n = 2k$, on a $A_{k+1} = 0$ si N est antisimple. Par dualité, on en déduit $FA_k = 0$. Comme il n'y a pas

d'anses d'indice k+1, on a aussi $\mathrm{tor}A_k = 0$, d'où $A_k = 0$.
Le cas n impair se démontre de manière analogue.
montre a).

La démonstration de b) utilise le lemme suivant dont le lecteur trouvera la preuve en fin de paragraphe.

Lemme 6.2 Pour un complexe

$$= \{ \ldots \longrightarrow C_i \longrightarrow C_{i-1} \longrightarrow \ldots \to C_o \longrightarrow 0\}$$

de Λ -modules libres de type fini, les cycles Z_i sont Λ -libres de type fini.

Par dualité de Poincaré les conditions de a) font que $H_{k-1}(\tilde{X})$ est sans $\mathbb{Z}$-torsion et est donc de dimension homologique 1 [L 2, Proposition 3.5]. On en déduit une présentation Λ-libre $0 \longrightarrow T \longrightarrow Z_{k-1} \longrightarrow H_{k-1}(\tilde{X})$ où Z_{k-1} est le Λ-module des (k-1)-cycles du complexe de chaînes $C(\tilde{X})$ (Z_{k-1} est Λ-libre par le lemme 6.2). On démontre alors b) comme suit :

1) k = 2 : alors $X \sim S^1$ et le noeud est antisimple.

2) k = 3 : Utilisant la Λ-présentation $0 \to T \to Z_2 \to H_2(\tilde{X}) \to 0$, on construit un 3-complexe

$$K = [\, S^1 \vee (\vee\, S^2)] \cup D^3$$

le bouquet de 2-sphères étant indexé par une base de Z_2 et l'attachement des 3-cellules étant déterminé par $T \to Z_2$. Ce complexe K admet un plongement $K \hookrightarrow X$ qui est $\frac{1}{2}(n+3)$-connexe. Comme dim $X \geqslant 6$, on peut appliquer le procédé classique d'élimination des anses [K 2] qui montre que le noeud est antisimple.

3) $k \geqslant 4$. On observe l'équivalence $\mathcal{C} \to \mathcal{C}(\tilde{X}) = \{C_i\}$ où $\mathcal{C}$ est le complexe

$$\ldots \to C_{k+3} \to Z_{k+2} \to 0 \to T \to Z_{k-1} \to C_{k-2} \to C_{k-3} \to \ldots \qquad \text{si } n = 2k$$

et

$$\ldots \to C_{k+4} \to Z_{k+3} \to 0 \to 0 \to Z_{k-1} \to C_{k-2} \to \ldots \qquad \text{si } n = 2k+1$$

Par [W 3, Théorème 2], $\mathcal{C} = \mathcal{C}(\tilde{Y})$ pour un CW-complexe fini Y équivalent à X. Comme pour le cas 2), on en déduit que X est antisimple.

Démonstration du lemme 6.2 :

Le fait que $Z_{-1} = 0$ permet de démarrer une récurrence sur i. Si Z_i est libre, la suite $C_{i+1} \longrightarrow Z_i \longrightarrow H_i(\ ) \longrightarrow 0$ est un début de résolution de $H_i(\mathcal{C})$.

Comme $\mathrm{Ext}^{\Lambda}_3(A;\Lambda) = 0$ pour tout Λ-module H(Syzygies), $Z_{i+1} = \mathrm{Ker}(C_{i+1} \longrightarrow Z_i)$ est projectif; Z_{i+1} est aussi de type fini puisque Λ est noetherien. Or un Λ-module projectif de type fini est libre (conséquence du théorème de Seshadri, cf[B IV, § 6]).

7 . Applications

La première application est une nouvelle démonstration du théorème de Levine [L 1].

Proposition 7.1 Un noeud N tel que $\pi_i(X) \cong \pi_i(S_1)$ pour $i \leqslant \frac{1}{2}(n+1)$ est isotope au noeud trivial ($n \geqslant 4$).

Démonstration Par dualité de Poincaré un tel noeud satisfait

$$H_i(\overline{X}) = 0 \text{ pour } i \leqslant \tfrac{1}{2}(n+3)$$

d'où N est antisimple par la proposition 6.1 et $N \in {}^{as}\Omega_n(S^1, id)$. Par théorie des obstructions $[E_{S^1}^{n+1}, E_{S^1}^{n+1}]_{rel\ \partial} = \{id\}$ d'où, par le théorème 3.3, $N \cong N(S^1, id)$ qui est le noeud trivial.

Soit $N_o \subset S^3$ un noeud classique fibré et $m_o : S^1 \longrightarrow X(N_o)$ un élément méridien. La paire $(X(N_o), m_o)$ représente un élément de $_n$ pour $n \geqslant 4$.

Proposition 7.2, Soit $N \subset S^{n+2}$ un n-noeud avec $n \geqslant 4$. Supposons qu'il existe un isomorphisme $h : \pi_1(X(N_o)) \longrightarrow \pi_1(X(N\ ))$ et que $\pi_i(X(N)) = 0$ pour $i \leqslant \frac{1}{2}(n+3)$. Alors N est isotope au noeud de Kervaire $N(X(N_o), h^{-1}(m))$.

Démonstration Comme $X(N_o)$ est de dimension homotopique 2, h donne application $\bar{h} : X(N_o) \longrightarrow X(N)$ qui induit un isomorphisme $\pi_i(X(N_o)) \longrightarrow \pi_i(X(N))$ pour $i \leq \frac{1}{2}(n+3)$. Donc N est antisimple et $N \in {}^{as}\Omega_n(X(N_o)h^{-1}(m))$. Comme N_o est fibré, $L_m(\pi_1(N_o)) \longrightarrow L_m(Z)$ est un isomorphisme [C] et, par le théorème 3.3 et la théorie des obstructions ($\pi_i(X(N_o) = 0$ pour $i \geq 2$) on a ${}^{as}\Omega(X(N_o), h^{-1}(m)) = \{ N(X(N_o), h^{-1}(m))\}$.

Proposition 7.3 Soit $(K, \alpha) \in \mathcal{K}_n$ et $N \subset S^{n+2}$ un noeud de méridien $m : S^1 \longrightarrow X(N)$. Supposons qu'il existe un homéomorphisme $h : X(N) \longrightarrow X(N(K, \alpha))$ tel que $\pi_1(h)(m) = \alpha$. Alors N est isotope à $N(K, \alpha)$.

En particulier, les noeuds de Kervaire sont déterminés à isotopie près par leurs complémentaires.

Démonstration (sous l'hypothèse que $Wh(\pi_1(k)) = 0$. Voir remarque ci-dessous) : Il est clair que N est antisimple et que $\rho(N) = (K, \alpha)$. Le théorème 3.2 nous dit que N représente un élément de $\mathrm{Aut}(E_K^{n+1})_{\mathrm{rel}\,\partial}$ qui est dans l'orbite de $\mathrm{id}(E_K^{n+1})$ par l'action de $\mathrm{Aut}(E_K^{n+1}, \alpha)$. Les actions de $\mathrm{Aut}(E_K^{n+1}, \bar{\alpha})$ et de $\mathrm{Aut}(E_K^{n+1}, \alpha)$ sur $\mathrm{Aut}(E_K^{n+1})_{\mathrm{rel}\,\partial}$ étant des actions par automorphismes, $\mathrm{id}(E_K^{n+1})$ est seule dans son orbite. Donc N est dans l'orbite de $\mathrm{id}(F_K^{n+1})$ par l'action de $\mathrm{Aut}(E_K^{n+1}, \bar{\alpha})$ce qui implique la conclusion de notre proposition.

Remarque : L'usage du théorème 3.2 requiert la condition $Wh(\pi_1(K)) = 0$. La proposition 7.3 est cependant vraie en général et se démontre de la même manière, en utilisant le fait plus fin que l'ensemble Γ^K_{n+2} de la proposition 5.1 porte une structure de groupe (c.f [H 2]) pour laquelle l'action de $Aut(E^{n+2}_K$, α ou $\overline{\alpha}$) est une action par automorphisme.

Pour la prochaine application, considérons un Λ-module ($\Lambda = Z[t, t^{-1}]$) de type fini A tel que (1-t) soit un isomorphisme (module de type K). Pour tout entier $k \geqslant 2$, il existe un CW-complexe K fini de dimension k+2 (k+1 si le sous-module torA des éléments de Z-torsion est trivial) tel que $\pi_1(K) = Z$, $H_*(K) = H_*(S^1)$, $H_k(\widetilde{K}) = A$ et $H_i(\widetilde{K}) = 0$ pour $i \neq k$. [L 2, lemme 9.4]. Un tel complexe K sera appelé un (A, k)-cercle homologique. Le n-noeud de Kervaire $N_n(K, \alpha)$ (α un générateur de $\pi_1(K)$) pour $n \geqslant 2\dim K$ satisfait alors les conditions suivantes

a) $\pi_1(X(N_n(K,\alpha))) = Z$

b) $H_k(\widetilde{X}(N_n(K, \alpha))) = A$

c) $H_i(\widetilde{X}(N_n(K, \alpha))) = 0$ pour $i \neq k$ et $i \leqslant \frac{n+3}{2}$

La fin de ce paragraphe est consacrée à des théorèmes de finitude des classes d'isotopie de noeud satisfaisant a), b) et c). Ceci nécessite une définition :

<u>Définition</u> : Soit A un module de type K et k, n deux entiers positifs. Le triple (A,k,n) est dit <u>admissible</u> si

i) $n \geq 2(k+2)$ ou
$n \geq 2(k+1)$ et $\mathrm{tor} A = 0$.

ii) Pour tout (A,k)-cercle homologique K, les groupes $\pi_i(K)$ sont finis pour $i = (n-k-1), n-k, n-k+1, n, n+1$, la dimension entre parenthèses n'étant à considérer que si $\mathrm{tor} A \neq 0$.

<u>Exemple</u> : Un triple (A,k,n) satisfaisant i) est admissible s'il remplit l'une des conditions suivantes :

1) A est fini;
2) $A \otimes \mathbb{Q} = \mathbb{Q}$ et k est impair ;
3) $A \otimes \mathbb{Q} = \mathbb{Q}$, k est pair et $n \neq 3k-2, 3k-1, (3k)$.

2) et 3) se démontrent en remarquant que $A \simeq \mathbb{Z}\,[\frac{1}{d}]$ pour un certain d (isomorphisme modulo la classe des groupes finis) et qu'alors, par théorie de la localisation, $\pi_i(K) \simeq \pi_i(S^k) \otimes \mathbb{Z}\,[\frac{1}{d}]$ (isomorphisme modulo les groupes finis).

4) Si A est un $\mathbb{Z}$-module de type fini et de rang r, la condition ii) est équivalente par théorie de la localisation à l'hypothèse

$$\pi_i(\underbrace{S^k \vee \ldots \vee S^k}_{r\ \text{copies}}) \otimes \mathbb{Q} = 0 \quad \text{pour}$$

$i = (n-k-1), n-k, n-k+1, n, n+1$ puisqu'alors $\pi_i(K)$ est de type fini et que $\pi_i(K) \otimes \mathbb{Q} \cong \pi_i(S^k \vee \ldots \vee S^k) \otimes \mathbb{Q}$.

Théorème 7.4. Soit (A,k,n) un triple admissible, $k \geq 3$. Alors les classes d'isotopie de n-noeuds N tels que

$$\begin{cases} \pi_1(X(N)) = Z \\ H_k(X(\tilde{N})) = A \\ H_i(\tilde{X}(N)) = 0 \text{ pour } i \neq k \text{ et } i \leq \frac{n+3}{2} \end{cases}$$

sont en nombre fini.

Par (6.1) de tels noeuds sont antisimples et leur image par ρ est un (A,k)-cercle homologique. Le théorème 7.4 découle donc directement du théorème 3.3 et des deux lemmes suivants :

Lemme 7.5 : Soit A un module de type K et k un entier ≥ 3. Alors il existe un nombre fini de types d'homotopie de (A,k)-cercles homologiques et leur dimension homotopique est $\leq k+2$. Si torA = 0, il y a exactement un type d'homotopie de (A,k)-cercles homologiques et sa dimension homotopique est k+1.

Lemme 7.6 : Soit (A,k,n) un triple admissible. Alors $[E_K^{n+1}, E_K^{n+1}]_{rel\,\partial}$ est un ensemble fini.

Démonstration du lemme 7.5 Soit K un (A , k)-cercle homologique. Considérons une résolution Λ-libre

$0 \longrightarrow L_2 \xrightarrow{\partial_1} L_1 \xrightarrow{\partial_0} L_0 \longrightarrow A \longrightarrow 0$ de A. On construit directement une équivalence entre le complexe C^o_* suivant

$$0 \longrightarrow C^o_{k+2} \xrightarrow{\partial_{k+2}} C^o_{k+1} \xrightarrow{\partial_{k+1}} C^o_k \longrightarrow 0$$

où $(C^o_{k+i}, \partial_{k+i}) = (L_i, \partial_i)$ et le complexe $C_*(\widetilde{K})$. Par [W 3 théorème 2] il existe un cercle homologique K_o homotopiquement équivalent à K tel que son (k-1)-squelette $K_o^{(k-1)}$ soit S^1 et tel que $C_*(\widetilde{K}_o) \simeq C^o_*$. Il suffit donc de montrer la finitude des types d'homotopie de complexes K tels que $K^{(k-1)} = S^1$ et $C_*(\widetilde{K}) = C^o_*$.

Il est clair que ces deux dernières conditions caractérisent le type d'homotopie du (k+1)-squelette de K. Ceci démontre déjà le lemme lorsque torA = 0, ce qui implique $C^o_{k+2} = 0$. La liberté dans la construction de K réside uniquement dans l'attachement des (k+2)-cellules, c'est-à-dire dans le choix d'une section $s : H_{k+1}(\widetilde{K}^{(k+1)}) \rightleftarrows \pi_{k+1}(K^{(k+1)})$ de l'homomorphisme d'Hurewicz.

Une telle section s déterminera un complexe K_s et une décomposition

$$\pi_{k+1}(K^{(k+1)}) = \pi_{k+1}(K_s) \oplus H_{k+1}(K^{(k+1)}).$$

Les suites exactes d'homotopie et d'homologie de l'application $\widetilde{K}_s \longrightarrow K(A, k)$ montrent que $\pi_{k+1}(K_s) \simeq H_{k+2}(K(A, k))$.

Affirmation Si B est un groupe abélien tel que torB est fini, alors $H_{k+2}(K(B, k))$ est fini pour $k \geqslant 3$.

Cette affirmation implique le lemme 7.5. En effet, un module d'Alexander satisfait aux hypothèses sur B; $H_{k+2}(K(A, k))$ fini implique la finitude des sections s et ainsi le lemme 7.5.

Démontrons maintenant l'affirmation, par récurrence sur r(B). Il suffit naturellement de considérer le cas où torB = 0. Si r(B) = 1, $B = \mathbb{Z}[\frac{1}{m}]$ et, comme $k \geqslant 3$:

$$H_{k+2}(K(\mathbb{Z}[\tfrac{1}{m}], k)) = H_{k+2}(K(\mathbb{Z}, k)) \otimes \mathbb{Z}[\tfrac{1}{m}] = \begin{cases} \mathbb{Z}/2\mathbb{Z} \text{ si } m \text{ est impair} \\ 0 \text{ sinon} \end{cases}$$

Si r(B) > 1, on choisit $0 \neq Q \in B$ et on considère le sous-groupe V(a) formé des éléments $x \in B$ pour lesquels il existe deux entiers $m, n \in \mathbb{Z}$ tels que $mx = na$. On a $r(V(a)) = 1$, d'où $r(B/V(a)) = r(B)-1$ et donc $H_{k+2}(K(B/V(a)), k)$ est fini par hypothèse de récurrence. On en déduit que H_{k+2} (K(B, k)) est fini, en utilisant la suite spectrale de Serre de la fibration

$$K(V(a),k) \longrightarrow K(B,k) \longrightarrow K(B/V(a), k)$$

Démonstration du lemme 7.6 La finitude de $[E_K^{n+1}, E_K^{n+1}]$ s'obtient par la théorie des obstructions dont les groupes à considérer sont $T_i = H^i(E_K^{n+1}, \partial E_K^{n+1}; \pi_i(K))$ (coefficients locaux). Comme K n'a que des cellules de dimension 0, 1, k, k+1, (k+2) (la dimension entre parenthèses n'étant à considérer que si torA $\neq 0$), on a

$T_i = 0$ si $i \neq (n-k-1), n-k, n-k+1, n, n+1$. Par l'hypothèse

ii) T_i est alors fini pour tout i d'où $[E_K^{n+1}, E_K^{n+1}]$ est un groupe fini.

8. Les groupes $\text{Aut}(E_k^n, \alpha)$ et $\text{Aut}(E_k^n, \overline{\alpha})$

Si (K, α) représente une classe de $\mathcal{K}_m$, on dénote par $\varepsilon(K, \alpha)$ le groupe des classes d'homotopie de self-équivalences d'homotopie simples $f : K \longrightarrow K$ telles que $f \cdot \alpha = \alpha$. Les homomorphismes naturels

$$\text{Aut}(E_k^n, \overline{\alpha}) \xrightarrow{\ j\ } \text{Aut}(E_k^n, \alpha) \xrightarrow{\ \mu\ } \varepsilon(K, \alpha)$$

vont nous permettre de comparer nos groupes d'automorphismes avec $\varepsilon(K, \alpha)$ qui est un objet de nature homotopique. Comme on a affaire à des équivalences d'homotopie simples, μ est surjective.

Proposition 8.1. $\text{Ker}\mu = \mathbb{Z}/2\mathbb{Z}$ et il y a une rétraction naturelle $t : \text{Aut}(E_k^n, \alpha) \longrightarrow \text{Ker}\mu$. Un élément de $\text{Aut}(E_k^n, \alpha)$ est dans Kert si et seulement si il préserve la trivialisation naturelle du fibré tangent de E_k^n .

On en déduit que $\text{Aut}(E_K^n, \alpha) = \mathbb{Z}/2\mathbb{Z} \oplus \varepsilon(K, \alpha)$ et l'existence d'automorphismes renversant la trivialisation du fibré tangent, fait utilisé au § 5.

Démonstration : Soit $f \in \text{Aut}(E_K^n, \alpha)$. On peut supposer que $t(S' \times D^{n-1}) = S' \times D^{n-1}$, d'où un homomorphisme $\text{Aut}(E_k^n, \alpha) \longrightarrow \text{Aut}(E_{S'}^n, \text{id})$.

Pour un polyèdre X, dénotons par $\tau(X)$ l'ensemble des épaississements stables de X. On a le diagramme suivant, dont les isomorphismes proviennent de [Ho, théorème 2.7], [W 1 Proposition 5] et de la théorie des obstructions:

$$\begin{array}{ccccc}
\mathrm{Ker}\mu & \xrightarrow{\ \cong\ } & \tau(\Sigma K) & \xrightarrow{\ \cong\ } & [\Sigma K, BO] \\
\Big\downarrow {\scriptstyle t|\mathrm{Ker}\mu} & & \Big\downarrow & & \Big\downarrow \cong \\
\mathrm{Aut}(E^n_{S'}, \mathrm{id}) & \xrightarrow{\ \cong\ } & \tau(\Sigma S') & \xrightarrow{\ \cong\ } & [\Sigma S', BO] \\
 & & & & \| \\
 & & & & \pi_2(BO) = \mathbb{Z}/2\mathbb{Z}
\end{array}$$

D'où $\mathrm{Ker}\mu = \mathrm{Aut}(E^n_{S'}, \mathrm{id}) \simeq \mathbb{Z}/2\mathbb{Z}$ et t est la rétraction annoncée. L'élément non-nul de $\mathrm{Aut}(E^n_{S'}, \mathrm{id})$ est donné par $S^1 \times D^{n-1} \ni (x,u) \longrightarrow (x, \varphi(x)(u))$ où $\varphi : S^1 \longrightarrow \mathrm{Aut}\, D^n$ représente le généraleur de $\pi_1(\mathrm{aut}\, D^n) \simeq \mathbb{Z}/2\mathbb{Z}$. Un tel élément inverse les trivialisations du fibré tangent.

<u>Proposition 8.2</u>. $\mu \circ j : \mathrm{Aut}(E^n_K, \overline{\alpha}) \longrightarrow \varepsilon(K)$ est un isomorphisme.

<u>Démonstration</u> : Ceci découle directement dù fait que $\mathrm{Aut}(E^n_K, \overline{\alpha}) = \mathrm{Ker}t$.

On a donc aussi la décomposition

$\mathrm{Aut}(E^n_K, \alpha) = \mathrm{Aut}(E^n_K, \overline{\alpha}) \oplus \mathbb{Z}/2\mathbb{Z}$.

BIBLIOGRAPHIE

[B] BASS H. Algebraic K-theory. Benjamin 1968.

[C] CAPPEL S. Mayer-Vietoris sequences in hermitian K-theory. Algebraic K-theory III, Springer Lect. Notes 343, 478-512.

[H 1] HAUSMANN J.-Cl. Groupes de sphères d'homologie entière. Thèse, Université de Genève, 1974.

[H 2] - Variétés sans anses au milieu de la dimension. En préparation.

[Ho] HODGSON J.P.E.- Obstruction to concordance for thickenings Inv. Math. 5(1968) 292-316.

[K 1] KERVAIRE M. Les noeuds de dimension supérieure. Bull. Soc. Math. de France 93 (1965) 225-271.

[K 2] KERVAIRE M. Le théorème de Barden-Mazur-Stallings. Comm. Math. Helv. 40(1965), 31-42.

[L 1] LEVINE J. Unknoting spheres in codimension 2. Topology 4(1966) 9-16.

[L 2] - Knot module. Trans. AMS. 229(1977) 1-50.

[L 3] - An algebraic classification of some knot in codimension 2. Comm. Math. Helv. 45 (1970) 185-198.

[L S] LASHOF R. and SHANESON S. - Classification of knots in codimension 2. BAMS 7 S (1969) 171-175.

[Wn] WALDHAUSEN F.- Whitehead groups of generalized free products. To appear.

[W 1] WALL C.T.C. Classification problems... IV (Thickenings) Topology 5 (1966) 73-94.

[W 2] - Surgery on compact manifolds. Academic Press 1970.

[W 3] - Finiteness conditions for CW-complexes II. Proc. of the Royal Soc. A. 295 (1966) 129-139.

[Wi] WINKELKEMPER H.E. - Manifolds as open books, BAMS 79(1973) 45-51 .

---

niversity of Geneva, Switzerland .

Signature of Branched Fibrations

by Louis H. Kauffman

I. Introduction

A branched fibration is a topological analog of a degenerating family of algebraic varieties that is parametrized over a manifold M, with the degenerate fibers lying over a codimension two submanifold $V \subset M$. This is a common situation in algebraic geometry, but there is a wide avenue of choice for the corresponding topological notion. We have chosen a definition that abstracts the main features associated with isolated (complex) hypersurface singularities. This means that the degeneration will be closely associated to a fibered knot; the knot plays the role of the link of the singularity.

In section 2 we review the definitions of knot, fibered knot and Seifert pairing. Theorem 2.9 shows that, over the complex numbers, the Seifert pairing of a fibered knot is non-trivial only on subspaces associated with unit-length eigenvalues of the monodromy.

In section 3 branched fibrations are defined and related to fibered knots and singularities. This is based on joint work of the author and Walter Neumann ([KN]). Theorem 3.2 states the main properties of the knot product construction of [KN]. This construction associates to a knot $\mathcal{K} = (S^k, K)$ and a fibered knot $\mathcal{L} = (S^\ell, L)$ a new product knot $\mathcal{K} \otimes \mathcal{L} = (S^{k+\ell+1}, K \otimes L)$. The product knot has a spanning manifold that is defined in terms of branched fibrations. In fact $K \otimes L = \partial M$ where $M = \tau(D^{k+1}, F)$, a branched fibration of D^{k+1} along a submanifold $F \subset D^{k+1}$ with $\partial F = K$. The branched fibration is obtained by a pull-back construction from a simpler branched fibration $\tau : D^{\ell+1} \to D^2$, branched over $0 \in D^2$. This branched fibration τ is directly associated with the fibered knot $\mathcal{L}$.

This situation leads to a signature problem: Let $\sigma_\tau(\mathcal{K})$ denote the signature of the branched fibration $\tau(D^{k+1}, F)$ when $4 \mid (k + \ell)$. Theorem 3.6 gives a formula for this signature in terms of the eigenvalues of the monodromy for $\mathcal{L}$ and the Seifert forms of $\mathcal{K}$ and $\mathcal{L}$. This result generalizes some computations of signatures of branched coverings (see [DK] and [N]).

In section 4, Theorem 3.6 is applied to some special cases involving Brieskorn singularities, cyclic branched coverings and concordance invariants of links in S^3.

In section 5 we show how to construct a more general class of branched fibrations by mimicking a method due to F. Hirzebruch ([H]) for ramified covers. This leads towards the question of a more general formula for signatures of branched fibrations. This seems to be a difficult problem and may, in fact, require a more general concept of branched fibration or a change in viewpoint.

In any case, I hope to have given an initial framework for these questions and to have shown some of the interesting connections among singularities, knot theory, and signatures.

Throughout the paper all manifolds are smooth; $\cong$ denotes isormorphism or diffeomorphism, while $\approx$ denotes homeomorphism.

II. Knots and Fibered Knots

This section will recall some standard notions in knot theory. The main result (Theorem 2.9) shows that, over the complex numbers, the Seifert pairing of a fibered knot is nontrivial only on subspaces associated with unit-length eigenvalues of the monodromy. This fact will be of use for the signature computations of section III.

Definition 2.1. A knot $\mathcal{K} = (S^n, K)$ is a pair consisting of an oriented n-sphere S^n and a codimension two, compact, closed, oriented submanifold $K \subset S^n$. If K is a homotopy sphere, then the knot is said to be spherical.

When $n = 3$, this definition is intended to include links. That is, a link $K \subset S^3$ is a collection of disjoint embedded oriented circles.

Definition 2.2. A spanning surface for a knot $\mathcal{K} = (S^n, K)$ is a compact oriented (n-1)-manifold F, embedded in S^n so that $\partial F = K$. Here the symbol, ∂, denotes oriented boundary.

It is worth remarking that knots always have spanning surfaces. Since the proof of this result is short and it motivates the definition of fibered knot, we include the argument in the next lemma.

Lemma 2.3. If $\mathcal{K} = (S^n, K)$ is any knot, then there exists a spanning surface $F \subset S^n$ for $\mathcal{K}$.

Proof. Let $E = S^n - N^\circ$ where N is a closed tubular neighborhood of K. Note that $H^1(E;\mathbb{Z}) = [E, S^1]$ where $[\ ,\ ]$ denotes homotopy classes of maps. Let $\alpha: E \to S^1$ represent a sum of the generators of $H^1(E;\mathbb{Z})$ with orientations specified by the orientation of $K \subset S^n$. We may assume that α is transverse regular to $* \in S^1$. It is not hard to see that N is diffeomorphic to $K \times D^2$ with $\partial\alpha^{-1}(*)$ corresponding to $K \times *$. Thus, by adding a collar to $\alpha^{-1}(*)$, one obtains $F \subset S^n$ with $\partial F = K$.

Remark. It may happen that the mapping $\alpha: E \to S^1$ described above can be chosen to be a smooth fibration. In this case one says that $\mathcal{K}$ is a fibered knot. The formal definition is as follows:

Definition 2.4. A knot $\mathcal{K} = (S^n, K)$ is fibered with fibered structure $b: S^n \to D^2$, if there is a smooth mapping $b: S^n \to D^2$, transverse to $0 \in D^2$ such that

i) $b^{-1}(0) = K \subset S^n$.

ii) $b/||b||: S^n - K \to S^1$ is a smooth fibration.

Here $||b||(x)$ denotes the distance from $b(x)$ to the origin in $\mathbb{R}^2$. A fibered knot with fibered structure b will sometimes be indicated by the notation $(S^n,K;b)$.

The first example of a fibered knot is the <u>empty knot of degree</u> <u>a</u>, $[a] = (S^1,\phi;a)$. Here $a:S^1 \to S^1$ is defined by the formula $a(x) = x^a$ (complex multiplication; a is an integer). The map is vacuously transverse to $0 \in D^2$ and a typical fiber is $F = a^{-1}(1) = \{1,\omega,\omega^2,\ldots,\omega^{a-1}\}$ where $\omega = \exp(2\pi i/a)$. Just as all of mathematics unfolds from the empty set, so do many interesting knots come from these empty knots. This comes about by the product construction discussed in the next section.

Another construction that gives rise to fibered knots involves the notion of the <u>link of a singularity</u>. Let $f:\mathbb{C}^n \to \mathbb{C}$ be a complex polynomial mapping such that $f(0) = 0$. One says that f has an <u>isolated singularity at 0</u> if the gradient $\nabla f = (\partial f/\partial z_1,\partial f/\partial z_2,\ldots,\partial f/\partial z_n)$ vanishes at $0 \in \mathbb{C}^n$ and is non-zero in some deleted neighborhood of $0 \in \mathbb{C}^n$. Under these conditions one can define a knot, the <u>link of f</u> by $L(f) = (S_\varepsilon^{2n-1},L(f))$ where $L(f) = f^{-1}(0) \cap S_\varepsilon^{2n-1}$ for $0 < \varepsilon << 1$. For ε sufficiently small, $L(f)$ is independent of the choice of ε. In ([M]) Milnor shows that $L(f)$ is a fibered knot. The fibration of the complement is given by the mapping $f/||f||:S_\varepsilon^{2n-1} - L(f) \to S^1$. The Brieskorn polynomials (see [B]) $f(z) = z_0^{a_0} + z_1^{a_1} + \ldots + z_n^{a_n}$ provide one such class of isolated singularities.

One method for studying a knot is to consider invariants for the embedding of a spanning surface. The simplest of these is the Seifert pairing:

<u>Definition 2.5.</u> Let $K = (S^n,K)$ be a knot with spanning surface $F \subset S^n$. The <u>Seifert pairing</u> $\theta_K:H_*(F) \times H_{n-*-1}(F) \to \mathbb{Z}$ is defined as follows: Let $i:F \to S^n - F$ be the mapping obtained by pushing F into its complement along the positive normal direction. Then $\theta_K(x,y) = \ell(i_*x,y)$ where ℓ denotes linking numbers in S^n,

Here H_* denotes the free part of the reduced homology. That is, $H_*(X) = \bar{H}_*(X)/T_*(X)$ where $T_*(X) \subset \bar{H}_*(X)$ is the torsion subgroup of the reduced integral homology group. While the Seifert pairing actually depends upon the choice of spanning surface F, we have chosen to omit this dependence in the notation.

If the submanifold F has a middle dimension, then there is a well-known relationship between the Seifert pairing and the intersection form on F. This is given by the following theorem of J.Levine ($[L_1]$).

<u>Theorem 2.6.</u> Let $F^{2n} \subset S^{2n+1}$ be a compact oriented 2n-dimensional manifold with boundary embedded in S^{2n+1}. Let $\theta:H_n(F) \times H_n(F) \to \mathbb{Z}$ be the middle-dimensional Seifert pairing, and let $\langle,\rangle:H_n(F) \times H_n(F) \to \mathbb{Z}$ be the intersection form on F. Then, for $x,y \in H_n(F)$, $\langle x,y\rangle = \theta(x,y) + (-1)^n\theta(y,x)$.

This is a key relationship for signature computations. If F^{2n} is given as above, and n is even, then $\langle,\rangle$ is a symmetric form and the <u>signature of F</u>

$\sigma(F)$, is defined to be the signature of the form $(H_n(F),\langle,\rangle)$. Thus the Seifert pairing determines the signature of F .

In the case of fibered knots, extra structure is provided by the monodromy $h_*:H_*(F) \to H_*(F)$ where F is a typical fiber. Here $h:F \to F$ is the diffeomorphism of the fiber that defines the structure for the corresponding fiber bundle over S^1 . The following proposition is well known.

Proposition 2.7. Let $L = (S^n,L)$ be a fibered knot with fiber G and monodromy h . Let Θ_L denote the Seifert pairing with respect to G . Then $\Theta_L(x,h_*y) = (-1)^{|x||y|}\Theta_L(y,x)$. Here if $x \in H_*(F)$, then $|x| = * + 1$.

Remark: In matrix terms this proposition becomes $H = \Theta^{-1}\Theta^T$ where H is a matrix of h_* with respect to some basis for $H_*(F)$ and Θ is a matrix of the Seifert pairing. The graded transpose Θ^T is defined by the equations $\Theta^T_p = (-1)^{(p-1)(n-p)}\Theta^t_p$ where Θ_p is that part of the Seifert matrix coming from the Seifert pairing on $H_p(F) \times H_{n-p-1}(F)$ and t denotes ordinary transpose.

In what follows, we write h instead of h_* .

Corollary 2.8. Let $K = (S^{2n+1},K)$ be a fibered knot with fiber F and monodromy h . Let $\langle,\rangle:H_n(F) \times H_n(F) \to \mathbb{Z}$ denote the intersection pairing and $\Theta:H_n(F) \times H_n(F) \to \mathbb{Z}$ denote the Seifert pairing in the middle dimension. Then for $x,y \in H_n(F)$, $\langle x,y\rangle = \Theta(x,(I-h)y)$.

Proof. By 2.7, $(-1)^{(n+1)(n+1)}\Theta(y,x) = \Theta(x,hy)$. By 2.6, $\langle x,y\rangle = \Theta(x,y) + (-1)^n\Theta(y,x) = \Theta(x,y) - \Theta(x,hy) = \Theta(x,(I-h)y)$. This proves the corollary.

In the rest of this section we will analyze the relationship between the eigenvalues of the monodromy and the structure of the Seifert pairing in the middle dimension. Thus we fix a fibered knot $K = (S^{2n+1},K)$ with fiber F , monodromy h , and Seifert and intersection forms as in Corollary 2.8. Let $A = H_n(F) \otimes \mathbb{C}$. Then $\langle,\rangle$ and Θ extend to forms on A via the formulas $\langle x \otimes \alpha, y \otimes \beta\rangle = \bar{\alpha}\beta\langle x,y\rangle$ and $\Theta(x \otimes \alpha, y \otimes \beta) = \bar{\alpha}\beta\Theta(x,y)$ for $\alpha,\beta \in \mathbb{C}$ and $x,y \in H_n(F)$. The bar denotes complex conjugation. Thus $\langle,\rangle$ be comes a Hermitian or skew-Hermitian form on A , and Θ a sesquilinear form. Let $h:A \to A$ denote the usual monodromy tensored with $1\mathbb{C}$.

Consider the Jordan normal form for $h:A \to A$. Then $A \cong \oplus_\lambda A_\lambda$ where λ is an eigenvalue for h and A_λ is the corresponding subspace. (A basis can be chosen for A_λ such that $h_\lambda = h|A_\lambda$ is a λ-Jordan block , and $h \cong \oplus h_\lambda$) .

Theorem 2.9. Let K be a fibered knot as above with monodromy h and complex Seifert pairing $\Theta:A \times A \to \mathbb{C}$. Then Θ is an orthogonal direct sum of its restrictions to the Jordan subspaces A_λ . That is, if $x \in A_\lambda$ and $y \in A_\mu$ with $\lambda \neq \mu$ then $\Theta(x,y) = 0$. Furthermore, if $\Theta|A_\lambda \not\equiv 0$ then $||\lambda|| = 1$. Thus, only unit length eigenvalues of the monodromy are relevant to the Seifert pairing.

Proof. To prove that $\Theta|A_\lambda \not\equiv 0 \Rightarrow ||\lambda|| = 1$ it will suffice to assume that $\lambda \neq 1$ and that $h|A_\lambda$ corresponds to a single Jordan block. That is, we assume that A_λ has a basis $\{e_0, e_1, \ldots, e_s\}$ such that $he_0 = \lambda e_0$ and $he_k = \lambda e_k + e_{k-1}$ for $k = 1,2,\ldots,s$. We shall let Θ denote $\Theta|A_\lambda$ with this basis. The proof will proceed by induction on s.

Note that $\overline{<x,y>} = (-1)^n <y,x>$. Thus we have $<e_0,e_0> = \Theta(e,(I-h)e_0) = \Theta(e_0,(1-\lambda)e_0)$. Hence $<e_0,e_0> = (1-\lambda)\Theta(e_0,e_0)$. Therefore $(1-\bar{\lambda})\overline{\Theta(e_0,e_0)} = (-1)^n(1-\lambda)\Theta(e_0,e_0)$. Since we also know that $<x,y> = \Theta(x,y) + (-1)^n\overline{\Theta(y,x)}$ and are assuming that $\lambda \neq 1$, this implies that $(1-\lambda)\Theta(e_0,e_0) = \Theta(e_0,e_0) + [(1-\lambda)/(1-\bar{\lambda})]\Theta(e_0,e_0)$. Hence, if $\Theta(e_0,e_0) \neq 0$, then $(1-\bar{\lambda})(1-\lambda) = (1-\bar{\lambda}) + (1-\lambda)$. Therefore $\lambda\bar{\lambda} = 1$ and $||\lambda|| = 1$. This completes the proof for $s = 0$.

Continuing by induction, we assume that the result has been shown for all Jordan blocks of size less than s. Let $B_{s-1} = [e_0, e_1, \ldots, e_{s-1}] \subsetneq A_\lambda$ denote the subspace spanned by these basis vectors. By induction, if $\Theta|B_{s-1} \not\equiv 0$ then $||\lambda|| = 1$. Thus we may assume that $\Theta|B_{s-1} \equiv 0$. Under this assumption we now make a second induction on s to show that $\Theta(e_s,e_k) \neq 0$ or $\Theta(e_k,e_s) \neq 0$ for any k satisfying $0 \leq k < s$ implies that $||\lambda|| = 1$.

To start this second induction, note that $<e_s,e_0> = \Theta(e_s,(I-h)e_0) = (1-\lambda)\Theta(e_s,e_0)$ while $<e_0,e_s> = \Theta(e_0,(I-h)e_s) = \Theta(e_0,(1-\lambda)e_s - e_{s-1}) = (1-\lambda)\Theta(e_0,e_s)$ (since $\Theta|B_{s-1} \equiv 0$, $\Theta(e_0,e_{s-1}) = 0$). The same argument as in the first induction at $s = 0$ now shows that $\Theta(e_0,e_s) \neq 0 \Rightarrow ||\lambda|| = 1$. To complete this second induction, suppose that $\Theta(e_s,e_k) = \Theta(e_k,e_s) = 0$ for $0 \leq k < \ell \leq s-1$. Then a similar computation shows that $\Theta(e_s,e_\ell) \neq 0$ or $\Theta(e_\ell,e_s) \neq 0 \Rightarrow ||\lambda|| = 1$. This completes the second induction.

We now may therefore assume that $\Theta(e_i,e_j) = 0$ for $i \neq s$ and $j \neq s$. Thus, since $\Theta \not\equiv 0$, $\Theta(e_s,e_s) \neq 0$. But $<e_s,e_s> = \Theta(e_s,(1-\lambda)e_s - e_{s-1}) = (1-\lambda)\Theta(e_s,e_s)$. Hence the same calculuation as before shows that $||\lambda|| = 1$. This completes the first induction and hence the proof that $\Theta|A_\lambda \not\equiv 0 \Rightarrow ||\lambda|| = 1$.

The rest of the proof is obtained by very similar induction arguments. The details will be omitted. Different Jordan blocks corresponding to the same eigenvalue λ are, in fact, orthogonal for Θ.

Remark. For eigenvectors there is a direct relationship between the eigenvalue and the value of the Seifert pairing. Suppose that $\Theta(x,x) = \alpha \neq 0$ and that $hx = \lambda x$. Then (as in the proof above) $(1-\bar{\lambda})\bar{\alpha} = (-1)^n(1-\lambda)\alpha$. If $\lambda \neq 1$, then $(1-\lambda)/(1-\bar{\lambda}) = -\lambda$. Hence $\lambda\alpha = (-1)^{n+1}\bar{\alpha}$. For example, if $\alpha = 1-\omega$ with $\omega \neq 1$, $||\omega|| = 1$ and $n = 0$ then $\lambda = -(1-\bar{\omega})/(1-\omega) = \bar{\omega}$. If $\alpha = (1-\omega_0)\ldots(1-\omega_n)$ with $\omega_i \neq 1$, $||\omega_i|| = 1$ then $\lambda = \bar{\omega}_0\bar{\omega}_1\ldots\bar{\omega}_n$.

An example. Consider the empty knot of degree a, $[a] = (S^1,\phi;a)$, where a is a positive integer. This knot has spanning surface $F = \{1,\omega,\omega^2,\ldots,\omega^{a-1}\}$ where $\omega = \exp(2\pi i/a)$. Thus $\tilde{H}_0(F) \cong \mathbb{Z}^{a-1}$ with basis $\{e_0,e_1,\ldots,e_{a-2}\}$ where

$e_k = [\omega^k] - [\omega^{k+1}]$ and $[p]$ denotes the integral homology class of the point $p \in F$. The monodromy acts via rotation by $2\pi/a$, hence $h[\omega^k] = [\omega^{k+1}]$ and therefore $he_k = e_{k+1}$ (but note that $1 + \omega + \omega^2 + \ldots + \omega^{a-1} = 0$ and $e_{a-1} = -(e_0 + e_1 + \ldots + e_{a-2}))$. Let $A = \bar{H}_0(F;\mathbb{C})$. The eigenvalues of the mondromy are $\bar{\omega}, \bar{\omega}^2, \ldots, \bar{\omega}^{a-1}$. A corresponding basis of eigenvectors is given by $E_k = e_0 + \omega^k e_1 + \omega^{2k} e_2 + \ldots + \omega^{(a-1)k} e_{a-1}$. The integral Seifert pairing has matrix Λ_a with respect to the basis $\{e_0, \ldots, e_{a-2}\}$, where

$$\Lambda_a = \begin{bmatrix} 1 & -1 & & & \\ & 1 & -1 & & 0 \\ & & \ddots & \ddots & \\ & 0 & & \ddots & -1 \\ & & & & 1 \end{bmatrix}$$ is an $(a-1) \times (a-1)$ matrix.

It is an easy calculation to see that the Seifert pairing over $\mathbb{C}$ is diagonal. Its matrix Δ_a, with respect to the basis $\{E_1, E_2, \ldots, E_{a-1}\}$ is given by the diagonal matrix

$$\Delta_a = \begin{bmatrix} 1-\omega & & & & \\ & 1-\omega^2 & & & 0 \\ & & 1-\omega^3 & & \\ & 0 & & \ddots & \\ & & & & 1-\omega^{a-1} \end{bmatrix}$$

III. Branched Fibrations and Knot Products

In this section we define branched fibrations and explain their relationship to singularities, fibered knots and knot products.

Definition 3.1. A branched fibration over D^2 is a smooth mapping $\tau: D^{n+1} \to D^2$ such that $\tau^{-1}(0) \simeq CL$ where $L \subset S^n$ is a knot, and CL denotes the cone on L. The mapping τ must satisfy the following two conditions:

i) $\tau|S^n: S^n \to D^2$ is a fibered structure for (S^n, L). (see 2.4).

ii) $\tau|D^{n+1} - CL: D^{n+1} - CL \to D^2 - \{0\}$ is a smooth fibration.

Any fibered knot gives rise to a branched fibration as follows: Let $L = (S^n, L; b)$ be a fibered knot. We may alter the structure map b so that it has only regular values in the interior of D^2. Then define $cb: D^{n+1} \to D^2$ by the equation $cb(rx) = rb(x)$ where $0 \leq r \leq 1$ and $x \in S^n$. Let $\tau: D^{n+1} \to D^2$ be the result of smoothing cb at the cone point. Then $\tau: D^{n+1} \to D^2$ is a branched fibration so that $\tau|S^n$ is the given fibered structure for L.

This correspondence between branched fibrations over the disk and fibered knots is an abstraction from the case of isolated singularities. If $f: (\mathbb{C}^n, 0) \to (\mathbb{C}, 0)$ is an isolated complex polynomial singularity with link $L(f) = (S_\varepsilon^{2n-1}, L(f))$, then we let $N_f = \{z \in D_\varepsilon^{2n} \mid \|f(z)\| \leq \delta\}$ where $0 < \delta << \varepsilon$. Then $f:N_f \to D^2$ is a branched fibration.

Note that the most elementary branched fibration over D^2 is simply a ramified covering of the disk: $\mu_a:D^2 \to D^2$, $\mu_a(z) = z^a$. This corresponds to the empty knot of degree a .

Just as $\mu_a:D^2 \to D^2$ is a local model for a cyclic branched covering, a branched fibration $\tau:D^{n+1} \to D^2$ may be used as a local model for more general branched fibrations.

Thus suppose we are given a manifold M containing a properly embedded codimension two submanifold $V \subset M$. Suppose that there exists a smooth mapping $\alpha:M \to D^2$, transverse to $0 \in D^2$, so that $V = \alpha^{-1}(0)$. Then a new branched fibration $\tau_M:\tau(M,V) \to M$ is formed by constructing the pull-back

$$\begin{array}{ccc} \tau(M,V) & \longrightarrow & D^{n+1} \\ \downarrow{\scriptstyle \tau_M} & & \downarrow{\scriptstyle \tau} \\ M & \xrightarrow{\alpha} & D^2 \end{array} .$$

That is, $\tau(M,V) = \{(m,x) \in M \times D^{n+1} \mid \alpha(m) = \tau(x)\}$ and $\tau_M(m,x) = m$. The mapping τ_M is a smooth fibration away from $V \subset M$ and $\tau_M^{-1}(v) \cong CL$ for each $v \in V$ ($L \subset S^n$ is the fibered knot corresponding to τ) .

If $\tau = \mu_a:D^2 \to D^2$, then this pull-back yields an a-fold cyclic branched covering with branch set V . In the general case, the restriction of τ_M to a normal disk to $v \in V$ recovers τ . The construction $\tau(M,V)$ will not always be independent of α ; this is the case when M is 2-connected (see [KN]) . See ([KN]) also for a general definition of a branched fibration of M along V .

Product Construction. Let $K \subset S^n$ be any knot and $L \subset S^n$ be a fibered knot with associated branched fibration $\tau:D^{m+1} \to D^2$. Then there is a well-defined knot product $K \otimes L = (S^{n+m+1}, K \otimes L)$. It is defined as follows:

Let $F^n \subset D^{n+1}$ be a properly embedded submanifold of D^{n+1} such that $\partial F = K \subset S^n$. Let $\alpha:D^{n+1} \to D^2$ be a mapping transverse to $0 \in D^2$ so that $\alpha^{-1}(0) = F$. We may then form the pull-back

$$\begin{array}{ccc} \tau(D^{n+1},F) & \longrightarrow & D^{m+1} \\ \downarrow & & \downarrow{\scriptstyle \tau} \\ D^{n+1} & \xrightarrow{\alpha} & D^2 \end{array} .$$

The knot product is defined by taking boundaries:

$$(S^{n+m+1}, K \otimes L) = (\partial(D^{n+1} \times D^{m+1}), \partial\tau(D^{n+1},F)) .$$

Note that when $L = [a]$, the empty knot of degree a , then $K \otimes [a]$ is the a-fold cyclic branched covering of S^n along K and the construction gives an embedding $K \otimes [a] \subset S^{n+2}$. This new knot, $K \otimes [a]$, is called the a-fold cyclic suspension of K .

This product construction enjoys a number of useful properties, as summarized in the next theorem (proved in [KN]).

Theorem 3.2. Let $K = (S^n, K)$ be a knot and $L = (S^m, L)$ a fibered knot. Then

i) If K and L are both fibered then so is $K \otimes L$, and $L \otimes K = (-1)^{(n-1)(m-1)} K \otimes L$. (Here $-(S,K) = (-S,-K)$.)

ii) The product operation is associative.

iii) Suppose that $F \subset D^{n+1}$ is obtained by pushing the interior of a spanning surface $\tilde{F}$ for K into the interior of D^{n+1} . Let $\tau : D^{m+1} \to D^2$ be the branched fibration corresponding to L; let G be a spanning manifold for L . Then $K \otimes L$ has a spanning manifold M that is diffeomorphic to $\tau(D^{n+1}, F)$. Furthermore, M has the homotopy type of the join $F * G$.

iv) With notation as in iii), note that $\bar{H}_*(M) \cong \bar{H}_*(F) \otimes \bar{H}_*(G)$ ($\bar{H}_* = H_{*-1}$, H as in section 2). If Θ_K and Θ_L are Seifert pairings of K and L with respect to the spannning surfaces $\tilde{F}$ and G , and $\Theta_{K \otimes L}$ the Seifert pairing of $K \otimes L$ with respect to M , then

$$\Theta_{K\otimes L} \simeq \Theta_K \otimes \Theta_L$$

using the above decomposition of the homology of M . That is, for elements of homogeneous degree

$$\Theta_{K\otimes L}(a \otimes a', b \otimes b') = (-1)^{|a'||b|} \Theta_K(a,b) \Theta_L(a',b')$$

(where $x \in \bar{H}_{|x|} = H_{|x|-1}$ defines the $\bar{H}_*$ grading).

v) If $f : (\mathbb{C}^n, 0) \to (\mathbb{C}, 0)$ and $g" (\mathbb{C}^m, 0) \to (\mathbb{C}, 0)$ are isolated complex hypersurface singularities, then $f + g : (\mathbb{C}^n \times \mathbb{C}^m, 0) \to (\mathbb{C}, 0)$, $(f+g)(x,y) = f(x) + g(y)$ is also an isolated singularity and $L(f+g) \simeq L(f) \otimes L(g)$.

Signature problems. Given a knot $K = (S^{4n+1}, K)$ with spanning surface F of dimension $4n$ one defines the signature of K , $\sigma(K)$, to be the signature of the spanning surface. Thus $\sigma(K) = \sigma(F)$. Standard arguments using the Novikov addition theorem (see [AS],[KT]) show that this signature is independent of the choice of spanning surface. In order to generalize this notion, let $\tau : D^{\ell+1} \to D^2$ be a branched fibration corresponding to a fibered knot $L = (S^\ell, L)$. Let $L = (S^k, K)$ be any knot and assume that $4 | (k + \ell)$. Define the τ-signature of K , $\sigma_\tau(K)$, by the formula $\sigma_\tau(K) = \sigma(K \otimes L)$.

Since we know from 3.2 that $K \otimes L$ has a spanning surface $M \simeq \tau(D^{k+1}, F)$ where F is a pushed-in spanning surface for K , the τ-signature is the signature of this branched fibration. Thus $\sigma_\tau(K) = \sigma(\tau(D^{k+1}, F))$. The rest of this section will be devoted to showing how to compute this signature. We shall use part iv) of Theorem 3.2 to reduce the problem to signatures of forms related to the Seifert pairings of the two knots.

Lemma 3.3. Let $K = (S^k, K)$ be any knot, $L = (S^\ell, L)$ be a fibered knot with associated branched fibration τ. Let $F \subset D^{k+1}$ be a pushed in spanning surface for K, and G be a fiber for L. Let $M = \tau(D^{k+1}, F)$. Assume that $4 | (k+\ell)$. Then $\sigma(M) = 0$ unless k and ℓ are both odd. Given that k and ℓ are odd, let $A = H_p(F) \otimes H_q(G)$ where $p = (k-1)/2$, $q = (\ell-1)/2$. Let $<,>: A \times A \to \mathbb{Z}$ denote the restriction of the intersection form on $H_n(M)$ to A (via 3.2, iv) with $n = (k+\ell)/2$. Then $\sigma_\tau(K) = \sigma(A, <,>)$.

Proof: We know that $H_n(M) \cong \bigoplus_{s+t=n-1} H_s(F) \otimes H_t(G)$. The intersection form decomposes as an orthogonal sum on parts $A(s,t) \oplus A(k-s-1, \ell-t-1) = B(s,t)$, $A(s,t) = H_s(F) \otimes H_t(G) \times H_{k-s-1}(F) \otimes H_{\ell-t-1}$. Let $\sigma(B(s,t))$ denote the signature of the form on $H_n(M) \times H_n(M)$ restrictied to $B(s,t)$. Now suppose that $s \neq k-s-1$ or $t \neq \ell-t-1$. Then the form on $B(s,t)$ will have a matrix $N = \begin{bmatrix} 0 & X \\ X^t & 0 \end{bmatrix}$ if a basis of $B(s,t)$ is chosen with respect to the tensor decomposition. But certainly $\sigma(N) = 0$ and this suffices to prove the lemma.

In the light of this lemma we let $K = (S^k, K)$, $L = (S^\ell, L)$ where $k = 2p+1$ and $\ell = 2q+1$. Assume that $4 | (k+\ell)$. Let Θ be the Seifert pairing for K on $H_p(F)$ and Ω be the Seifert pairing for L on $H_q(G)$. We will continue to use this notation for the rest of the section.

Definition 3.4. Let $\Theta_\lambda^\varepsilon$ be the complex hermitian or skew-hermitian form defined by the equation $\Theta_\lambda^\varepsilon = (1-\bar{\lambda})\Theta + \varepsilon(1-\lambda)\Theta^*$ where $\varepsilon = \pm 1$ and λ is any complex number of unit length. (* denotes conjugate transpose here.) Define $\sigma(K;\lambda,\varepsilon)$ by the formula $\sigma(K;\lambda,\varepsilon) = \sigma(\Theta_\lambda^\varepsilon)$. Note that if a form $\Theta_\lambda^\varepsilon$ is skew-hermitian then $\sigma(\Theta_\lambda^\varepsilon)$ is, by definition, the signature of the hermitian form $(-i)\Theta_\lambda^\varepsilon$.

The notation that follows will be useful in formulating the next theorem. Let F and G be spanning surfaces for K and L as described above. Let $A = \bar{H}_p(F;\mathbb{C}) \otimes \bar{H}_q(G;\mathbb{C})$ and write $\bar{H}_q(G;\mathbb{C}) = \bigoplus_\lambda B_\lambda$ where B_λ is the Jordan subspace of $\bar{H}_q(G;\mathbb{C})$ corresponding to an eigenvalue λ of the monodromy $h: \bar{H}_q(G;\mathbb{C}) \to \bar{H}_q(G;\mathbb{C})$. Finally, let $\Omega(\lambda)$ denote the restriction of the Seifert pairing Ω to the subspace B_λ.

Definition 3.5. With notation as above, let $\varepsilon = (-1)^q$ and $\mu = (-1)^{(p+1)(q+1)}$. Define $\sigma(\tau;\lambda)$ by the formula $\sigma(\tau;\lambda) = \sigma(\Omega(\lambda) + \varepsilon\Omega(\lambda)^*)$. Define $\sigma_\tau^\lambda(K)$ by the formula $\sigma_\tau^\lambda(K) = \mu\sigma(\Theta \otimes \Omega(\lambda) + \Theta^* \otimes \Omega(\lambda)^*)$.

Theorem 3.6. Let $K = (S^{2p+1}, K)$ be any knot and τ a branched fibration corresponding to the fibered knot $L = (S^{2q+1}, L)$. Let $\varepsilon = (-1)^q$ and $\mu = (-1)^{(p+1)(q+1)}$. Then, using the notation developed above, one has the following formulas for $\sigma_\tau(K)$.

1) $\sigma_\tau(K) = \sum_\lambda \sigma_\tau^\lambda(K)$ where λ runs over all eigenvalues of the monodromy h for τ satisfying $||\lambda|| = 1$.

2) If each Jordan subspace B_λ for $||\lambda|| = 1$, $\lambda \neq 1$ is in fact the λ-eigenspace (i.e., $h|B_\lambda$ has no nilpotent part), then $\sigma_\tau^\lambda(K) = \mu\sigma(\tau;\lambda)\sigma(K;\lambda,\varepsilon)$. Hence

$$\sigma_\tau(K) = \mu\Sigma_{\lambda \in E} \sigma(\tau;\lambda)\sigma(K;\lambda,\varepsilon) + \sigma_\tau^1(K)$$

where $E = \{\lambda \,|\, ||\lambda|| = 1,\ \lambda \neq 1,\ \lambda$ an eigenvalue of $h\}$.

Proof. First note that the sign $\mu = (-1)^{(p+1)(q+1)}$ comes from the grading convention on the Seifert pairing for $K \otimes L$ as given in 3.2(iv). By 3.3, we need only consider $A = H_p(F) \otimes H_q(G)$ for signature computations. Since by Theorem 2.9 the decomposition $A = \bigoplus_\lambda B_\lambda$ gives an orthogonal decomposition of Θ_L , we conclude that the signature $\sigma_\tau(K) = \sigma(K \otimes L)$ is the sum of the signatures obtained from tensoring $\overline{H}_p(F;\mathbb{C})$ with B_λ . This gives part 1).

To see the reduction in part 2) it is convenient to use matrix notation. Let V denote a matrix for $\Omega(\lambda)$. Let H be the monodromy on this subspace. Let W be a matrix for Θ_k on $\overline{H}_p(F;\mathbb{C})$. Then by the remark after Proposition 2.7, $H = -\varepsilon V^{-1}V^*$ where $\varepsilon = (-1)^q$ and $*$ denotes conjugate transpose. For part 2) we are given that $H = \lambda I$, I an identity matrix, and $\lambda \neq 1$, $||\lambda|| = 1$. Thus $V^* = -\varepsilon\lambda V$. Let $X = W \otimes V + W^* \otimes V^*$. Then we know that $\sigma_\tau^\lambda(K) = \mu\sigma(X)$. But $X = W \otimes V - \varepsilon\lambda W^* \otimes V$ and it is easy to see from this that $X = [(1-\overline{\lambda})W + \varepsilon(1-\lambda)W^*] \otimes [((1-\lambda)(1-\overline{\lambda}))^{-1}(V + \varepsilon V^*)]$. Since the signature of a tensor product of (skew)hermitian forms is the product of the signatures, this shows that $\sigma(X) = \sigma(K;\lambda,\varepsilon)\sigma(\tau;\lambda)$. This completes the proof of the theorem.

Remark. The signatures $\sigma(K;\lambda,\varepsilon)$ are well-known (see [DK],[L2] and [T]) . It is also worth remarking that computing $\sigma_\tau(K)$ actually amounts to finding $\sigma(\tau(D^{k+1},F))$ where $F \subset D^{k+1}$ is any properly embedded codimension-two submanifold with $\partial F = K \subset S^k$. We defined $\sigma_\tau(K)$ when F was a pushed-in spanning surface. A standard (see [KT]) Novikov addition argument shows that $\sigma_\tau(K) = \sigma(\tau(D^{k+1},F))$ for an arbitrary surface as above.

IV. Applications and Examples

Brieskorn manifolds. Here τ is the branched fibration corresponding to the singularity $f(z) = z_0^{a_0} + a_1^{a_1} + ... + z_n^{a_n}$ and $L = (S^{2n+1},L)$ where $L = L(f)$. By Theorem 3.2, we know that $\Theta_L \simeq (-1)^{\frac{n(n+1)}{2}} \Lambda_{a_0} \otimes \Lambda_{a_1} \otimes ... \otimes \Lambda_{a_n}$ where Λ_a is the Seifert pairing for the empty knot of degree a . Since (as in section 2) $\Lambda_a \simeq$ Diag$(1 - \omega, 1 - \omega^2,...,1 - \omega^{a-1})$ where $\omega = \exp(2\pi i/a)$, we let $\omega_k = \exp(2\pi i/a_k)$. Then Θ is isomorphic to a diagonal matrix with entries $e(\vec{\kappa}) = (-1)^{\frac{n(n+1)}{2}} (1-\omega_0^{\kappa_0})(1-\omega_1^{\kappa_1})...(1-\omega_n^{\kappa_n})$ where $\vec{\kappa} = (\kappa_0,...,\kappa_n)$ and $0 < \kappa_j < a_j$ for $j = 0,1,...,n$. The monodromy matrix H is also diagonal with eigenvalues $\lambda_{\vec{\kappa}} = \overline{\omega}_0^{\kappa_0}\overline{\omega}_1^{\kappa_1}...\overline{\omega}_n^{\kappa_n}$. Thus $\sigma(\tau;\lambda_{\vec{\kappa}}) = \begin{cases} \text{Sign}(\text{Re}(e(\vec{\kappa})) ,n \text{ even} \\ \text{Sign}(\text{Im}(e(\vec{\kappa})) ,n \text{ odd} \end{cases}$

where Re and Im stand for real and imaginary parts respectively. It then follows (as in [B]) that if $\varepsilon(\vec{\kappa}) = \sigma(\tau;\lambda\vec{\kappa})$ then

$$\varepsilon(\vec{\kappa}) = \begin{cases} 1 & \text{if } 0 < \sum_s \kappa_s/a_s < 1 \pmod 2 \\ -1 & \text{if } 1 < \sum_s \kappa_s/a_s < 2 \pmod 2 \\ 0 & \text{otherwise} \end{cases}.$$

Thus, given $K = (S^{2p+1},K)$ with $4|(p+n)$ we conclude that

$$\sigma_\tau(K) = (-1)^{(n+1)(p+1)} \sum_{\vec{0}<\vec{\kappa}<\vec{a}} \varepsilon(\vec{\kappa})\sigma(K;\lambda\vec{\kappa},\varepsilon)$$

where $\varepsilon = (-1)^n$ and $\vec{\kappa} < \vec{a}$ means $\kappa_i < a_i$ for all i.

Cyclic Covers. A case of special interest is $f(z_0) = z_0^{a_0}$. This corresponds to a standard cyclic cover. Thus $\sigma_{[a]}(K) = \sum_{s=1}^{a-1} \sigma(K;\lambda^s,1)$ where $K = (S^{2q+1},K)$, q is odd, and $\lambda = \exp(2\pi i/a)$.

Note that $\sigma_{[a]}(K) = \sigma([a](D^{2q+2},F)) = \mathrm{Sign}(F_a)$ where $F \subset D^{2q+1}$ is a spanning manifold for K and F_a is the a-fold cyclic branched covering of D^{2q+2} along F. In this case the decomposition corresponds to an eigenspace decomposition of $\bar{H}_{q+1}(F_a;\mathbb{C})$ with respect to the covering translation for the branched covering $F_a \to D^{2q+2}$. The monodromy of the empty knot [a] corresponds directly to this deck transformation.

Links in S^3. Recall that two oriented links L and $L' \subset S^3$ are concordant if there is an embedding $j: C \times [0,1] \hookrightarrow S^3 \times [0,1]$ such that $(S^3,L) = (S^3 \times 0, j(C \times 0))$ and $(S^3,L') = (S^3 \times 1, j(C \times 1))$ where C is a disjoint collection of (oriented) circles. If a is a power of a prime number, then it follows as in [GD] that $\sigma_{[a]}(L)$ is a concordance invariant of $L \subset S^3$. Similarly, $\sigma(L;\omega,1)$ is a concordance invariant for $\omega = \exp(2\pi i\kappa/a)$, $0 < \kappa < a$, $a = P^s$, p a prime. Let $\sigma_\omega(L) = \sigma(L;\omega,1)$. By using our approach for cyclic covers, we can extract information about these signatures. For example, we shall give a well-known (see [T]) relationship among signature, genus and nullity.

First recall some further definitions: The multiplicity, $\mu(L)$, of a link $L \subset S^3$ is the number of components of the link. Given $L \subset S^3$, let $F' \subset D^4$ be a pushed-in connected spanning surface for L, and let F'_a denote the a-fold cyclic cover of D^4 branched along E', and L_a the a-fold cyclic cover of S^3 branched along L. With ω as above, let $\langle,\rangle_\omega = (1-\bar{\omega})\Theta + (1-\omega)\Theta^*$ where $\Theta: \bar{H}_1(F';\mathbb{C}) \times \bar{H}_1(F';\mathbb{C}) \to \mathbb{C}$ is the Seifert pairing. The ω-nullity is defined by the formula, $\eta_\omega(L) = (\text{nullity } \langle,\rangle_\omega) + 1$. It is easy to see that $\eta_\omega(L) - 1$ equals the rank of the ω-eigenspace of $H_1(L_a;\mathbb{C})$ with respect to the covering transformation.

Theorem 4.1. Let $F \subset D^4$ be any connected orientable, properly embedded surface so that $\partial F = L \subset S^3$. Let $g(F)$ denote the genus of F. Let ω be a primitive a^{th} root of unity where $a = p^s$, p a prime. Then
$|\sigma_\omega(L)| \leq g(F) + \mu(L) - \eta_\omega(L)$.

Proof. Let N denote the a-fold cyclic branched cover of D^4 along F. It is easy to see that $H_1(N) = 0$, and the relative sequence for $(N,\partial N)$ then shows that $\beta_1(N,\partial N) = 0$ (β_1 = first Betti number). Hence $\beta_3(N) = 0$ by Lefschetz duality. Let X denote Euler characteristic. Then $X(N) = a - (a-1)X(F)$ and $X(F) = 1 - \beta_1(F) = 1 - (2g + \mu - 1) = (2 - 2g) - \mu$ where $g = g(F)$ and $\mu = \mu(L)$. Hence $\beta_2(N) = (a - 1)\beta_1(F)$. Let $H_2(N;\mathbb{C}) = \oplus_{\kappa=1}^{a-1} H(\kappa)$ where $H(\kappa)$ denotes the eigenspace corresponding to ω^κ. Then $\dim H(\kappa) = (\frac{1}{a-1})\dim H_2(N;\mathbb{C}) = \beta_1(F)$. Let $\langle,\rangle = \langle,\rangle_\omega$ = the intersection pairing restricted to $H(1)$. Thus

$$\begin{aligned}|\sigma_\omega(L)| &\leq \dim H(1) - \text{nullity} \langle,\rangle \\ &= \beta_1(F) - \text{nullity} \langle,\rangle \\ &= 2g + \mu - 1 - \eta_\omega(L) + 1 \\ &= 2g + \mu - \eta_\omega .\end{aligned}$$

Hence $|\sigma_\omega(L)| \leq 2g(F) + \mu(L) - \eta_\omega(L)$. This proves the theorem.

G-signatures. For cyclic covers these methods also let us calculate g-signatures where g is some power of the covering translation. Let $K = (S^{2q+1},K)$ with q odd, and let $F \subset D^{2q+2}$ be a spanning manifold in the disk so that $\partial F = K$. Let $g:F_a \to F_a$ be a covering translation so that $g_*:H_{n+1}(F_a;\mathbb{C}) \to H_{n+1}(F_a;\mathbb{C})$ is represented on the eigenspaces $H(\kappa)$ by multiplication by ξ^κ for ξ an ath root of unity. The usual definition of g-signature (see [AS]) becomes $\sigma(g,F_a) = \Sigma_{\kappa=1}^{a-1} \xi^\kappa(H(\kappa))$ in this context. Since our method gives $\sigma(H(\kappa))$ in terms of the Seifert pairing, this formula may be computed explicitly.

An interesting special case is given by $V \xrightarrow{\pi} S^2$, a d-fold cyclic cover branched along d points. The branched covering space V is the completion of the covering of $X = S^2 - \{d \text{ points}\}$ which corresponds to the representation $\pi_1(X) \xrightarrow{\alpha} \mathbb{Z} \twoheadrightarrow \mathbb{Z}/d\mathbb{Z}$ where α takes each standardly oriented generator to $1 \in \mathbb{Z}$. It is easy to see that $V - \{d\text{-disks}\} \cong F_d$ where F_d is the fiber of the branched fibration corresponding to $z_0^d + z_1^d$. From this one computes directly that $\sigma(g,V) = d \coth(i\theta/2)$ where g corresponds to $\xi = e^{i\theta}$. This is, of course, a very special case of the Atiyah-Singer G-signature theorem. This approach has been used by Patrick Gilmer in [G]. It is also instructive to form V directly by a cut-and-paste construction and then do the calculations from a direct geometric base.

V. A Generalized Hirzebruch Construction

So far we have restricted ourselves to branched fibrations closely related to knot theory. In order to form more general branched fibrations, we generalize a construction due to Hirzebruch ([H]).

Let $V^{m-2} \subset M^m$ be an inclusion of closed manifolds. Assume that the homology class of V, $[V] \in H_{m-2}(M;\mathbb{Z})$, is Poincare dual to dx for $x \in H^2(M;\mathbb{Z})$ where d is a (non-zero) integer. Under these conditions there is a well-defined d-fold cyclic branched cover of M with branch set V. We assert the corresponding statement for a certain class of branched fibrations.

<u>Definition 5.1.</u> Call a branched fibration $\tau : D^{2k} \to D^2$ <u>d-equivariant</u> if it satisfies the following conditions: Let S^1 act on $D^{2k} = \{z \in \mathbb{C} | \, ||z|| = 1\}$ by the formula $\lambda \cdot (z_1, z_2, \ldots, z_k) = (\lambda^{\alpha_1} z_1, \lambda^{\alpha_2} z_2, \ldots, \lambda^{\alpha_k} z_k)$ for a chosen set of positive integers $\alpha_1, \alpha_2, \ldots, \alpha_k$. Then, with respect to this action, $\tau(\lambda \cdot z) = \lambda^d \tau(z)$.

For example, the branched fibration corresponding to a Brieskorn polynomial $f(z_1, \ldots, z_k) = z_1^{a_1} + \ldots + z_k^{a_k}$ is d-equivariant for d = least common multiple $(a_1, a_2, \ldots, a_k)$.

<u>Proposition 5.2.</u> Let $\tau : D^{2k} \to D^2$ be a d-equivariant branched fibration. Let $V^{m-2} \subset M^m$ be an embedding of closed manifolds so that the Poincare dual of $[V] \in H_{m-2}(M;\mathbb{Z})$ is divisible by d. Then there is a manifold $\hat{M}$ and a smooth mapping $\pi : \hat{M} \to M$ forming a τ-branched fibration of M along V.

<u>Proof.</u> Let $\bar{\Lambda} \to M$ be the complex disk bundle with first Chern class $c_1(\bar{\Lambda}) = x$ where dx = Poincare dual of $[V]$. Let D_0^2 denote D^2 with the circle action $\lambda \cdot z = \lambda^d z$. Note that if X and Y are S^1-spaces, then one defines $X \times_{S^1} Y = \{[x,y] \, | (\lambda x, y) \sim (x, \lambda y)$ for $\lambda \in S^1$; $[x,y]$ = equivalence class of $(x,y)\}$ Thus $D_0^2 \times_{S^1} \Lambda = \bar{\Lambda}^d$ where Λ denotes the circle bundle associated with $\bar{\Lambda}$ and $\bar{\Lambda}^d$ is the d-fold tensor product of $\bar{\Lambda}$ with itself.

Let $E = D^{2k} \times_{S^1} \Lambda$. Then τ induces a mapping $\mathcal{T} : E \to \bar{\Lambda}^d$ by the formula $\mathcal{T}[z,x] = [\tau(z), x]$. Note that $\mathcal{T}$ is a τ-branched fibration of $\bar{\Lambda}^d$ along its zero section $M \subset \bar{\Lambda}^d$.

Since $c_1(\bar{\Lambda}^d) = dx$ is Poincaré dual to $V \subset M$, there exists a section $s : M \to \bar{\Lambda}^d$ so that s is transverse to M and $s(M) \cap M = V$.

Let $\hat{M} = \mathcal{T}^{-1} s(M)$ and $\pi : \hat{M} \to M$ be the restriction of $a : E \to M$ where $a[z,x] = p(x)$, and $p : \Lambda \to M$ is the bundle projection. This constructs the desired branched fibration.

<u>Remark.</u> For branched coverings this construction appears in an article of Hirzebruch ([H]). The construction is summarized by the following diagram:

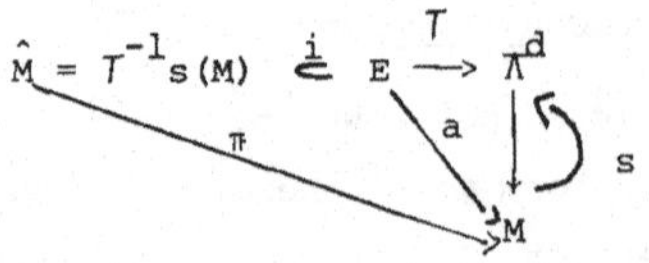

The next theorem generalizes a result obtained by P.Gilmer for cyclic coverings ([G]) .

Theorem 5.3. With notation as above, we may obtain information on the tangent bundle of $\hat{M}$ as follows: Let T( ) denote the real tangent bundle. Let Λ^d denote the underlying real vector bundle of $\overline{\Lambda}^d$ and $\mathbb{E}$ the underlying real vector bundle of $E \to M$. Then

$$T(\hat{M}) \oplus \pi^*(\Lambda^d) \simeq \pi^*(T(M) \oplus \mathbb{E}) .$$

Proof. The proof is identical in form to that given by Gilmer. We have

$$\begin{aligned} T(\hat{M}) \oplus \nu^E_{\hat{M}} &= i^*T(E) \\ &= i^*(a^*T(M) \oplus a^*\mathbb{E}) \\ &= \pi^*(T(M) \oplus \mathbb{E}) \end{aligned}$$

while $\nu^E_{\hat{M}} = i^*T^*\nu^{\Lambda^d}_{S(M)} = \pi^*(\Lambda^d)$. Here ν^Y_X denotes the normal bundle of X in Y .

Remark. For cyclic branched covers, this result is extraordinarily useful, since $\dim \hat{M} = \dim M$ and $\hat{M}$ is closed if M is closed. Thus the formula of 5.3 in conjunction with the Hirzebruch Index Theorem gives formulas for the signature of cyclic branched covers.

The situation is not all so fortunate for branched fibrations. Now $\hat{M}$ need not be closed and it will have higher dimension than M . Thus 5.3 is only a first step towards the signature of branched fibrations. We hope to continue this study in another paper.

References

[AS] M.F. Atiyah and I.M. Singer, The index of elliptic operators III; Annals of Math. (2) 87 (1968), 546-604.

[B] E. Brieskorn, Beispiele zur Differential Topologie von Singularitäten; Invent. Math. 2 (1966), 1-14.

[DK] A. Durfee and L. Kauffman, Periodicity of branched cyclic covers; Math. Ann. 218 (1975), 157-174.

[G] P. Gilmer, Equivariant signatures of ramified coverings; (preprint).

[GD] D. Goldsmith, Some new invariants of link concordance; (to appear).

[H] F. Hirzebruch, The signature of ramified coverings; In: Global Analysis. Papers in honor of Kodaira, 253-265, Princeton University Press (1969).

[K1] L. Kauffman, Branched coverings open books and knot periodicity; Topology 13 (1974), 143-160.

[K2] L. Kauffman, Products of knots; Bull. Am. Math. Soc. 80 (1974), 1104-1107.

[KN] L. Kauffman and W. Neumann, Products of knots, branched fibrations and sums of singularities; (to appear).

[KT] L. Kauffman and L. Taylor, Signature of links; Trans. Amer. Math. Soc. 216 (1976), 351-365.

[L1] J. Levine, Polynomial invariants of knots of codimension two; Ann. of Math. 84 (1966), 537-554.

[L2] J. Levine, Knot cobordism groups in codimension two; Comm. Math. Helv. 44 (1969), 229-244.

[M] J. Milnor, Singular points of complex hypersurfaces; Princeton University Press (1968).

[N] W. Neumann, Cyclic suspension of knots and periodicity of signature for singularities; Bull. Amer. Math. Soc. 80 (1974), 977-981.

[T] A.G. Tristram, Some cobordism invariants of links; Proc. Camb. Phil. Soc. 66 (1969), 257-264.

Department of Mathematics
University of Michigan
Ann Arbor, Michigan 48109

COBORDISMUS VON KNOTEN

Renate Vogt

In der vorliegenden Arbeit verstehen wir unter einem Knoten eine glatte orientierbare Untermannigfaltigkeit K^q von S^{q+2}, wobei K eine geschlossene, einfach zusammenhängende q-dimensionale Mannigfaltigkeit ist. Der Knoten heißt sphärisch, wenn K^q homotopieäquivalent zu S^q ist.

Zwei Knoten K_0^q, $K_1^q \subset S^{q+2}$ heißen cobordant, wenn es eine glatte orientierte Untermannigfaltigkeit X^{q+1} von $S^{q+2} \times I$ gibt, so daß gilt:

(i) $\partial X = X \cap (S^{q+2} \times \partial I) = K_0^q \times \{0\} \cup K_1^q \times \{1\}$,

(ii) Die Inklusionen $K_0^q \to X$, $K_1^q \to X$ sind Homotopieäquivalenzen.

Cobordismus ist eine Äquivalenzrelation auf der Menge aller q-dimensionalen Knoten ($q \geq 0$). Auf der Menge $\mathcal{C}_q$ der Cobordismusklassen von q-dimensionalen Knoten wird durch die zusammenhängende Summe (s. [3]) die Struktur einer abelschen Halbgruppe induziert. Wir bezeichnen mit $\mathcal{C}_q^s$ die Untergruppe der Cobordismusklassen von sphärischen Knoten.

Die Gruppen $\mathcal{C}_q^s$ wurden für gerade q von Kervaire [2] und für ungerade $q \geq 5$ von Levine [5,6] vollständig bestimmt als

$$\mathcal{C}_q^s = 0 \qquad \text{für q gerade ,}$$

$$\mathcal{C}_q^s \cong Z^\infty + Z_2^\infty + Z_4^\infty \qquad \text{für q ungerade, } q \geq 5 \text{ .}$$

In dieser Arbeit sollen nun die Halbgruppen $\mathcal{C}_q$ untersucht werden, wobei weitgehend auf die Methoden von Kervaire und Levine zurückgegriffen wird.

§ 1. Hilfsmittel

Die Existenz der Seifertfläche ist nach Erle [1] auch für nichtsphärische Knoten gesichert:

Lemma 1: Jeder Knoten $K^q \subset S^{q+2}$ ist der Rand einer orientierbaren Mannigfaltigkeit $V^{q+1} \subset S^{q+2}$. V heißt Seifertfläche für K.

Um zu entscheiden, ob zwei Knoten cobordant sind, erweist sich das folgende Kriterium als nützlich:

Lemma 2: Zwei Knoten $K_o^q, K_1^q \subset S^{q+2}$ mit $q \geq 5$ sind cobordant genau dann, wenn es eine Mannigfaltigkeit $W^{q+1} \subset D^{q+3}$ gibt mit

(i) $W^{q+1} \cap S^{q+2} = \partial W = K_o \# (-K_1)$,

(ii) Die Inklusionen $K_o^q \setminus \mathring{D}^q \to W^{q+1}$, $K_1^q \setminus \mathring{D}^q \to W^{q+1}$ sind Homotopieäquivalenzen.

§ 2. Knoten gerader Dimension

Seien $K_o^{2n}, K_1^{2n} \subset S^{2n+2}$ zwei (n-1)-zusammenhängende, 2n-dimensionale Knoten mit $n \geq 3$, und seien V_o^{2n+1}, V_1^{2n+1} Seifertflächen für K_o^{2n}, K_1^{2n}. Dann ist $V := V_o \natural (-V_1)$ eine Seifertfläche für $K := K_o \# (-K_1)$.

Wir nehmen an, daß K_o und K_1 dieselbe Homologie haben, d.h.

$$H_n(K_o) \cong H_n(K_1) \cong Z^k .$$

Dabei ist k gerade, weil K_o und K_1 gerade Eulercharakteristik haben.

Durch Surgery können wir die Seifertfläche V zunächst (n-1)-zusammenhängend machen (s. [4]). Indem wir Surgery entlang der "primitiven" Elemente im Sinne von Kervaire-Milnor [4] durchführen, können wir im nächsten Schritt erreichen, daß $rg\, H_n(V) = k$ ist. Um die Torsion von $H_n(V)$ zu töten, muß man nach der Methode von Wall [9] vorgehen. Man hat dann die Seifertfläche V von K abgeändert zu einer (n-1)-zusammenhängenden Mannigfaltigkeit $\tilde{V}$ mit

$$H_n(\tilde{V}) \cong Z^k .$$

Durch weitere $\frac{k}{2}$ Surgeries kann man schließlich erreichen, daß die Inklusionen $K_o \to \tilde{V}$, $K_1 \to \tilde{V}$ Isomorphismen in der Homologie erzeugen, also Homotopieäquivalenzen sind.

Da sich alle diese Surgeries auf V innerhalb der Kugel D^{2n+3} durchführen lassen (s. [2]), ist $\tilde{V}$ eine Untermannigfaltigkeit von D^{2n+3} und erfüllt damit die Bedingungen von Lemma 2.

Wir haben den folgenden Satz bewiesen:

Satz 1: Sei $\mathcal{C}_{2n}^{(n-1)}$ die Halbgruppe der Cobordismusklassen von (n-1)-zusammenhängenden Knoten. Für alle $n \geq 3$ ist die Abbildung

$$\Phi : \mathcal{C}_{2n}^{(n-1)} \longrightarrow N$$
$$K \longmapsto \tfrac{1}{2}\, rg\, H_n(K)$$

ein Isomorphismus. Das Erzeugende von $\mathcal{C}_{2n}^{(n-1)}$ wird repräsentiert durch den kanonisch eingebettenen Torus

$$S^n \times S^n \subset S^{2n+2} .$$

Der Beweis des Satzes beruht im wesentlichen darauf, daß sich die Homologie der Seifertfläche, auch in der mittleren Dimension, durch Surgery vereinfachen läßt. Mit diesen Methoden kann daher eine Verallgemeinerung des Satzes auf Knoten, die nicht (n-1)-zusammenhängend sind, nicht gelingen.

§ 3. Knoten ungerader Dimension

Bei sphärischen Knoten enthält die Seifertform genügend Informationen, um zu entscheiden, ob zwei Knoten cobordant sind. Wir werden zeigen, daß auch bei hochzusammenhängenden einfachen Knoten mit freier Homologie eine Cobordismusklassifikation mit Hilfe der Seifertform möglich ist. Dabei heißt ein (2n-1)-dimensionaler Knoten *einfach*, wenn er eine (n-1)-zusammenhängende Seifertfläche besitzt.

Sei $K^{2n-1} \subset S^{2n+1}$ ein Knoten mit Seifertfläche V^{2n}. Sei

$$i^{+}: V \longrightarrow S^{2n+1} \setminus V$$

die Verschiebung von V in positiver Normalenrichtung. Dann ist die *Seifertform* von K bezüglich V definiert als die folgende Bilinearform

$$\begin{aligned} \theta : H_n^f(V) \otimes H_n^f(V) &\longrightarrow \quad Z \\ x \otimes y \quad &\longmapsto L(x, i_*^{+}(y)) , \end{aligned}$$

wobei $H_n^f(V) := H_n(V)/\mathrm{Torsion}$ und L die Linking-Form in S^{2n+1} ist.

Die Seifertform hat die folgende wichtige Eigenschaft: ist σ die Schnittform auf V, so ist

$$\theta + (-1)^n \theta^t = \sigma .$$

Seien K_0^{2n-1}, K_1^{2n-1} zwei cobordante (n-2)-zusammenhängende einfache Knoten mit freier Homologie. Für $i \in \{0,1\}$ sei V_i eine (n-1)-zusammenhängende Seifertfläche für K_i, θ_i die zugehörige Seifertform, σ_i die Schnittform auf V_i.

Wir haben die folgende Situation:

$$\begin{array}{ccccccccccc} 0 & \longrightarrow & H_n(K_i) & \longrightarrow & H_n(V_i) & \longrightarrow & H_n(V_i,K_i) & \longrightarrow & H_{n-1}(K_i) & \longrightarrow & 0 \\ 0 & \longrightarrow & Z^{c_i} & \xrightarrow{\text{inj.}} & Z^{r_i} & \longrightarrow & Z^{r_i} & \longrightarrow & Z^{c_i} & \longrightarrow & 0 \end{array}$$

Da K_0 und K_1 dieselbe Homologie haben, ist natürlich $c_0 = c_1 =: c$.

Die Basis $x_1,\ldots,x_c$ von $H_n(K_0)$ läßt sich erweitern zu einer Basis $x_1,\ldots,x_{r_0}$ von $H_n(V_0)$, ebenso läßt sich die Basis $y_1,\ldots,y_c$

von $H_n(K_1)$ zu einer Basis $y_1,\dots,y_{r_1}$ von $H_n(V_1)$ erweitern.

Sei π_i für $i\in\{0,1\}$ der folgende Homomorphismus

$$\begin{aligned}\pi_i\colon H_n(V_i) &\longrightarrow [H_n(V_i)]^* \\ x &\longmapsto (\pi_i(x)\colon y\mapsto \sigma_i(y,x)) \ .\end{aligned}$$

Die Homologie von K_i berechnet sich als

$$\begin{aligned}H_n(K_i) &\cong \operatorname{Ker}\ \pi_i \\ H_{n-1}(K_i) &\cong \operatorname{Coker}\ \pi_i \ .\end{aligned}$$

Es folgt also:

(1) $$c = r_0 - \operatorname{rg}\sigma_0 = r_1 - \operatorname{rg}\sigma_1 \ .$$

Genauer kann man sagen, daß die Schnittmatrizen bezüglich der Basen $x_1,\dots,x_{r_0}$ bzw. $y_1,\dots,y_{r_1}$ die Form haben

$$S_0 = \begin{pmatrix} 0 & 0 \\ 0 & \tilde S_0\end{pmatrix} \begin{matrix}\}\ c \\ \}\ r_0-c\end{matrix} \quad , \quad |\det \tilde S_0| = 1 \ ,$$

$$S_1 = \begin{pmatrix} 0 & 0 \\ 0 & \tilde S_1\end{pmatrix} \begin{matrix}\}\ c \\ \}\ r_1-c\end{matrix} \quad , \quad |\det \tilde S_1| = 1 \ .$$

Die Homologieklassen $x_1,\dots,x_c$ bzw. $y_1,\dots,y_c$ können durch eingebettete n-Sphären in K_0 bzw. K_1 repräsentiert werden. Der Cobordismus $X^{2n}\subset S^{2n+1}\times I$, $X\cong K_0\times I$, induziert Einbettungen $S^n_j\times I\to S^{2n+1}\times I$ $(j=1,\dots,c)$, so daß x_j durch $S^n_j\times\{0\}$, y_j durch $S^n_j\times\{1\}$ repräsentiert wird, wobei die Basen $x_1,\dots,x_c$ und $y_1,\dots,y_c$ geeignet gewählt sein müssen. Aus geometrischen Gründen sind die Verschlingungszahlen der entsprechenden n-Sphären in $S^{2n+1}\times\{0\}$ und in $S^{2n+1}\times\{1\}$ gleich.

Wenn wir $H_n(K_i)$ als Untermodul von $H_n(V_i)$ auffassen, so haben wir gezeigt:

(2) $$\theta_0\big|H_n(K_0)\otimes H_n(K_0) = \theta_1\big|H_n(K_1)\otimes H_n(K_1) \ ,$$

d.h. die Seifertmatrizen haben bezüglich der Basen $x_1,\dots,x_{r_0}$ bzw. $y_1,\dots,y_{r_1}$ die Gestalt

$$M_0 = \begin{pmatrix} A & B_0 \\ \pm B_0^t & C_0\end{pmatrix} \begin{matrix}\}\ c \\ \}\ r_0-c\end{matrix} \quad ,$$

$$M_1 = \begin{pmatrix} A & B_1 \\ \pm B_1^t & C_1 \end{pmatrix} \begin{matrix} \} \; c \\ \} \; r_1 - c \end{matrix} \quad .$$

Wir leiten als nächstes eine Bedingung an die Seifertformen θ_0 und θ_1 ab, die genau der von Levine in [5] definierten Cobordismusrelation für Matrizen entspricht.

Sei $X \cong K_0 \times I$ ein Cobordismus zwischen K_0 und K_1. Es existiert eine Untermannigfaltigkeit G^{2n+1} von $S^{2n+1} \times I$ mit $\partial G = V_0 \cup X \cup V_1$ (s. [7]). Betrachte das Diagramm

$$\begin{array}{ccc} H_n(V_o) \oplus H_n(V_1) & \xrightarrow{\;i_o + i_1\;} & H_n(\partial G) \\ & \searrow^{\varphi} & \downarrow j \\ & & H_n(G) \end{array}$$

wobei alle Abbildungen durch Inklusionen induziert sind. Es gilt:

<u>Lemma 3</u>: Die Bilinearform $\theta_0 + (-\theta_1)$ verschwindet auf $\operatorname{Ker} \varphi$ (siehe dazu [7]).

Aus der Mayer-Vietoris-Sequenz bestimmt man

$$H_n(\partial G) \cong Z^{r_0 + r_1} \quad .$$

Nach derselben Methode wie Kervaire [3] schließt man, daß

$$\operatorname{rg} \operatorname{Ker} j = \tfrac{1}{2} \operatorname{rg} H_n(\partial G) = \tfrac{1}{2}(r_0 + r_1) =: r$$

ist. Daraus folgt, daß $\operatorname{rg} \operatorname{Ker} \varphi \geq r$ ist. Wir haben damit die dritte Bedingung an die Seifertformen θ_0, θ_1 abgeleitet:

(3) Es gibt einen Untermodul H der Dimension r von $H_n(V_0) \oplus H_n(V_1)$, auf dem die Bilinearform $\theta_0 + (-\theta_1)$ verschwindet.

Man kann noch genauere Angaben darüber machen, welche Klassen aus $H_n(V_0) \oplus H_n(V_1)$ zu $\operatorname{Ker} \varphi$ gehören. Zunächst ist klar, daß $(x_1 - y_1), \dots, (x_c - y_c)$ in $\operatorname{Ker}(i_0 + i_1)$, also auch in $\operatorname{Ker} \varphi$ liegen:

(4) $(x_1 - y_1), \dots, (x_c - y_c) \in H_n(V_0) \oplus H_n(V_1)$ lassen sich zu einer Basis von H ergänzen.

Schließlich läßt sich noch beweisen, daß $x_1, \dots, x_c$ nicht in H liegen. Dazu zeigen wir, daß $i_0(x_1), \dots, i_0(x_c) \notin \operatorname{Ker} j$ sind. Wir ändern die Mannigfaltigkeit G durch Surgery wie in § 2 zu einer Mannigfaltigkeit $\tilde{G}$ ab, so daß

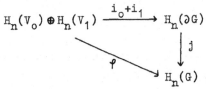

$$\tilde{G} \ \text{(n-1)-zusammenhängend},$$
$$H_n(\tilde{G}) \cong Z^r ,$$
$$\partial\tilde{G} = \partial G$$

ist. Dabei nehmen wir keine Rücksicht auf die Einbettung von G in $S^{2n+1}\times I$. Die Inklusion $\partial G \to \tilde{G}$ induziert eine Abbildung $\tilde{j}: H_n(\partial G) \to H_n(\tilde{G})$, und es gilt

Lemma 4: Ker j ist ein Untermodul der Dimension r von Ker $\tilde{j}$.

Zum Beweis sieht man sich die Surgeries, die von G zu $\tilde{G}$ führen, genau an.

Man wählt nun eine Basis $v_1,\dots,v_r$ von Ker $\tilde{j}$, die sich zu einer Basis von $H_n(\partial G)$ erweitern läßt, und beweist:

Lemma 5: Die Schnittzahl von v_i und v_j $(i,j \in \{1,\dots,r\})$ in ∂G ist gerade.

Dabei benutzt man wesentlich die Arbeiten von Whitney [10,11].

Andererseits wissen wir, daß die Matrizen der Schnittformen auf V_0 und V_1 bezüglich der Basen $x_1,\dots,x_{r_0}$ und $y_1,\dots,y_{r_1}$ gegeben sind durch

$$S_0 = \begin{pmatrix} 0 & 0 \\ 0 & \tilde{S}_0 \end{pmatrix} \quad \text{und} \quad S_1 = \begin{pmatrix} 0 & 0 \\ 0 & \tilde{S}_1 \end{pmatrix} .$$

Da sich $x_1,\dots,x_{r_0},y_{c+1},\dots,y_{r_1}$ durch c Klassen $z_1,\dots,z_c$ zu einer Basis von $H_n(\partial G)$ erweitern läßt, erhält man bei geeigneter Wahl der $z_1,\dots,z_c$ als Schnittzahlmatrix von ∂G:

$$\begin{pmatrix} 0 & 0 & 0 & E \\ 0 & \tilde{S}_o & 0 & 0 \\ 0 & 0 & \tilde{S}_1 & 0 \\ \pm E & 0 & 0 & Z \end{pmatrix}$$

Man kann die $v_1,\dots,v_r$ o.B.d.A. so wählen, daß

$v_1,\dots,v_k$ Linearkombinationen aus $x_1,\dots,x_c$,

$v_{k+1},\dots,v_{k+p}$ Linearkombinationen aus $x_1,\dots,x_{r_0},y_{c+1},\dots,y_{r_1}$,

$v_{k+p+1},\dots,v_{k+p+q}$ Linearkombinationen aus $x_1,\dots,x_{r_0},y_{c+1},\dots,y_{r_1}$, $z_1,\dots,z_c$

sind mit $k \leq c$, $p \leq 2(r-c)$, $q \leq c$, $k+p+q = r$.

Mit Hilfe von Lemma 5 und der Schnittmatrix von ∂G läßt sich dann leicht zeigen, daß $k+q = c$ und $p = r-c$ ist. Die $v_{k+1},\dots,v_{k+p}$ bilden nun zusammen mit $x_1-y_1,\dots,x_c-y_c$ eine Basis von H. Daraus folgt, daß $x_1,\dots,x_c \notin H$ ist. Es läßt sich sogar zeigen

(5) $x_1,\dots,x_c$ läßt sich zu einer Basis von $(H_n^f(V_0)+H_n^f(V_1))/H$ ergänzen.

Wir haben damit 5 Bedingungen an die Seifertformen der Knoten K_0, K_1 abgeleitet, die erfüllt sind, wenn K_0 und K_1 cobordant sind. Umgekehrt sind diese Bedingungen für $n \geq 3$ auch hinreichend dafür, daß die Knoten cobordant sind.

Angenommen, wir haben zwei $(n-2)$-zusammenhängende $(2n-1)$-dimensionale Knoten K_0 und K_1 mit $(n-1)$-zusammenhängenden Seifertflächen V_0 und V_1, wobei $n \geq 3$ ist. Zunächst können wir erreichen, daß $\operatorname{rg} H_n(V_0) = \operatorname{rg} H_n(V_1) = r$ ist, indem wir für $i = 0$ oder $i = 1$ V_i ersetzen durch

$$V_i \# (S^n \times S^n) \# \dots \# (S^n \times S^n) .$$

Bezüglich geeigneter Basen $x_1,\dots,x_r$ von $H_n^f(V_0)$ und $y_1,\dots,y_r$ von $H_n^f(V_1)$ sollen die Seifertmatrizen M_0 und M_1 die Eigenschaft haben, daß

$$M_0 + (-1)^n M_0^t = S_0 = \begin{pmatrix} 0 & 0 \\ 0 & \tilde{S}_0 \end{pmatrix} \begin{matrix} \}\, c \\ \}\, r-c \end{matrix} \quad , \quad |\det \tilde{S}_0| = 1 ,$$

$$M_1 + (-1)^n M_1^t = S_1 = \begin{pmatrix} 0 & 0 \\ 0 & \tilde{S}_1 \end{pmatrix} \begin{matrix} \}\, c \\ \}\, r-c \end{matrix} \quad , \quad |\det \tilde{S}_1| = 1 ,$$

und

$$M_0 = \begin{pmatrix} A & B_0 \\ \pm B_0^t & C_0 \end{pmatrix} \begin{matrix} \}\, c \\ \}\, r-c \end{matrix} \quad ,$$

$$M_1 = \begin{pmatrix} A & B_1 \\ \pm B_1^t & C_1 \end{pmatrix} \begin{matrix} \}\, c \\ \}\, r-c \end{matrix} \quad .$$

Wenn außerdem die Bedingungen (3), (4), (5) erfüllt sind, so läßt sich die Schnittmatrix von $V_0 \natural (-V_1)$ bezüglich einer geeignet gewählten Basis $s_1,\dots,s_r,t_1,\dots,t_r$ von $H_n^f(V_0) \oplus H_n^f(V_1)$ in der Form

$$\left(\begin{array}{cc|cc} \multicolumn{2}{c|}{0} & E & 0 \\ & & 0 & 0 \\ \hline E & 0 & D & 0 \\ 0 & 0 & 0 & 0 \end{array}\right) \begin{array}{l} \}\ r-c \\ \}\ c \\ \}\ r-c \\ \}\ c \end{array}$$

schreiben.

Auf der Seifertfläche $V_0 \natural (-V_1)$ des Knotens $K_0 \# (-K_1)$ führen wir Surgery entlang der Klassen $s_1,\ldots,s_r$ durch (siehe [8]), und zwar wieder innerhalb der Kugel D^{2n+2}. Dadurch erhalten wir eine Mannigfaltigkeit $W^{2n} \subset D^{2n+2}$ mit

$$\partial W = W \cap S^{2n+1} = K_0 \# (-K_1) ,$$

$$H_*(W) = H_*(K_0 \setminus \mathring{D}^{2n-1}) = H_*(K_1 \setminus \mathring{D}^{2n-1}) .$$

Indem man die Wirkung der einzelnen Surgeries genau beobachtet, zeigt man, daß die Inklusionen

$$K_0 \setminus \mathring{D}^{2n-1} \longrightarrow W$$

$$K_1 \setminus \mathring{D}^{2n-1} \longrightarrow W$$

Homotopieäquivalenzen sind. Wir können wieder Lemma 2 anwenden und haben bewiesen, daß die Knoten cobordant sind.

Wir können das folgende Ergebnis formulieren:

<u>Satz 2</u>: Für $n \geq 3$ gilt: zwei $(n-2)$-zusammenhängende, einfache, $(2n-1)$-dimensionale Knoten mit freier Homologie sind genau dann cobordant, wenn ihre Seifertformen bezüglich $(n-1)$-zusammenhängender Seifertflächen die oben formulierten Bedingungen (1) - (5) erfüllen.

Wenn man auf die Voraussetzung verzichtet, daß die Knoten freie Homologie haben, lassen sich die Bedingungen (1) - (4) ohne Schwierigkeit, die Bedingung (5) nur in abgeschwächter Form beweisen. Die Umkehrung bereitet erhebliche Schwierigkeiten.

Literatur

1. Dieter Erle, Quadratische Formen als Invarianten von Einbettungen der Kodimension 2, Topology 8 (1969), S. 99-114.

2. Michel A. Kervaire, Les noeds de dimensions superieures, Bull. Soc. Math. de France 93 (1965), S. 225-271.

3. Michel A. Kervaire, Knot cobordism in codimension two, Lecture Notes 197, S. 82-105.

4. Michel A. Kervaire, John W. Milnor, Groups of homotopy spheres, Ann. of Math. 77 (1968), S. 504-537.

5. Jerome Levine, Knot cobordism groups in codimension two, Comm. Math. Helv. 44 (1969), S. 229-244.

6. Jerome Levine, Invariants of knot cobordism, Invent. math. 8 (1969), S. 98-110.

7. Jerome Levine, An algebraic classification of some knots in co-dimension two, Comm. Math. Helv. 45 (1970), S. 185-198.

8. John W. Milnor, A procedure for killing homotopy groups of differen-tiable manifolds, Symposia in Pure Math. AMS vol. 3 (1961), S. 39-55.

9. C. T. C. Wall, Killing the middle homotopy groups of odd dimensional manifolds, Trans. Amer. Math. Soc. 103 (1962), S. 421-433.

10. Hassler Whitney, The singularities of a smooth n-manifold in (2n-1)-space, Ann. of Math. 45 (1944), S. 247-293.

11. Hassler Whitney, The general type of singularity of a set of 2n-1 smooth functions of n variables, Duke Math. Journal 10 (1943), S. 161-172.

Renate Vogt,
Mathematisches Institut
der Universität Bonn,
Wegelerstraße 10,
53 Bonn,
West Germany

ATTEMPTING TO CLASSIFY KNOT MODULES AND THEIR HERMITIAN PAIRINGS

C. Kearton

0. Introduction

This is a preliminary report of some research which is still in progress. Some of the hypotheses are more restrictive than is necessary, but indications are given of what I believe to be the appropriate generalisations.

A knot is an oriented pair (S^3, S^1), smooth or piecewise-linear. Let K be the closed complement of a tubular neighbourhood of S^1; by Alexander duality

$$H_1(K) \cong H^1(S^3 - K) \cong H^1(S^1)$$

and so the orientations determine, via the duality isomorphisms, a natural choice of generator t for the infinite cyclic group $H_1(K)$, written multiplicatively.

The Hurewicz theorem gives an epimorphism

$$h : \pi_1(K) \twoheadrightarrow H_1(K) ;$$

let $\tilde{K} \to K$ be the covering corresponding to ker h. Then $H_1(K) = (t : \ )$ acts on K as the group of covering transformations, and this action extends linearly to make $H_*(\tilde{K})$ into a module over $\Lambda = \mathbb{Z}[t, t^{-1}]$.

It is well known [6, 8, 11] that $H_1(\tilde{K})$ is presented as a Λ-module by $tA - A'$, where A is a Seifert matrix of the knot and A' denotes the transpose of A. Thus there is a short exact sequence

$$F \rightarrowtail^{tA - A'} G \overset{\psi}{\twoheadrightarrow} H_1(\tilde{K})$$

where F and G are free Λ-modules of rank 2k say.

If Λ_0 denotes the field of fractions of Λ, regarded as a Λ-module, then there is a pairing

$$\langle\ ,\ \rangle : \quad H_1(\tilde{K}) \times H_1(\tilde{K}) \longrightarrow \Lambda_0/\Lambda$$

defined by

$$\langle \psi\underline{x}, \psi\underline{y} \rangle = (1 - t)\, \underline{x}'\, (tA - A')^{-1}\bar{\underline{y}} \quad .$$

Here $\underline{x}$, $\underline{y}$ are column vectors in $G \cong \Lambda^{2k}$.

Conjugation in Λ is defined as the linear extension of $t \mapsto t^{-1}$, and is denoted by $\bar{\ }$. $\langle\ ,\ \rangle$ is clearly Hermitian, ie it is linear in the first variable, conjugate linear in the second, and conjugate symmetric.

It is not hard to show that $(H_1(\tilde{K}), \langle\ ,\ \rangle)$ is an invariant of the knot [8, 12]. The pairing is due to R.C. Blanchfield [1] (cf. [3, 4]).

The pairs $(H_1(\tilde{K}), \langle\ ,\ \rangle)$ have been characterised as follows [3, 4, 8] :

Theorem 0.1 Let M be a Λ-module,
$\{\ ,\ \} : M \times M \longrightarrow \Lambda_0/\Lambda$ a pairing. Then $(M, \{\ ,\ \}) \cong (H_1(\tilde{K}), \langle\ ,\ \rangle)$ for some knot iff the following are satisfied.

(i) M is a finitely - generated Λ- torsion - module.

(ii) $(t - 1) : M \longrightarrow M$ is an isomorphism.

(iii) $\{\ ,\ \}$ is a non-singular Hermitian pairing.

Non-singular means that the associated map

$$M \longrightarrow \overline{\mathrm{Hom}}\ (M, \Lambda_0/\Lambda)$$

is an isomorphosm.

The classification of knot modules and their pairings thus reduces to the study of modules and pairings satisfying (i) - (iii) above.

1. Hypotheses and notation

Let $(M, \langle\ ,\ \rangle)$ be a pair satisfying the three conditions of theorem 0.1. Assume that M is annihilated by $\varphi^k \in \Lambda$ for some integer k and some irreducible polynomial φ which is symmetric, i.e. $\varphi = \bar{\varphi}$. Since M is $\mathbb{Z}$- torsion - free [4 ; p 155], $M \subset M \otimes_{\mathbb{Z}} \mathbb{Q} \cong \bigoplus_i F_i$, where $F_i \cong \bigoplus \mathbb{Q}[t, t^{-1}]/(\varphi^i)$; this is a consequence of the structure theorem for modules over a PID. Assume that $F_i = 0$ except for $i = n$; in the terminology of Levine [7], M is homogeneous of degree n.

Let τ be the image of t in $R = \Lambda/(\varphi)$, so that $R = \mathbb{Z}[\tau, \tau^{-1}]$. Assume that R is a Dedekind domain in which 2 is a unit.

If A is a Λ-module, let $A^* = \overline{\mathrm{Hom}}\ (A, \Lambda_0/\Lambda)$, the Λ-module of conjugate linear maps $A \to \Lambda_0/\Lambda$. For $x \in A^*$, $y \in A$, define $[x, y] = x(y)$.

Suppose that A, B are Λ-modules; then

$\lambda : A \longrightarrow B^*$
$\lambda^* : B \longrightarrow A^*$ are related by $[\lambda^* b, a] = \overline{[\lambda a, b]}$

$\mu : A \longrightarrow B$
$\tilde{\mu} : B^* \longrightarrow A^*$ are related by $[\tilde{\mu} b^*, a] = [b^*, \mu a]$

A pairing

$$\langle\ ,\ \rangle : A \times B \longrightarrow \Lambda_0/\Lambda$$

which is linear on A, conjugate linear on B, defines a map $\lambda : A \longrightarrow B^*$ by

$$[\lambda a, b] = \langle a, b\rangle$$

The map $\theta : M \longrightarrow M^*$ corresponding to the pairing $\langle\ ,\ \rangle$ on M is an isomorphism by condition (iii). Furthermore,

$$[\theta x, y] = \langle x, y\rangle = \overline{\langle y, x\rangle} = \overline{[\theta y, x]} = [\theta^* x, y],$$

and so $\theta = \theta^*$

In the sequel, $\ker \varphi^i = \ker (M \xrightarrow{\varphi^i} M)$; in particular, $\ker \varphi^n = M$.

One can define a Λ-module $\bar{\Lambda}$ in the following way: $\bar{\Lambda} = \Lambda$ as an additive group, and for $x \in \bar{\Lambda}$, $\lambda \in \Lambda$, let $\lambda . x = \bar{\lambda} x$ (where . denotes the action of Λ on $\bar{\Lambda}$). This extends to $\bar{\Lambda}_o$, $\overline{\Lambda_o/\Lambda} = \bar{\Lambda}_o/\bar{\Lambda}$ in the obvious way.

Clearly, $\overline{\text{Hom}}\ (A, \Lambda_o/\Lambda) \cong \text{Hom}\ (A, \overline{\Lambda_o/\Lambda})$ in a natural way.

Define $\overline{\text{Ext}}\ (A, \Lambda_o/\Lambda) = \text{Ext}\ (A, \overline{\Lambda_o/\Lambda})$, so that, for example, applying the functor $\overline{\text{Hom}}\ (\ , \Lambda_o/\Lambda)$ to the short exact sequence $A \rightarrowtail B \twoheadrightarrow C$ yields the exact sequence

$$C^* \rightarrowtail B^* \longrightarrow A^* \longrightarrow \overline{\text{Ext}}\ (C, \Lambda_o/\Lambda) \longrightarrow \overline{\text{Ext}}\ (B, \Lambda_o/\Lambda) \longrightarrow \ldots\ldots$$

2. Some technical results

Lemma 2.1 Let A, C be finitely - generated Λ-torsion-modules, with C a free $\Lambda/(\varphi)$ -module of rank m. Then $\text{Ext}\ (C, A) \cong \oplus_1^m\ A/\varphi A$.

Proof: Let $K \xrightarrow{\alpha} P \twoheadrightarrow C$ be short exact, with K and P free Λ-modules of rank m, $\alpha = \varphi$ times the identity. By [9 ; theorem 3.6],

$$\text{Ext}\ (C, A) \cong \text{Hom}\ (K, A)\ /\ \alpha'\ \text{Hom}\ (P, A)$$

where $\alpha' = \text{Hom}\ (\alpha, A)$ is the dual of α.

But $P \cong \oplus_1^m \Lambda$, so $\text{Hom}\ (P, A) \cong \oplus_1^m\ \text{Hom}\ (\Lambda, A) \cong \oplus_1^m A$.

Similarly, $K \cong \oplus_1^m\ A$, and α' is φ times the identity, whence the result.

Remark The isomorphism depends on a choice of basis for C, and may be described as follows.

Let $c_1, \ldots, c_m$ be a Λ-basis for C, and $A \xrightarrow{\sigma} B \xrightarrow{\pi} C$ a short exact sequence representing an element of Ext (C, A). Then $\exists\ b_i \in B$ with $\pi b_i = c_i$, $1 \leq i \leq m$; and $\exists\ a_i \in A$ with $\sigma a_i = \varphi b_i$, $1 \leq i \leq m$. The a_i determine an element of $\oplus_1^m\ A/\varphi A$.

The following result occurs as Lemma II.12 in [5]

Lemma 2.2 Let A be a finitely - generated $\mathbb{Z}$-torsion-free Λ-module, such that $(t - 1) : A \rightarrow A$ is surjective. Then A has a presentation by a matrix ta - b, where a and b are square non-singular integer matrices.

Lemma 2.3 Let A be a Λ-module presented by a square non-singular matrix S. Then $\overline{\text{Ext}}\ (A, \Lambda_o/\Lambda) = 0$.

Proof There is a short exact sequence $P \xrightarrow{S} Q \twoheadrightarrow A$ with P, Q free Λ-modules of rank k. [By [9 ; theorem 3.6],

$$\overline{\text{Ext}}\ (A, \Lambda_o) = \text{Ext}\ (A, \bar{\Lambda}_o) \cong \frac{\text{Hom}\ (P, \bar{\Lambda}_o)}{S'\ \text{Hom}\ (Q, \bar{\Lambda}_o)} \cong \frac{\oplus_1^k \bar{\Lambda}_o}{S' \oplus_1^k \bar{\Lambda}_o} = 0$$

as S is invertible over the field Λ_0.

Applying $\overline{\mathrm{Hom}}$ (A,) to the short exact sequence $\Lambda \rightarrowtail \Lambda_0 \twoheadrightarrow \Lambda_0/\Lambda$, one obtains

$$\to \overline{\mathrm{Ext}}\,(A, \Lambda) \to \overline{\mathrm{Ext}}\,(A, \Lambda_0) \to \overline{\mathrm{Ext}}\,(A, \Lambda_0/\Lambda) \to 0,$$

the sequence terminating in zero because of the free resolution $P \rightarrowtail Q \twoheadrightarrow A$.

$$\text{Whence} \quad \overline{\mathrm{Ext}}\,(A, \Lambda_0/\Lambda) = 0.$$

3. The main diagram

The first part of the diagram consists of the following commutative quadrilateral : all the maps are inclusions or quotients, and every row is a short exact sequence.

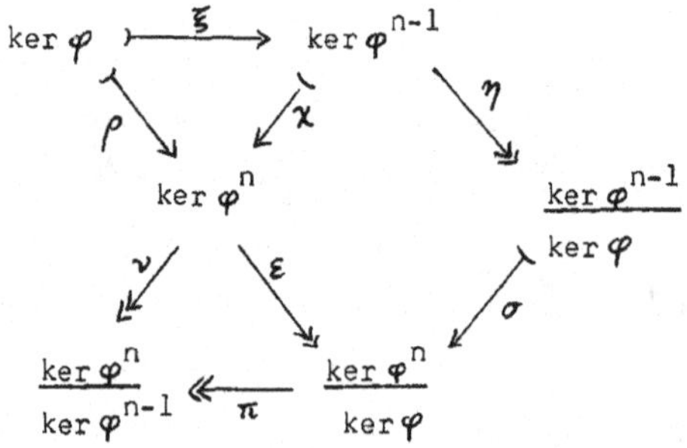

Define a map $\lambda : \ker\varphi \to \left(\dfrac{\ker\varphi^n}{\ker\varphi^{n-1}}\right)^*$ by the equation

$$[\lambda x, \nu y] = \langle \rho x, y \rangle \quad ; \quad x \in \ker\varphi, \quad y \in \ker\varphi^n .$$

This is well-defined, for $\langle \rho x, y\rangle = 0$ if $y \in \ker\varphi^{n-1}$.

The following maps are similarly defined.

$\mu : \ker\varphi^{n-1} \to \left(\dfrac{\ker\varphi^n}{\ker\varphi}\right)^*$ by $[\mu x, \varepsilon y] = \langle \chi x, y \rangle$; $x \in \ker\varphi^{n-1}$, $y \in \ker\varphi^n$.

$\lambda_0 : \dfrac{\ker\varphi^n}{\ker\varphi^{n-1}} \to (\ker\varphi)^*$ by $[\lambda_0 \nu x, y] = \langle x, \rho y\rangle$; $x \in \ker\varphi^n$, $y \in \ker\varphi$.

$\mu_0 : \dfrac{\ker\varphi^n}{\ker} \to (\ker\varphi^{n-1})^*$ by $[\mu_0 \varepsilon x, y] = \langle x, \chi y\rangle$; $x \in \ker\varphi^n$, $y \in \ker\varphi^{n-1}$

$\theta_0 : \dfrac{\ker\varphi^{n-1}}{\ker\varphi} \to \left(\dfrac{\ker\varphi^{n-1}}{\ker\varphi}\right)^*$ by $[\theta_0 \eta x, \eta y] = \langle \chi x, \chi y\rangle$; $x, y \in \ker\varphi^{n-1}$

Applying $\overline{\mathrm{Hom}}$ (, Λ_0/Λ) to the diagram above, one obtains the commutative diagram

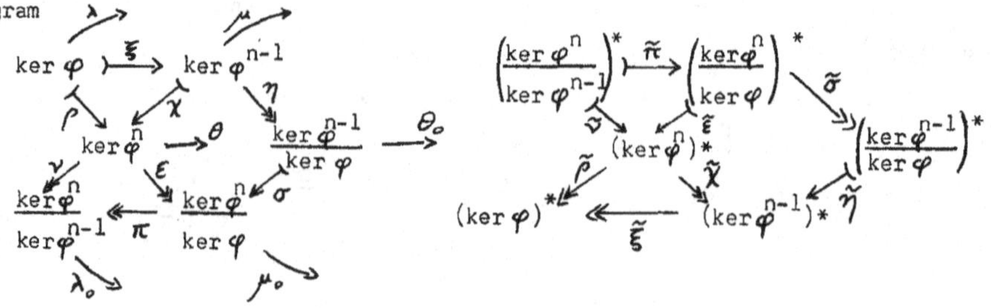

To see that the diagram commutes is just a matter of checking definitions. For example $[\tilde{\nu}\lambda x, y] = [\lambda x, \nu y] = \langle \rho x, y\rangle = [\theta\rho x, y]$ for all $x \in \ker\varphi$, $y \in \ker\varphi^n$.

Therefore $\tilde{\nu}\lambda = \theta\rho$.

Note that every module on the left-hand side of the diagram satisfies the hypotheses of Lemma 2.2, and therefore the conclusion of Lemma 2.3. It follows that the maps $\tilde{\sigma}, \tilde{\chi}, \tilde{\rho}$ and $\tilde{\xi}$ are all surjections, and so the right-hand side of the diagram is also a quadrilateral of short exact sequences.

(i) λ is an isomorphism.

For since $\tilde{\nu}\lambda = \theta\rho$, and $\theta\rho$ is monomorphic, so is λ.

If $z^* \in \left(\frac{\ker\varphi^n}{\ker\varphi^{n-1}}\right)^*$, then since θ is an isomorphism $\exists\, z \in \ker\varphi^n$ with $\theta z = \tilde{\nu}z^*$

But $\varphi z^* = 0$, so $\varphi z = 0$, and hence $\exists\, x \in \ker\varphi$ with $\rho x = z$. Then $\tilde{\nu}\lambda x = \theta\rho x = \tilde{\nu}z^*$, and so $\lambda x = z^*$. Therefore λ is surjective.

(ii) μ is an isomorphism.

The proof is similar to (i).

(iii) λ_0 is an isomorphism.

For if $\lambda_0\nu x = 0$ for $x \in \ker\varphi^n$, then $\tilde{\rho}\theta x = 0$ and so $\theta x \in \operatorname{Im}\tilde{\varepsilon}$. Since θ, μ are isomorphisms and $\chi, \tilde{\varepsilon}$ are monomorphisms, it follows that $x \in \operatorname{Im}\chi$, and so $\nu x = 0$. Therefore λ_0 is injective.

λ_0 is surjective because $\lambda_0\nu = \tilde{\rho}\theta$, and $\tilde{\rho}\theta$ is surjective.

(iv) μ_0 is an isomorphism.

The proof is similar to (iii)

(v) θ_0 is an isomorphism.

θ_0 is injective because $\tilde{\eta}\theta_0 = \mu_0\sigma$, and $\mu_0\sigma$ is injective. θ_0 is surjective because $\theta_0\eta = \tilde{\sigma}\mu$ and $\tilde{\sigma}\mu$ is surjective.

(vi) $\lambda_0 = \lambda^*$

For if $x \in \ker\varphi^n$, $y \in \ker\varphi$;

$$[\lambda_0\nu x, y] = \langle x, \rho y\rangle = \overline{\langle \rho y, x\rangle} = \overline{[\lambda y, \nu x]} = [\lambda^*\nu x, y]$$

(vii) $\mu_0 = \mu^*$

The proof is similar to (vi)

(viii) $\theta_0 = \theta_0^*$

For if $x, y \in \ker\varphi^{n-1}$;

$$[\theta_0\eta x, \eta y] = \langle \chi x, \chi y\rangle = \overline{\langle \chi y, \chi x\rangle} = \overline{[\theta_0\eta y, \eta x]} = [\theta_0^*\eta x, \eta y]$$

Thus from the pair (M, θ) one obtains the pair $(\ker\varphi^{n-1}/\ker\varphi, \theta_0)$, homogeneous of degree n-2 and satisfying the conditions of theorem 0.1. Iterating this process leads to the case n = 1 or n = 2, depending on the parity of the degree of M. The case n = 1 is simply the theory of Hermitian forms over a Dedekind domain; for the case n = 2 the reader should consult Levine [7] (but note the caveat in the introduction to [8]).

The rest of this paper will concentrate on the inductive step, i.e. on passing from $(\ker \varphi^{n-1}/\ker \varphi\ ,\ \theta_0)$ back to $(\ker \varphi^n,\ \theta)$.

4. The inductive step.

The main diagram can be written as

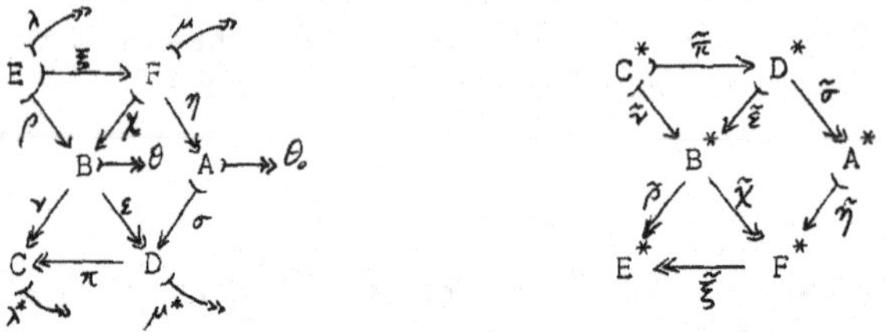

Suppose that a pair $(A,\ \theta_0)$ is given, satisfying the conditions for a knot module and pairing, A being homogeneous of degree $n - 2$ and rank m. The plan is to construct a pair $(B,\ \theta)$ related to $(A,\ \theta_0)$ as in the diagram above.

Let C be a free R-module of rank m, regarded as a Λ-module via the quotient map $\Lambda \twoheadrightarrow R$. Choosing a basis $c_1, \ldots, c_m$ of C gives an isomorphism $\mathrm{Ext}\ (C,\ A) \cong \oplus_1^m A/\varphi A$; let $\underline{a} = (a_1, \ldots, a_m) \in \oplus_1^m A$ determine an element

$$A \overset{\sigma}{\rightarrowtail} D \overset{\pi}{\twoheadrightarrow} C$$

of $\mathrm{Ext}\ (C,\ A)$. The a_i and c_i are related by $\pi d_i = c_i,\ \varphi d_i = \sigma a_i$.

Applying $\overline{\mathrm{Hom}}\ (\ \ ,\ \Lambda_0/\Lambda\ )$ to this short exact sequence yields

$$C^* \overset{\tilde{\pi}}{\rightarrowtail} D^* \overset{\tilde{\sigma}}{\twoheadrightarrow} A^*\ ,$$

as $\overline{\mathrm{Ext}}\ (C,\ \Lambda_0/\Lambda\ ) = 0$ by Lemma 2.3.

Choose a short exact sequence $E \overset{\xi}{\rightarrowtail} F \overset{\eta}{\twoheadrightarrow} A$ which is isomorphic to this by $(\ \lambda\ ,\ \mu\ ,\ \theta_0\ )$; thus

$$\begin{array}{ccccc} E & \overset{\xi}{\rightarrowtail} & F & \overset{\eta}{\twoheadrightarrow} & A \\ \lambda \downarrow \wr & & \mu \downarrow \wr & & \theta_0 \downarrow \wr \\ C^* & \overset{\tilde{\pi}}{\rightarrowtail} & D^* & \overset{\tilde{\sigma}}{\twoheadrightarrow} & A^* \end{array}$$

commutes. Applying $\overline{\mathrm{Hom}}\ (\ \ ,\ \Lambda_0/\Lambda\ )$ to $E \overset{\xi}{\rightarrowtail} F \overset{\eta}{\twoheadrightarrow} A$ yields

$$A^* \overset{\tilde{\eta}}{\rightarrowtail} F^* \overset{\tilde{\xi}}{\twoheadrightarrow} E^*$$

as $\overline{\mathrm{Ext}}\ (A,\ \ \Lambda_0/\Lambda\ \ ) = 0$ by Lemma 2.3.

Thus one obtains a commutative diagram

Now choose $f_1, \ldots, f_m \in F$ such that $\eta f_i = a_i$; then $\underline{f} = (f_1, \ldots, f_m)$ determines a short exact sequence

$$F \overset{\chi}{\rightarrowtail} B \overset{\nu}{\twoheadrightarrow} C$$

such that

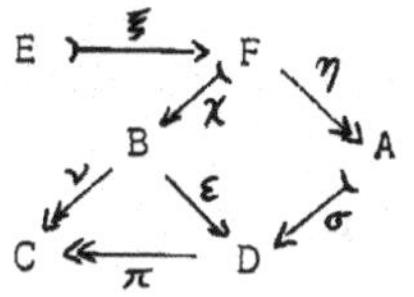

commutes.

To show that ε is onto, let $d \in D$. Then $\exists\, b \in B$ such that $\nu b = \pi d$.

$$\therefore \pi(d - \varepsilon b) = 0$$

$\therefore \exists\, a \in A$ such that $\sigma a = d - \varepsilon b$, and so $\exists\, f \in F$ such that $\sigma\eta f = d - \varepsilon b$. But $\varepsilon\chi f = \sigma\eta f = d - \varepsilon b$, so $d = \varepsilon(\chi f + b)$.

Define $\rho = \chi\xi$; then $E \overset{\rho}{\rightarrowtail} B \overset{\varepsilon}{\twoheadrightarrow} D$ is a short exact sequence. For $\varepsilon\rho = \varepsilon\chi\xi = \sigma\eta\xi = 0$; and if $\varepsilon b = 0$, then $\nu b = \pi\varepsilon b = 0$, so $\exists\, f \in F$ such that $\chi f = b$. But $\sigma\eta f = \varepsilon\chi f = 0$, so $\eta f = 0$, so $f = \xi e$. Thus $b = \chi f = \chi\xi e = \rho e$; and so the sequence is exact.

Thus there is a commutative diagram

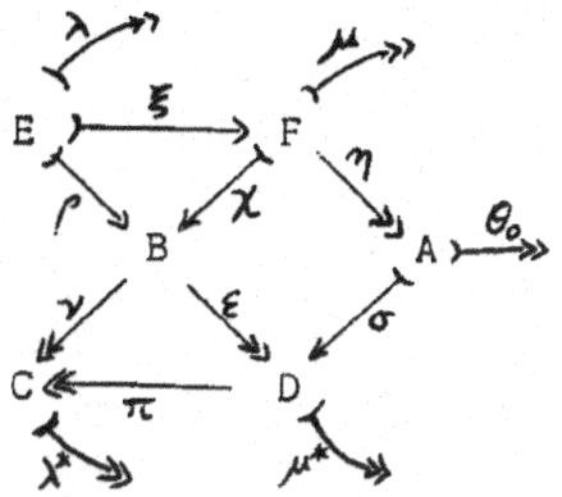

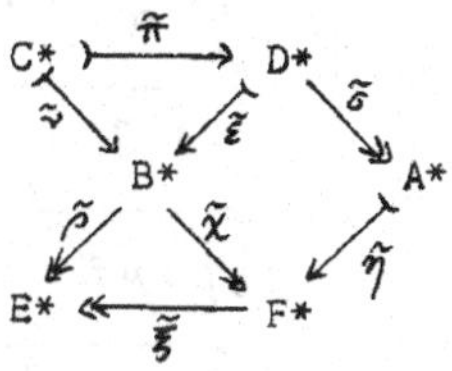

$\tilde{\chi}$ is onto because $\overline{\text{Ext}}\,(C, \Lambda_0/\Lambda) = 0$, and so $\tilde{\rho}$ is also onto.

$(A/\varphi A) \otimes \mathbb{Q}$ is a vector space over the field of fractions R_0 of $R = \Lambda/(\varphi)$, of dimension m. Assume now that the image of $a_1, \ldots, a_m$ is a basis of this vector space; this is clearly equivalent to assuming that D is homogeneous of degree n-1.

Recall how a_i, f_i, c_i are related:

$\pi d_i = c_i$, $\varphi d_i = \sigma a_i$.

$$\eta f_i = a_i.$$

$\nu b_i = c_i$, $\varphi b_i = \chi f_i$.

Choose $b_i^* \in B^*$ such that $\chi b_i^* = \mu^* \varepsilon b_i$, and let $\tilde{\varepsilon} d_i^* = \varphi b_i^*$. I claim that d_i^* is determined by the choice of f_i. For let b_i' be another choice of b_i, $b_i^{*\prime}$

another choice of b_i^*. Then $\varphi b_i = \varphi b_i' = \chi f_i$, so $b_i - b_i' \in \operatorname{Im}\rho = \ker\varepsilon$, and $\varepsilon b_i = \varepsilon b_i'$.

If $\tilde{\chi} b_i^* = \tilde{\chi} b_i^{*\prime} = \mu^* \varepsilon b_i$, then $b_i^* - b_i^{*\prime} \in \operatorname{Im}\tilde{\nu}$, and so $\varphi b_i^{*\prime} = \varphi b_i^* = \tilde{\varepsilon} d_i^*$. $\tilde{\varepsilon}$ is a monomorphism, so d_i^* is determined by f_i.

Consider the statement

$$[\mu f_i, \varepsilon b_j] = \overline{[\mu f_j, \varepsilon b_i]}, \quad \forall i,j \qquad (*)$$

I claim that the truth of (*) depends only on the class which $\underline{f} = (f_1,\dots,f_m)$ represents in $\oplus_1^m F/\varphi F$, not on the f_i or b_i.

For suppose that $f_i' = f_i + \varphi f_{oi}$, and $\nu b_i' = c_i$, $\varphi b_i' = \chi f_i'$.

Then $\varphi(b_i' - b_i) = \chi f_i' - \chi f_i = \chi\varphi f_{oi}$, and so $b_i' - b_i = \chi f_{oi} + \delta_i$ where $\varphi\delta_i = 0$.

$$\begin{aligned}[\mu f_i', \varepsilon b_j'] &= [\mu f_i + \varphi\mu f_{oi}, \varepsilon b_j + \varepsilon\chi f_{oj}] \qquad \text{as } \varepsilon\delta_j = 0 \\ &= [\mu f_i, \varepsilon b_j] + [\mu f_i, \varepsilon\chi f_{oj}] + [\varphi\mu f_{oi}, \varepsilon b_j] + [\varphi\mu f_{oi}, \varepsilon\chi f_{oj}]\end{aligned}$$

$$\begin{aligned}\text{Now } [\varphi\mu f_{oi}, \varepsilon b_j] &= [\mu f_{oi}, \varepsilon\varphi b_j] \\ &= [\mu f_{oi}, \varepsilon\chi f_j] \\ &= [\mu f_{oi}, \sigma\eta f_j] \\ &= \overline{[\mu^*\sigma\eta f_j, f_{oi}]} \\ &= \overline{[\tilde{\eta}\tilde{\sigma}\mu f_j, f_{oi}]} \\ &= \overline{[\mu f_j, \sigma\eta f_{oi}]} \\ &= \overline{[\mu f_j, \varepsilon\chi f_{oi}]}\end{aligned}$$

Similarly, $[\varphi\mu f_{oi}, \varepsilon\chi f_{oj}] = \varphi[\mu f_{oi}, \varepsilon\chi f_{oj}] = \varphi\overline{[\mu f_{oj}, \varepsilon\chi f_{oi}]}$

$$\begin{aligned}\text{Thus } [\mu f_i', \varepsilon b_j'] &= [\mu f_i, \varepsilon b_j] + [\mu f_i, \varepsilon\chi f_{oj}] + [\varphi\mu f_{oi}, \varepsilon b_j] + [\varphi\mu f_{oi}, \varepsilon\chi f_{oj}] \\ &= [\mu f_i, \varepsilon b_j] + \overline{[\mu f_j, \varepsilon\chi f_{oi}]} + \overline{[\varphi\mu f_{oj}, \varepsilon b_i]} + \overline{[\varphi\mu f_{oj}, \varepsilon\chi f_{oi}]} \\ &= \overline{[\mu f_j', \varepsilon b_i']} + [\mu f_i, \varepsilon b_j] - \overline{[\mu f_j, \varepsilon b_i]}\end{aligned}$$

$$\therefore [\mu f_i', \varepsilon b_j'] = \overline{[\mu f_j', \varepsilon b_i']} \iff [\mu f_i, \varepsilon b_j] = \overline{[\mu f_j, \varepsilon b_i]}$$

Assume for the moment that (*) is satisfied. Then I claim that $\mu f_i = d_i^*$ for each i.

For $\tilde{\eta}\tilde{\sigma}\mu f_i = \tilde{\eta}\theta_0\eta f_i = \mu^*\sigma\eta f_i = \mu^*\varepsilon\chi f_i$

$$= \mu^* \varepsilon \varphi b_i = \varphi \tilde{\chi} b_i^* = \tilde{\chi} \tilde{\varepsilon} d_i^* = \tilde{\eta} \tilde{\sigma} d_i^* .$$

$\tilde{\eta}$ is mono, so $\tilde{\sigma} \mu f_i = \tilde{\sigma} d_i^*$. Thus $\mu f_i - d_i^* = \gamma_i \in \ker \tilde{\sigma} = \operatorname{Im} \tilde{\pi}$.

Now
$$\begin{aligned}
\overline{[\mu f_j, \varepsilon b_i]} &= [\mu^* \varepsilon b_i, f_j] \\
&= [\tilde{\chi} b_i^*, f_j] \\
&= [b_i^*, \chi f_j] \\
&= [b_i^*, \varphi b_j] \\
&= [\varphi b_i^*, b_j] \\
&= [\tilde{\varepsilon} d_i^*, b_j] \\
&= [d_i^*, \varepsilon b_j]
\end{aligned}$$

So
$$\begin{aligned}
[\mu f_i, \varepsilon b_j] &= [d_i^*, \varepsilon b_j] + [\gamma_i, \varepsilon b_j] \\
&= \overline{[\mu f_j, \varepsilon b_i]} + [\gamma_i, \varepsilon b_j] \\
&= [\mu f_i, \varepsilon b_j] + [\gamma_i, \varepsilon b_j]
\end{aligned}$$

$\therefore [\gamma_i, \varepsilon b_j] = 0$ for all i, j ; and so $\gamma_i = 0$.

To sum up, the diagram above commutes, with

$$\nu b_i = c_i , \quad \varphi b_i = \chi f_i$$

$$\tilde{\rho} b_i^* = \lambda^* c_i , \quad \varphi b_i^* = \tilde{\varepsilon} \mu f_i$$

Thus it is possible to define a map $\theta : B \longrightarrow B^*$ by $\theta b_i = b_i^*$

such that

$$\begin{array}{ccccc}
F & \overset{\chi}{\rightarrowtail} & B & \overset{\nu}{\twoheadrightarrow} & C \\
\mu \downarrow \wr & & \downarrow \theta & & \wr \downarrow \lambda^* \\
D^* & \overset{\tilde{\varepsilon}}{\rightarrowtail} & B^* & \overset{\tilde{\rho}}{\twoheadrightarrow} & E^*
\end{array}$$

commutes; by the five-lemma, θ is an isomorphism.

By its construction, $\tilde{\chi} \theta = \mu \varepsilon$, and so the whole diagram commutes.

Dualising, $\theta^* : B \longrightarrow B^*$ has the same properties; so if $\theta \neq \theta^*$ one can replace it by $\frac{1}{2}(\theta + \theta^*)$ to obtain a Hermitian map.

Suppose that $\theta = \theta^*$ and $\psi = \psi^*$ are two such maps. Then

$$\tilde{\chi}(\theta - \psi) = \mu^* \varepsilon - \mu^* \varepsilon = 0$$

and so $\kappa = \theta - \psi : B \longrightarrow B^*$ is a map with $\operatorname{Im} \kappa \subset \operatorname{Im} \tilde{\nu} = \ker \tilde{\chi}$.

Suppose that $h : B \to B$ is a map which makes the following diagram commute.

Then h is an isomorphism and $\operatorname{Im}(h-1) \subset \ker \varepsilon$. The map $\tilde{h}\theta h : B \to B^*$ is Hermitian, and is equivalent to θ for our purposes. Given θ, ψ as above, is there an h such that $\psi = \tilde{h}\theta h$? Writing $h = 1 + \alpha$, where $\alpha : B \to \ker$, we need

$$\psi = \tilde{h}\theta h = \theta + \kappa$$

$$\text{i.e.} \quad (1 + \tilde{\alpha})\,\theta\,(1 + \alpha) = \theta + \alpha$$

$$\theta + \tilde{\alpha}\theta + \theta\alpha + \tilde{\alpha}\theta\alpha = \theta + \kappa$$

Since $\tilde{\alpha}\theta\alpha = 0$, this reduces to

$$\kappa = \tilde{\alpha}\theta + \theta\alpha$$

Setting $\alpha = \frac{1}{2}\theta^{-1}\kappa$ gives the desired result, for then $\tilde{\alpha} = \frac{1}{2}\kappa^*\theta^{*-1} = \frac{1}{2}\kappa\theta^{-1}$ as $\theta = \theta^*$, $\kappa = \kappa^*$.

5. The condition (*)

Now return to the requirement that

$$[\mu f_i, \varepsilon b_j] = \overline{[\mu f_j, \varepsilon b_i]} \quad ; \quad \forall i, j \, .$$

First note that the map ε is not uniquely determined by the f_i. Indeed, if ε is one such map, define $\varepsilon' : B \to D$ by

$$\varepsilon' b_i = \varepsilon b_i + \delta_i \quad ; \quad 1 \leq i \leq m,$$

where $\varphi\delta_i = 0$. Then the diagram

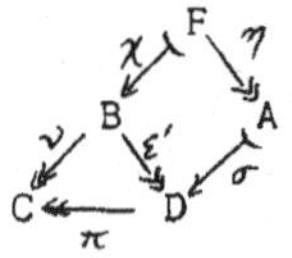

commutes, for $\varepsilon'\chi f_i = \varepsilon\chi f_i$ since $\varphi\delta_i = 0$, and $\pi\varepsilon' b_i = \pi\varepsilon b_i$ for the same reason. Of course, the diagram is equivalent to the original one (with ε) by an isomorphism which is the identity on A, C, F, and B.

Furthermore:

$$\begin{aligned} \varphi[\mu f_i, \varepsilon b_j] &= [\mu f_i, \varepsilon\varphi b_j] \\ &= [\mu f_i, \varepsilon\chi f_j] \\ &= [\mu f_i, \sigma\eta f_j] \\ &= [\tilde{\sigma}\mu f_i, \eta f_j] \end{aligned}$$

$$= [\theta_0 \eta f_i, \ \eta f_j]$$

$$= \overline{[\theta_0 \eta f_j, \ \eta f_i]} \qquad \text{since } \theta_0 = \theta_0^*$$

$$= \varphi \overline{[\mu f_j, \ \varepsilon b_i]}$$

and so $\varphi\left([\mu f_i, \varepsilon b_j] - \overline{[\mu f_j, \varepsilon b_i]}\right) = 0$

Let $f_i' = f_i + \xi e_i$, $1 \leq i \leq m$; then $\eta f_i' = \eta f_i = a_i$; of course, the f_i' will in general determine a different element of Ext(C, F).

If

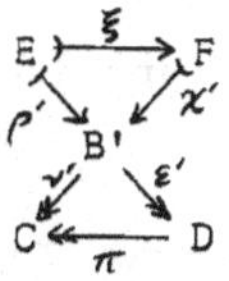

are the corresponding maps for the f_i', then

$$[\mu f_i', \ \varepsilon' b_j'] = [\mu f_i, \ \varepsilon' b_j'] + [\mu \xi e_i, \ \varepsilon' b_j']$$

But $\varphi(\varepsilon' b_j' - \varepsilon b_j) = \varepsilon' \chi' f_j' - \varepsilon \chi f_j = \sigma \eta f_j' - \sigma \eta f_j = 0.$

Since it is possible to alter ε' by a map $B' \longrightarrow \ker(\varphi | D)$, ε' can be chosen so that $\varepsilon' b_j' = \varepsilon b_j$ for each j. Then

$$[\mu f_i', \ \varepsilon' b_j'] = [\mu f_i, \varepsilon b_j] + [\mu \xi e_i, \varepsilon b_j]$$

Setting $x_{ij} = [\mu \xi e_i, \ \varepsilon b_j]$, one obtains

$$[\mu f_i', \ \varepsilon' b_j'] - \overline{[\mu f_j', \ \varepsilon' b_i']} = [\mu f_i, \ \varepsilon b_j] - \overline{[\mu f_j, \varepsilon b_i]} + x_{ij} - \bar{x}_{ji} .$$

The left-hand side will be zero if we can solve the equations

$$x_{ij} - \bar{x}_{ji} = c_{ij} \quad ; \quad 1 \leq i, j \leq m .$$

where $c_{ij} \in R$ and $c_{ij} = -\bar{c}_{ji}$.

If these equations have a solution $x_{ij} \in R$, then there exist $e_i \in E$ with

$$x_{ij} = [\mu \xi e_i, \ \varepsilon b_j]$$

since $\mu : E \xrightarrow{\approx} C^*$.

Recall that R_0 is the field of fractions of $R = \mathbb{Z}[\tau, \tau^{-1}] = \Lambda/(\varphi)$. Now

$$c_{ij} = \sum_{k=-\ell}^{\ell} \alpha_{ij}^{k} \tau^{k}$$

where $2\ell + 2$ is the degree of φ, $\alpha_{ij}^{k} \in \mathbb{Z}$, and the k in α_{ij}^{k} is a superscript.

$$c_{ij} = -\bar{c}_{ji} \Rightarrow \sum_{k=-\ell}^{\ell} \alpha_{ij}^{k} \tau^{k} = - \sum_{k=-\ell}^{\ell} \alpha_{ji}^{-k} \tau^{k}$$

whence $\alpha_{ij}^{k} = -\alpha_{ji}^{-k}$ for all i, j, k.

If $x_{ij} = \sum_{k=-\ell}^{\ell} \beta_{ij}^{k}\tau^{k}$, then

$$x_{ij} - \bar{x}_{ji} = \sum_{k=-\ell}^{\ell} (\beta_{ij}^{k} - \beta_{ji}^{-k})\tau^{k}$$

and so we need $\beta_{ij}^{k} - \beta_{ji}^{-k} = \alpha_{ij}^{k}$ for all i, j, k.

This can be achieved by taking $\beta_{ij}^{k} = \frac{1}{2}\alpha_{ij}^{k}$ for each i, j, k.

More generally, take $\beta_{ii}^{0} = 0$ for each i, assign $\beta_{ij}^{0} \in \mathbb{Z}$ arbitrarily for $i < j$, and assign $\beta_{ij}^{k} \in \mathbb{Z}$ arbitrarily for $k > 0$. The other values of β_{ij}^{k} are then determined by the equations above.

Finally, note that $\gamma(f_i) = f_i' = f_i + \xi e_i$ defines a map $\gamma: F \to F$ which makes the diagram

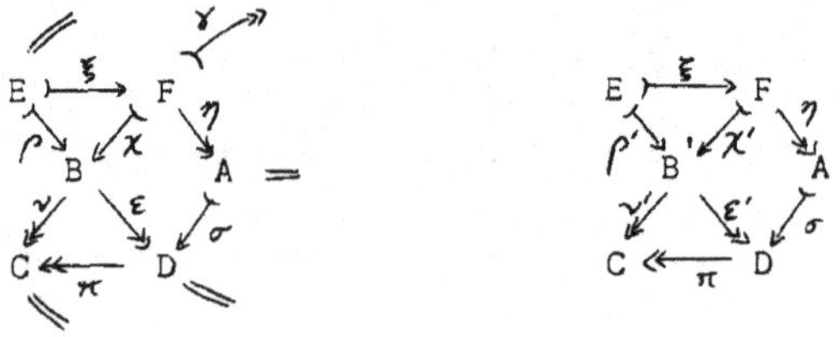

commute. Defining $\delta: B \to B$ by $\delta(b_i) = b_i'$ yields a map (since $\varepsilon b_i = \varepsilon' b_i'$) which keeps the diagram commutative. Thus up to isomorphism, the left-hand side of the main diagram is determined by the elements $(a_i) \in A/\varphi A$. Coupling this with the fact that θ is unique up to conjugation, it follows that the main diagram is determined up to isomorphism by $(a_i) \in A/\varphi A$, $1 \leq i \leq m$.

6. Ext (C, A)

Recall how $\mathcal{E}: A \overset{\sigma}{\rightarrowtail} D \overset{\pi}{\twoheadrightarrow} C$ determines an element of Ext $(C, A) \cong \oplus_1^m A/\varphi A$

Given a basis $c_1, \ldots, c_m$ of C, there exist $d_i \in D$, $a_i \in A$ such that $\pi d_i = c_i$ and $\varphi d_i = \sigma a_i$. Suppose that $\beta: C \to C$ is an automorphism. Then $\beta c_i = \sum_{j=1}^{m} \beta_{ij} c_j$ for some $\beta_{ij} \in \Lambda$.

$$\sigma \sum_{j=1}^{m} \beta_{ij} a_j = \sum_{j=1}^{m} \beta_{ij}\, \sigma a_j = \sum_{j=1}^{m} \beta_{ij}\, \varphi d_j = \varphi \sum_{j=1}^{m} \beta_{ij} d_j$$

$$\text{and} \quad \pi \sum_{j=1}^{m} \beta_{ij} d_j = \sum_{j=1}^{m} \beta_{ij} c_j = \beta c_i$$

Thus $(a_i) \in A/\varphi A$, $1 \leq i \leq m$ classify $\mathcal{E} \in$ Ext (C, A), and

$(\sum_{j=1}^{m} \beta_{ij} a_j) \in A/\varphi A$, $1 \leq i \leq m$ classify $\mathcal{E}\beta \in$ Ext (C, A).

$$\begin{array}{lccccc} \mathcal{E}\beta & : & A & \overset{\sigma}{\rightarrowtail} & D & \overset{\beta^{-1}\pi}{\twoheadrightarrow} & C \\ & & \| & & \| & & \downarrow \wr \beta \\ \mathcal{E} & : & A & \overset{\sigma}{\rightarrowtail} & D & \overset{\pi}{\twoheadrightarrow} & C \end{array}$$

The matrix (β_{ij}) is invertible over $R = \Lambda/(\varphi)$, and so $\{(a_i) : 1 \leq i \leq m\}$ and $\{(\sum_{j=1}^{m} \beta_{ij} a_j) : 1 \leq i \leq m\}$ are bases of the same R-module in $\Lambda/\varphi A$.

Suppose conversely that $\{(a_i) : 1 \leq i \leq m\}$ and $\{(a_i') : 1 \leq i \leq m\}$ are bases of the same R-submodule of $\Lambda/\varphi A$. The $(a_i') = \sum_{j=1}^{m} \beta_{ij}(a_j)$ for an invertible matrix (β_{ij}) over R.

Define $c_i' = \sum_{j=1}^{m} \beta_{ij} c_j$, and let $\pi d_i' = c_i'$, $\sigma a_i'' = \varphi d_i'$.

As above, $(a_i'') = \sum_{j=1}^{m} \beta_{ij}(a_i) = (a_i')$ in $\Lambda/\varphi A$, and so $\{(a_i') : 1 \leq i \leq m\}$ determines the element $\mathcal{E}\beta \in \mathrm{Ext}(C, A)$, where $\beta : C \xrightarrow{\cong} C$ has matrix (β_{ij}).

Define $\mathcal{E} \sim \mathcal{E}'$ iff $\mathcal{E}\beta = \mathcal{E}'$ for some $\beta : C \xrightarrow{\cong} C$. Then we have proved

<u>Lemma 6.1</u> Let $\mathcal{E}, \mathcal{E}' \in \mathrm{Ext}(C, A)$. $\mathcal{E} \sim \mathcal{E}'$ iff they determine the same R-submodule of $\Lambda/\varphi A$.

Consider the diagram

$$\begin{array}{lccccccc} \mathcal{E} & : & A & \overset{\sigma}{\rightarrowtail} & D & \overset{\pi}{\twoheadrightarrow} & C \\ & & \alpha \downarrow \wr & & \wr \downarrow \delta & & \wr \downarrow \gamma \\ \alpha\mathcal{E}\gamma^{-1} & : & A & \overset{\sigma'}{\rightarrowtail} & D' & \overset{\pi'}{\twoheadrightarrow} & C \end{array}$$

where γ is an automorphism, α an isometry. Clearly $\mathcal{E}$ and $\alpha\mathcal{E}\gamma^{-1}$ are equivalent for our purposes. If L, L' are the submodules of $\Lambda/\varphi A$ corresponding to $\mathcal{E}$, $\alpha\mathcal{E}\gamma^{-1}$, then clearly $\alpha_* L = L'$, where $\alpha_* : \Lambda/\varphi A \longrightarrow \Lambda/\varphi A$ is induced by α. Thus it is necessary to classify R-submodules of $\Lambda/\varphi A$ up to equivalence given by automorphisms of $\Lambda/\varphi A$ induced by isometries of A. This seems a difficult problem.

7. <u>Localisation</u>

Consider now the assumption that $R = \Lambda/(\varphi) = \mathbb{Z}[\tau, \tau^{-1}]$ is a Dedekind domain. If R is a principal ideal domain, then since $C = \ker\varphi^n/\ker\varphi^{n-1}$ is $\mathbb{Z}$-torsion-free it will also be a free R-module and the foregoing discussion will apply. Otherwise, let R be the T-integers of the algebraic number field $\mathbb{Q}(\tau)$, where T fully divides the set S of prime spots on $\mathbb{Q}$. Thus $|\alpha|_p \leq 1$ for all $\alpha \in R$ and all $p \in T$.

The polynomial φ is irreducible over $\mathbb{Q}$, but may factorise over $\mathbb{Q}_p$ where $p \in S$. Say $\varphi = \varphi_1 \varphi_2 \cdots \varphi_r$, each φ_i corresponding to a prime $p_i \in T$ which divides p. Here $\mathbb{Q}_p$ denotes p-adic numbers, i.e. the completion of $\mathbb{Q}$ at the prime spot p.

If $\tau_1, \ldots, \tau_s$ are the roots of φ, then the disiriminant $\Delta(\varphi)$ of φ is given by

$$\Delta(\varphi) = \prod_{i<j} (\tau_i - \tau_j)^2 .$$

Recall that the resultant R(f, g) of two polynomials f, g with roots $\alpha_1, \ldots, \alpha_m, \beta_1, \ldots, \beta_n$ is given by

$$R(f, g) = a_0^n b_0^m \prod_{i=1}^m \prod_{j=1}^n (\alpha_i - \beta_j)$$

where

$$f(x) = a_0 x^m + a_1 x^{m-1} + \ldots + a_m$$

$$g(x) = b_0 x^n + b_1 x^{n-1} + \ldots + b_n$$

It follows easily that if $p \in S$

$$|\Delta(\varphi)|_p = |\Delta(\varphi_1)|_p \ldots . |\Delta(\varphi_r)|_p \cdot \prod_{i<j} |R(\varphi_i, \varphi_j)|_p^2$$

Let Δ be the discriminant of $\mathbb{Q}(\tau)/\mathbb{Q}$, and Δ_i the discriminant of $\mathbb{Q}(\tau)_{pi}/\mathbb{Q}_p$; then for $p \in S$

$$|\Delta|_p = |\Delta(\varphi)|_p \quad \text{and} \quad |\Delta_i|_p = |\Delta(\varphi_i)|_p .$$

Moreover, $|\Delta|_p = \prod_{i=1}^r |\Delta_i|_p$.

Since $|R(\varphi_i, \varphi_j)|_p \leqslant 1$ for all i, j, it follows that

$$|R(\varphi_i, \varphi_j)|_p = 1 \quad \text{for all } i, j \quad (i \neq j).$$

Now let (B, θ) be a knot module and pairing, homogeneous of degree n over φ. Let (B_p, θ_p) denote (B, θ) localised at $p \in S$, and $B_i = \{b \in B_p : \varphi_i^n b = 0\}$.

<u>Proposition 7.1</u> $\qquad B_p = B_1 \perp B_2 \perp \ldots \perp B_r$

<u>Proof</u> $\qquad$ Let $\hat{\varphi}_i = \varphi / \varphi_i = \prod_{j \neq i} \varphi_j$.

Now $|R(\varphi_i, \hat{\varphi}_i)|_p = \prod_{j \neq i} |R(\varphi_i, \varphi_j)|_p = 1$

and so $|R(\varphi_i^n, \hat{\varphi}_i^n)|_p = 1$

There exist polynomials $\alpha, \beta \in \mathbb{Z}_p[t, t^{-1}]$ such that $\alpha\varphi_i^n + \beta\hat{\varphi}_i^n = 1$.

If $b \in B_p$, then $b = \alpha\varphi_1^n b + \beta\hat{\varphi}_1^n b$, and of course $\varphi_1^n \beta\hat{\varphi}_1^n b = \beta\varphi^n b = 0$, so $\beta\hat{\varphi}_1^n b \in B_1$. Furthermore, if $b_1 \in B_1$ then

$$\langle \alpha\varphi_1^n b, b_1 \rangle = \langle \alpha b, \varphi_1^n b_1 \rangle = 0$$

and so $\alpha\varphi_1^n b \in B_1^{\perp}$, the orthogonal complement of B_1.

Thus

$$B_p = B_1 \perp B_1^{\perp}$$

and the argument continues easily to produce the desired result.

One advantage of localising is that for a prime p_i dividing $p \in S$, the corresponding ring of integers R_i is a P I D. Hence the module C is always free over R_i.

8. Generalisation

It should be possible to dispense with the assumption that M is homogeneous. Assume instead that M is annihilated by φ^n but by no lower power. Then

$$M \otimes_{\mathbb{Z}} \mathbb{Q} \cong \bigoplus_{i=1}^{n} F_i, \qquad F_i = \oplus \frac{\mathbb{Q}[t, t^{-1}]}{(\varphi^i)}$$

Let K_{n-1} be the kernel of the map $\ker\varphi^n \to (\ker\varphi)^*$ induced by $\langle\, ,\,\rangle$; thus K_{n-1} is the kernel of the composition

$$\ker\varphi^n \xrightarrow{\theta} (\ker\varphi^n)^* \xrightarrow{\tilde{\rho}} (\ker\varphi)^*$$

where $\rho : \ker\varphi \to \ker\varphi^n$ is the inclusion map.

Let $\ker'\varphi = \ker\varphi \cap K_{n-1}$; then $\ker'\varphi$ is the radical of $\langle\, ,\,\rangle \big|_{\ker\varphi} = \ker(\theta|\ker\varphi)$. Thus $\ker\varphi$ splits as $\ker'\varphi \perp K'$, using the theory of hermitian forms over Dedekind domains [2], and so $\ker\varphi^n$ has K' as an orthogonal summand. Splitting off K', we may assume that $\ker\varphi \subset K_{n-1}$.

The main diagram becomes

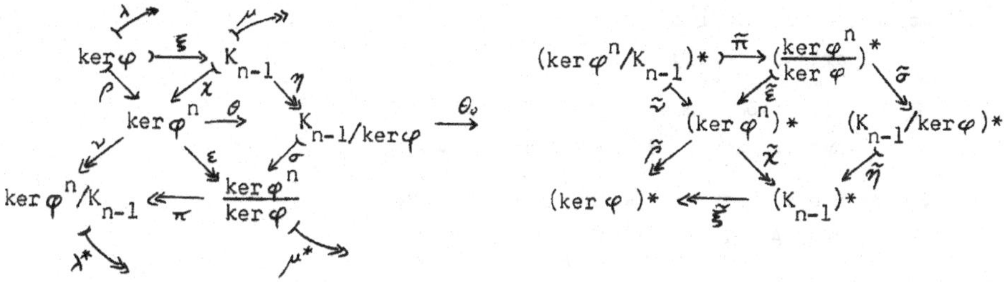

Setting $A = K_{n-1}/\ker\varphi$, $C = \ker\varphi^n/K_{n-1}$, assume that C has rank m as an R-module. Then $\mathrm{Ext}(C,A) \cong \oplus_1^m A/\varphi A$; but the dimension of $(A/\varphi A)\otimes_{\mathbb{Z}}\mathbb{Q}$ as a vector space over R_0 is no longer m : in general it will be less than m. Clearly this makes life more difficult.

REFERENCES

1. Blanchfield, R.C.; Intersection theory of manifolds with operators with applications to knot theory.
Ann. of Math, 65 (1957), 340 - 356.

2. Jacobowitz, R.; Hermitian forms over local fields.
Amer. Jour. Math. LXXXIV (1962), 441 - 465.

3. Kearton, C.; Classification of simple knots by Blanchfield duality.
Bull. Amer. Math. Soc. 79(1973), 952 - 955.

4. Kearton, C.; Blanchfield duality and simple knots.
Trans. Amer. Math. Soc. 202 (1975), 141 - 160.

5. Kervaire, M.A.; Les noeuds de dimensions superieures.
Bull. Soc. math. France, 93 (1965), 225 - 271.

6. Levine, J.; Polynomial invariants of knots of codimension two.
Ann. of Math. 84 (1966), 537 - 554.

7. Levine, J.; Knot modules.
Annals of Mathematics, Study 84. Knots, Groups, and 3-Manifolds; edited by L.P. Neuwirth.

8. Levine, J.; Knot modules, I.
Trans. Amer. Math. Soc. (1977), 1-50.

9. MacLane, S.; Homology.
Springer-Verlag. (1975)

10. Milnor, J.; On isometries of inner product spaces.
Invent, math., 8 (1969), 83 - 97.

11. Seifert, H.; Uber das Geschlect von Knoten.
Math. Ann., 110 (1934), 571 - 592.

12. Trotter, H.F.; On S-equivalence of Seifert matrices.
Invent. math., 20 (1973), 173 - 207.

SOME RESULTS ON HIGHER DIMENSIONAL KNOT GROUPS

J. LEVINE

Introduction :

A. This paper presents a number of contributions to the study of knot groups in dimensions above the classical case - particularly for 2-dimensional knots.

An <u>n -knot</u> is understood to denote a closed smooth submanifold K of a closed manifold Σ, where K is homeomorphic to S^n and Σ is homotopy equivalent to S^{n+2}. If Σ is homeomorphic to S^{n+2}, the knot is called <u>ordinary</u>. Of course, if $n \geq 3$, all n-knots are ordinary (see [S]). <u>The group</u> of the knot is the fundamental group of the complement $\Sigma - K$; we sometimes denote it by $\pi(K)$.

In [K : (I.1)], Kervaire showed that the group π of any knot has the following properties :

(i) π is finitely-presented

(ii) $H_1(\pi) = \pi/_{[\pi,\pi]}$ is infinite cyclic

(iii) $H_2(\pi) = 0$

(iv) There exists $\mu \in \pi$ (which we shall call a <u>meridian</u> for π) such that π is generated by the conjugates of μ .

Conversely, Kervaire shows that, for any $n \geq 3$, any group π with properties (i) - (iv) is the group of some n-knot. We will refer to these groups as <u>3-knot groups.</u> For n = 2, π is the group of some 2-knot if it satisfies the additional property:

(iii)' deficiency $\pi = 1$.

The deficiency of a group is the maximum of the deficiences of its finite presentations, where the deficiency of a presentation with g generators and r relators is $g - r$. Since deficiency $\pi \leq \text{rank } H_1(\pi)$, an easy observation, we see that deficiency $\pi \leq 1$ for any 3-knot group. A characterization of 2-knot groups (i.e. groups of 2-dimensional knots) or ordinary 2-knot groups remains unsolved since :

a) There are ordinary 2-knot groups which do not satisfy (iii)' - (see [K : (I.2)]).

b) There are 3-knot groups which are not 2-knot groups (see e.g. [L 1], [G] or [F]).

B. An important tool in the study of knot groups is the <u>knot module</u>. If π is any group, then $H_1(\pi)$ operates on the groups $H_q([\pi,\pi])$ by conjugation - this is well-defined since inner automorphisms of $[\pi,\pi]$ induce the identity on homology (see [HS : (IV, 16)]). If $H_1(\pi)$ is written multiplicatively and $H_q([\pi,\pi])$ additively, we have a (left) group action. Denoting by Λ the integral group ring of $H_1(\pi)$, we may extend this action linearly to make $H_q([\pi,\pi])$ a left Λ-module. Alternatively, these modules can be defined directly as the homology groups $H_q(\pi; \Lambda)$, where Λ serves as local coefficients under the action of π defined by the epimorphism $\pi \longrightarrow H_1(\pi)$ (see [C]). In our present situation, $H_1(\pi)$ is always infinite cyclic ; if we choose a generator t ("orientation" of the knot), we may write $\Lambda = Z[t, t^{-1}]$.

These modules are most useful when q = 1 or 2. We write $A(\pi) = H_1(\pi; \Lambda)$, $B(\pi) = H_2(\pi; \Lambda)$, and call $A(\pi)$ the <u>module of π</u>,

$B(\pi)$ the <u>2nd module of π.</u> If $\pi = \pi(K)$, then $A(\pi) \simeq H_1(\tilde{X})$, where $X = \Sigma - K$ and $\tilde{X}$ is the infinite cyclic covering of X associated to $[\pi,\pi]$. By Hopf's theorem $B(\pi)$ is a quotient module of $H_2(\tilde{X})$. The Λ -action on $H_q(\tilde{X})$ is defined by the covering transformations of $\tilde{X}$.

A problem subsidiary to that of characterizing knot groups is the characterization of knot modules. Even though the characterization of 3-knot groups is accomplished by Kervaire's theorem, there is no satisfactory general description of 3-knot modules. The only condition known on 3-knot modules is that they be of <u>type K</u> (see [L 1 : § 8]). One can conjecture that any such Λ -module is a 3-knot module. This is known to be true for Z-torsion free Λ -modules (see [L 1 : (11.1)]), which are, in fact, all realizable as 2-knot modules. The Z-torsion submodule of a 2-knot module must satisfy a duality property (see [L 1], [F] and theorem (III.1) below).

C. The aim of the present work is as follows :

I. The 2-knots constructed by Kervaire, for 3-knot groups satisfying (iii)', are not necessarily ordinary. Using a notion defined by Andrews-Curtis [AC], we give a condition on π sufficient to make the Kervaire knots ordinary.

II. We show that there exist ordinary 2-knot groups with any prescribed deficiency $\leqslant 1$. This answers a question posed by Lomonaco in [Ki].

III. The proofs that some 3-knot groups cannot be 2-knot groups relies on the duality property of $A(\pi)$ mentioned above. We derive a relation between $A(\pi)$ and $B(\pi)$, when π is a 2-knot group, which we show, by example, to be independent of the duality property on $A(\pi)$.

IV. Using these properties of 2-knot groups, we give a complete determination of all 2-knot groups whose commutator subgroups are finitely-generated abelian.

V. Turning to 3-knot groups, we give an effective criterion that a group π, with $H_1(\pi)$ infinite cyclic and $[\pi,\pi]$ finitely-generated abelian, be a 3-knot group. The criterion uses certain polynomial invariants of π. This gives, easily, examples of 3-knot groups which are not 2-knot groups.

VI. As another application of V, we give a sufficient criterion for a Z-torsion Λ -module to be a 3-knot module. This criterion is clearly not necessary, as one can show easily by twist-spinning constructions.

I. Let $\{ x_1, \ldots, x_n : r_1(x_1, \ldots, x_n) = 1, \ldots, r_m(x_1, \ldots, x_n) = 1\}$ be a presentation of a group. Consider the following transformations (writing $r_i = r_i(x_1, \ldots, x_n)$) :

a) replace r_i by r_i^{-1},

b) replace r_i by wr_iw^{-1}, where w is any word in $x_1, \ldots, x_n$,

c) replace r_i by r_ir_j, for any $j \neq i$,

d) add a generator x and a relation $xw^{-1} = 1$, where w is any word in $x_1, \ldots, x_n$,

e) inverse transformation of (d).

We will say two presentations are AC - equivalent if a sequence of these transformations will convert one presentation to the other.

Obviously, AC-equivalent presentations give the same group and have the same deficiency. Andrews-Curtis [AC] conjecture that, conversely, any two presentations of a group with the same deficiency are AC-equivalent. In fact, this conjecture is false, in general, (see [D]) but the most important case of presentations of the trivial group with deficiency 0 remains unresolved.

Let $\{ x_o, \ldots, x_n : r_1 = r_n = 1 \}$, where $r_i = r_i(x_o, \ldots, x_n)$, be a presentation of a 3-knot group π. Let $\mu = \mu(x_o, \ldots, x_n)$ represent a meridian element of π; then the presentation : $\{ x_o, \ldots, x_n : r_1 = r_n = 1, \mu = 1\}$ presents the trivial group. We will say that this presentation is <u>induced</u> from the original presentation of π.

<u>Theorem I</u> : Suppose π is a 3-knot group satisfying (iii)' which admits a presentation of deficiency one whose induced presentation of the trivial group (using any meridian) is AC- equivalent to the trivial presentation. Then π is an ordinary 2-knot group.

<u>Proof</u> : Recall the construction of Kervaire [K : (I.1)]. Let X_o be the connected sum of (n+1) copies of $S^1 \times D^4$; then $\pi_1(X_o)$ is the free group on $x_o, \ldots, x_n$, where x_i is represented by $S^1 \times 0$ in the (i+1)-st copy of $S^1 \times D^4$. $\pi_1(\partial X_o) \approx \pi_1(X_o)$ and the relations r_i may be represented by disjoint imbeddings of S^1 in ∂X_o. Since the normal bundles are trivial, these imbeddings extend to disjoint imbeddings $\phi_i : S^1 \times D^3 \longrightarrow \partial X_o$. The ϕ_i are uniquely, up to isotopy, determined by r_i if we insist that the canonical framing on X_o (induced from R^5 by the natural inclusion $X_o \subseteq R^5$), restricted to $\phi_i(S^1 \times D^3)$, agree via ϕ_i with the framing inherited from $D^2 \times D^3$. This is equivalent to the requirement that the canonical framing of

X_0 extend to a framing of X_1, where X_1 is obtained from X_0 by adding handles of index 2 via the $\{\emptyset_1\}$. We may also represent

a meridian μ by an imbedding $\emptyset : S^1 \times D^3 \to \partial X_0$ disjoint from $\{\emptyset_1\}$ satisfying the same conditions : the manifold X obtained by adding a final handle via $\emptyset$ depends only on the induced presentation. ∂X is a homotopy 4-sphere and the knot $\emptyset(* \times S^2) \subseteq \partial X$ represents the desired knot.

X is contractible, since it collapses to the 2-dimensional complex constructed in [AC] from the induced presentation of the trivial group. If the induced presentation were trivial, it is clear that $X = D^5$, since the handles just fill in the holes in X_0: $S^1 \times D^4 \longmapsto S^1 \times D^4 \cup D^2 \times D^3 = D^5$. Now it follows from the arguments in [AC], using the handle-body arguments of Smale [S], that the transformations (a) - (d) do not change X ; since X depends only on the presentation (e) will also not change X. Theorem I follows.

II. Let K be the 2-twist-spin (see [Z]) of the trefoil knot, and π_m the group of the m-fold connected sum of K with itself.

Theorem II : π_m has deficiency 1 - m .

Note that classical 1-knots and the 2-knot groups of Kervaire (§ I) all have deficiency 1. A question which remains is whether a prime 2-knot can have deficiency < 0 .

Proof of theorem II : The group $\pi = \pi_1$ of K has presentation $\{ x, y : xyx = yxy, x^2y = yx^2\}$ (see [Z : § 4]). A simple computation shows that the Λ -module $A(\pi)$ is cyclic with annihilator ideal :

$(t^2 - t + 1,\ t^2 - 1) = (3,\ t + 1)$. In other words, $A(\pi)$ is the cyclic group of order 3, with $t = -1$. Now $A(\pi_m) = A(\pi) \oplus \ldots \oplus A(\pi)$ (m summands) ; so $A(\pi_m)$ is the vector space of rank m over $Z/3$, with $t = -1$. In particular, notice that $A(\pi_m)$ cannot be generated, as a a Λ -module, by fewer than m elements.

Lemma (II.1) : Let A be a finitely generated Λ -module ($\Lambda = Z[t,t^{-1}]$) with rank r and deficiency d. Then $Ext^2_\Lambda(A,\Lambda)$ can be generated, as a Λ -module, by r - d elements.

Proof : Suppose $0 \longrightarrow F_2 \longrightarrow F_1 \longrightarrow F_0 \longrightarrow A \longrightarrow 0$ is a free resolution of A with $r_0 - r_1 = d$, where r_i = rank F_i . The existence of such a resolution follows from the facts that Λ has homological dimension 2 and projective modules over Λ are free(see [L 1] for background). Since the alternating sum of the ranks of the terms of an exact sequence is zero, we have $r_2 = r + r_1 - r_0 = r - d$. Now $Ext^2_\Lambda(A,\Lambda)$ is the second cohomology module of the dual cochain complex :

$$Hom_\Lambda\ (F_0,\Lambda) \longrightarrow Hom_\Lambda(F_1,\Lambda) \longrightarrow Hom_\Lambda(F_2,\Lambda)$$

and, therefore, is a quotient of $Hom_\Lambda(F_2,\Lambda)$, which is free of rank r_2. This prove Lemma (II.1).

We now prove that deficiency $\pi_m \leqslant 1 - m$. If not, then deficiency $A(\pi_m) > -m$, since a presentation of $A(\pi_m)$ can be derived from a presentation of π_m, with one less generator and the same number of relations (see [C]). By Lemma (II.1), $Ext^2_\Lambda\ (A(\pi_m),\Lambda)$ can then be generated by less than m generators, since $A(\pi_m)$ has zero rank. But $Ext^2_\Lambda\ (A(\pi_m),\Lambda) \approx Hom_Z(A(\pi_m),\ Q/Z)$ - (see [L 1 : (4.2)]) - is isomorphic to $A(\pi_m)$ and, therefore, as noticed above, requires m generators.

It remains to prove deficiency $\pi_m \geqslant 1-m$.

<u>Lemma (II.2)</u> : If π' and π'' are groups of n-knots K', K'', and π the group of the connected sum K' # K'' , then :

$$\text{deficiency } \pi \geqslant (\text{deficiency } \pi') + (\text{deficiency } \pi') - 1 .$$

<u>Proof</u> : π is the free product of π' and π'' with amalgamation of a meridian of π' with one of π'' .

Let $\{ x_1, \ldots, x_n = r_1 = \ldots = r_k = 1 \}$, $\{ y_1, \ldots, y_m : s_1 = \ldots = s_l = 1\}$

be presentations of π' and π'' of maximum deficiency. If $M = M(x_1, \ldots, x_n)$, $\nu = \nu(y_1, \ldots, y_m)$ represent meridians, then :

$\{ x_1, \ldots, x_n, y_1, \ldots, y_m : r_1 = \ldots = r_k = 1 = s_1 = \ldots = s_l, \mu = \nu$

is a presentation of π. Lemma (II.2) follows ;

Since we have written down a presentation of π with deficiency 0, a simple recursive argument using Lemma (II.2) proves deficiency $\pi_m \geqslant 1 - m$.

This completes the proof of theorem II.

III. In addition to the properties (i) - (iv) of 3-knots groups enumerated by Kervaire, 2-knot groups also have the following property :

<u>Theorem (III.1)</u> : If π is a 2-knot group, and $T(\pi)$ the Z-torsion submodule of $A(\pi)$, then $T(\pi)$ supports a bilinear, symmetric, non-

singular pairing $<\ ,\ >$, with values in Q/Z , and the property : $< t\alpha, t\beta > = < \alpha,\beta >$ for any $\alpha,\beta \in T(\pi)$.

We refer the reader to [F] or [L 1 : (6.5)] for a proof. We now present another restriction on 2-knot groups.

Theorem (III.2) : If π is a 2-knot group, then the 2nd knot module $B(\pi)$ is a quotient of $\mathrm{Ext}^1_\Lambda(\overline{A(\pi)},\Lambda)$.

Recall that, for any Λ -module A, $\bar{A}$ denotes the conjugate module : $\bar{A}$ is additively isomorphic to A, but t acts on $\bar{A}$ identically with t^{-1} on A.

Corollary III : If π is a 2-knot group and $A(\pi)$ is finite then $B(\pi) = 0$.

Proof of Corollary III : For any 3-knot group π, $A(\pi)$ is a Λ -module of type K (see [L 1 : § 8]), i.e. $A(\pi)$ is finitely-generated and the element $t - 1 \in \Lambda$ defines an automorphism of $A(\pi)$. By [L 1 : (3.3)] $\mathrm{Ext}^1_\Lambda(A,\Lambda) = 0$ if A is Z-torsion and of type K. Now corollary III follows directly from theorem (III.2).

In fact, corollary III is a consequence also of theorem (III.1) in the special case of $[\pi,\pi]$ abelian (we omit the proof). In the general case, however, this is not true, as shown by the following examples. Consider the group $\pi = \pi_{p,q}$ given by the presentation $\{ u, x_1, x_2, y_1, y_2 : x_1x_2 = x_2x_1, y_1y_2 = y_2y_1, x_i^p = 1 = y_i^p, ux_iu^{-1} = x_i^q, u^{-1}y_iu = y_i^q\ (i = 1,2)\}$ we assume p,q are relatively prime positive integers. An easy computation shows that $H_1(\pi)$ is infinite cyclic if p, q-1 are relatively prime, and then u is a meridian. Using the Reidemeister-Schreier theorem [KMS], we compute that $[\pi,\pi]$ is a free product of two groups, each a direct product of two cyclic groups of order p generated by x_1, x_2 and y_1, y_2, respectively. Then $A(\pi)$ is a direct sum of four copies of Z/p with generators X_1, X_2, Y_1, Y_2 , and the Λ -module structure is given by $t\,X_i = q\,X_i$, $t\,Y_i = q'Y_i$ $(i = 1,2)$, where $qq' \equiv 1 \bmod p$. Now $H_2([\pi,\pi])$ is a direct product of two copies of Z/p since :

1) $H_2(G) \approx \Lambda^2 G$, the exterior power, for any finitely-generated abelian group G, and

2) $H_2(G * G') \approx H_2(G) \oplus H_2(G')$, for any groups G, G' . (see [HS : (VI.14,15)]).

As generators of $H_2([\pi,\pi])$ we may take $X_1 \wedge X_2$ and $Y_1 \wedge Y_2$. The Λ -module structure on $B(\pi) = H_2([\pi,\pi])$ is given by :

$$t(X_1 \wedge X_2) = t\,X_1 \wedge t\,X_2 = q^2 X_1 \wedge X_2$$

$$t(Y_1 \wedge Y_2) = t\,Y_1 \wedge t\,Y_2 = (q')^2\, Y_1 \wedge Y_2 .$$

Thus $B(\pi)$ is of type K if and only if p, q+1 are relatively prime (since $q^2-1 = (q-1)(q+1)$, and q-1 is already relatively prime to p). Applying Lemma III (to follow), we conclude that $\pi_{p,q}$ is a 3-knot group if p is relatively prime to q, q-1 and q+1. However by corollary III, $\pi_{p,q}$ is never a 2-knot group.

Notice that theorem (III.1) does not help here since the formulae $< X_i,Y_j > = \delta_{ij}/p$, $< X_i,X_j > = 0 = < Y_i,Y_j >$ defines the required pairing on $A(\pi) = T(\pi)$.

A useful criterion for recognizing 3-knot groups is the following :

<u>Lemma III</u> : Let π be a finitely-presented group with $H_1(\pi)$ infinite cyclic. Then $B(\pi)$ is finitely generated over Λ , and $H_2(\pi) = 0$ if and only if $B(\pi)$ is of type K.

<u>Proof</u> : Let X be a finite complex with $\pi_1(X) \approx \pi$, and $\tilde{X}$ the infinite

cyclic covering of X corresponding to $[\pi,\pi] \subseteq \pi$. The cellular chain complex $C_*(\tilde{X})$ is free, over Λ , of finite type. Since Λ is noetherian, the homology modules $H_q(\tilde{X})$ are finitely generated. Therefore $A(\pi) = H_1(\tilde{X})$, and $B(\pi)$, which is a quotient of $H_2(\tilde{X})$ by Hopf's theorem [H], are finitely generated.

Let $E = K(\pi, 1)$ be the Eilenberg-MacLane complex. If $\tilde{E}$ is the infinite cyclic covering of E associated to $[\pi,\pi] \subseteq \pi$, then $\tilde{E}$ is also an Eilenberg-MacLane complex. The Λ -module structure on $H_q([\pi,\pi]) = H_q(\tilde{E})$ coincides with the action of the group of covering transformations, if $t \in \Lambda$ is associated to a suitable generator. Consider the short exact sequence of chain maps :

$$0 \longrightarrow C_*(\tilde{E}) \xrightarrow{t-1} C_*(\tilde{E}) \longrightarrow C_*(E) \longrightarrow 0$$

the associated exact homology sequence is :

$$(*) \quad \ldots \longrightarrow H_{q+1}(\pi) \longrightarrow H_q([\pi,\pi]) \xrightarrow{t-1} H_q([\pi,\pi]) \longrightarrow H_q(\pi) \longrightarrow \ldots$$

Since $H_1(\pi)$ is infinite cyclic and $t-1 = 0$ on $H_0([\pi,\pi])$, we see $A(\pi) = (t-1)A(\pi)$. Since Λ is noetherian and $A(\pi)$ is finitely generated, we conclude $A(\pi)$ is of type K, as we already know. So $H_2(\pi) \approx B(\pi) / (t-1) B(\pi)$ and Lemma III follows.

IV. We use the results of § III to determine all 2-knot groups whose commutator subgroup is finitely-generated abelian. We first construct some knots.

(a) Consider the following 2-bridge knots :

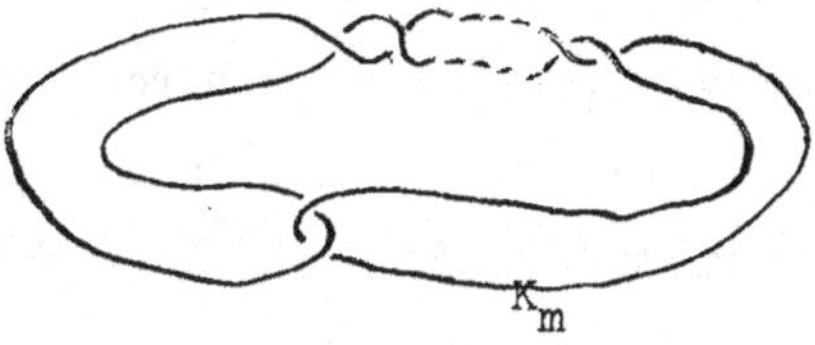

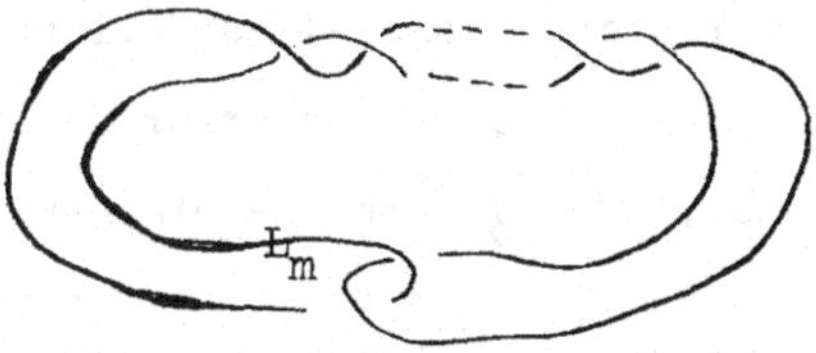

where K_m and L_m are defined to have m full twists. The 2-twist-spin (see [Z]) of these knots have groups with presentation :

$$\pi_q = \{ x,y : xyx^{-1} = y^{-1},\ y^q = 1 \}$$

where $q = 4m-1$ for K_m, and $q = 4m+1$ for L_m . This can be computed directly, or follows from Zeeman's result [Z] that the k-twist spin of a knot K is fibred with fibre the k-fold branched cover of K, monodromy given by the usual covering transformation, and Schubert's results [Sch] identifying the double branched coverings of 2-bridge knots as certain lens space.

(b) Let $M = (M_{ij})$ be a 3×3-integer matrix with determinant $M = +1$, determinant $(1-M) = \pm 1$. M determines a linear diffeomorphism $\bar{h}$ of R^3 which, in turn, determines a diffeomorphism h of the 3-torus $T = R^3/Z^3$. Note that determinant $M = +1$ insures that h is orientation-preserving ; thus, after an isotopic deformation, we may assume h is stationary on a 3-disk $D \subseteq T$. Let $T_o = \overline{T - D}$ and consider the "mapping torus" X of h/T_o, i.e. $X = T_o \times I$ with the identifications $(x, o) = (h(x), 1)$, for all $x \in T_o$. Then $\partial X = S^2 \times S^1$ and $\pi_1(X) \approx H \times_{\phi} Z$, the semi-direct product (the usual Cartesian product, with multiplication given by the formula :

$(h,m) . (h', n) = (h\, \phi^m (h'), m+n)$; $h, h' \in H$; $m, n \in Z$), where $H = \pi_1(T)$ is free abelian of rank 3, and $\phi = h_*$ is the linear automorphism of H defined by the matrix M. Note that X is a homology circle - $H_1(X)$ is infinite cyclic because $1 - \phi$ is an automorphism, and $H_2(X) = 0$ because $1 - h_*$ is an automorphism of $H_2(T)$. This latter fact is proved as follows. $H_2(T) \approx \Lambda^2 H_1(T) = \Lambda^2 H$, where Λ^2 denotes the 2nd exterior power, and h_* corresponds to $\Lambda^2 \phi$. If $\alpha_1, \alpha_2, \alpha_3$ are the eigen-values of M, then $\det M = \alpha_1\alpha_2\alpha_3 = +1$

and $\det(1 - M) = \prod_i (1 - \alpha_i) = \pm 1$, by assumption. The eigenvalves of $\Lambda^2 M$ are $\{ \alpha_i\alpha_j \}$, $i < j$, and so $1 - \Lambda^2 M$ is unimodular if $\prod_{i<j} (1 - \alpha_i\alpha_j) = \pm 1$. But

$$\prod_{i<j} (1 -\alpha_i\alpha_j) = \pm \prod_i (1 -\alpha_i^{-1}) = \pm (\alpha_1\alpha_2\alpha_3)^{-1} \prod_i (\alpha_i - 1) = \pm 1 .$$

Now form $\Sigma = X \cup (S^2 \times D^2)$, attached along their boundaries. Since X is a homology circle and $H_1(\partial X) \approx H_1(X)$, it follows that Σ is a homology 4-sphere. Moreover $\pi_1(\Sigma) \approx \pi_1(X)/\langle \xi \rangle$, where ξ is a generator of the Z factor in $\pi_1(X) = H \times_\emptyset Z$. Since $\emptyset - 1$ is an automorphism of H, it follows that Σ is simply-connected and so Σ is a homotopy sphere. The 2-knot $S^2 \times 0 \subseteq \Sigma$ is a fibred knot with group $\pi = H \times_\emptyset Z$.

This construction appears in [CS], and an obvious problem is the precise identification of Σ .

<u>Theorem IV</u> : If π is a 2-knot group with finitely-generated abelian commutator subgroup, then π must be one of the groups of (a) and (b).

These groups are all realized by fibred knots.

The knots constructed in (a) and (b) are all fibred. Those in (b) are so by construction, while those in (a) are twist-spin knots and, according to Zeeman[Z], all twist-spin knots are fibred.

Any group π with $H_1(\pi)$ infinite cyclic can be written as a semi-direct product $\pi = G \times_\emptyset Z$ where $G = [\pi,\pi]$, or, equivalently, G is the normal closure of [G,G] and the set S of elements of the form $g^{-1} \emptyset (g)$, $g \in G$. Then it is easy to check that π has a meridian element (Kervaire's property (iv)-see introduction) if and only if G is the normal closure of S alone. Finally $H_2(\pi) = 0$ if and only if the Λ-module $H_2(G)$, where $t = \emptyset_*$, has the property that t-1 is an automorphism. This follows from the proof of Lemma III. As a consequence, the following Lemma is immediate.

<u>Lemma IV</u> : A finitely-presented group $\pi = G \times_{\emptyset} Z$ is a 3-knot group if and only if :

(i) G is the normal closure of S, and

(ii) $H_2(G)$ is a Λ -module of type K.

In case G is finitely generated abelian, these criterion can be expressed in the following somewhat simpler form :

(i) G is a Λ -module of type K, and

(ii) $\Lambda^2 G$ is a Λ -module of type K.

The Λ-module structure is, of course, defined by $t = \emptyset$ on G, and $t = \Lambda^2\emptyset$ on $\Lambda^2 G$. This follows from the observation that $S = (t-1)G$ and $H_2(G) \approx \Lambda^2 G$ ([HS : VI.14]).

The knots of (a) and (b) produce the following examples of $(G, \emptyset)$ which define 2-knot groups.

(a) G cyclic of order q, any odd integer, and $\emptyset = -1$,

(b) G free abelian of rank 3 and $\emptyset$ any linear automorphism satisfying : $\mathrm{DET}\ \emptyset = +1$, $\mathrm{DET}(1-\emptyset) = \pm 1$.

We must show these are the only possibilities. Obviously $A(\pi) = G$, with $t = \emptyset$, and $B(\pi) = \Lambda^2 G$, with $t = \Lambda^2\emptyset$. To apply theorem (III.2) we must identify $\mathrm{Ext}^1_\Lambda(\overline{G},\Lambda)$. Let $F = \mathrm{Hom}_z(G, Z)$ with Λ-module structure defined by $t = \mathrm{Hom}_z(\emptyset^{-1}, 1)$.

<u>Claim</u> : $\mathrm{Ext}^1_\Lambda(\overline{G},\Lambda) \approx$ F as Λ-module.

By [L 1 : (3.2)], we may assume G is free abelian. A free resolution of G is given by $0 \to \hat{G} \xrightarrow{t-\phi} \hat{G} \to G \to 0$ where $\hat{G} = G \otimes_Z \Lambda$ and $t - \phi$ is a matrix over Λ . Taking the dual $\mathrm{Hom}_\Lambda(\cdot , \Lambda)$ of the sequence gives :

$$0 \to \mathrm{Hom}_\Lambda(\hat{G},\Lambda) \xrightarrow{t-\phi^*} \mathrm{Hom}_\Lambda(\hat{G}, \Lambda) \longrightarrow \mathrm{Ext}^1_\Lambda(G,\Lambda) \longrightarrow 0$$

where ϕ^* is the dual of ϕ. But there is a standard Λ-module isomorphism : $F \otimes_Z \Lambda \xrightarrow{\approx} \mathrm{Hom}_\Lambda(\hat{G} , \Lambda)$ defined by $h \otimes 1 \longmapsto ((g,t^i) \longmapsto h(g)t^i)$. Under this isomorphism $t - \phi^*$ corresponds to $1 \otimes t - \mathrm{Hom}_Z(\phi, 1) \otimes 1$. Thus we conclude that $\mathrm{Ext}^1_\Lambda(G,\Lambda) \approx$ F with $t = \mathrm{Hom}_Z(\phi, 1)$. Since $\overline{G}$ coincides with G, additively, but $t = \phi^{-1}$, the claim follows.

Now we may apply theorem (III.2). One consequence is rank F $\geq$ rank $\Lambda^2 G$ = rank $\Lambda^2 F$, where rank is an abelian group. If rank F = rank G = r, we have $r \geq \frac{r(r-1)}{2}$, or $r \leq 3$. If r = 1 or 2, then F or $\Lambda^2 F$ is a Λ-module of type K isomorphic, additively, to Z. Since this is obviously impossible we conclude r = 0 or 3.

If r = 3, then G must be Z-torsion free. In fact, if G contained elements of finite order, then $\Lambda^2 G$ would require more than 3 elements to generate additively. But $F \approx \mathrm{Ext}^1_\Lambda(\overline{G}, \Lambda)$ is free abelian of rank 3 and, therefore, could not map epimorphically onto $\Lambda^2 G$, as required by theorem (III.2).

To complete the reduction of the case r = 3 to (b) we must show det $\phi = +1$. Since F and $\Lambda^2 G$ are free abelian groups of the same rank, the epimorphism $F \longrightarrow \Lambda^2 G$ is an isomorphism. In particular, the eigenvalues $\{\alpha_1^{-1},\alpha_2^{-1}, \alpha_3^{-1}\}$ of $\mathrm{Hom}_Z(\phi^{-1}, 1)$ and $\{\alpha_1\alpha_2,\alpha_2\alpha_3,\alpha_1\alpha_3\}$ must coincide. Thus $\alpha_1^{-1}\alpha_2^{-1}\alpha_3^{-1} = (\alpha_1\alpha_2)(\alpha_2\alpha_3)(\alpha_1\alpha_3)$, which implies $(\alpha_1\alpha_2\alpha_3)^3 = +1$ or $\alpha_1\alpha_2\alpha_3 = +1$. Since det $\phi = \alpha_1\alpha_2\alpha_3$, the reduction is complete.

It remains to prove that G is finite only when $(G,\emptyset)$ are as in (a). When G is finite, though, corollary III implies $\Lambda^2 G = 0$. Thus G is cyclic. To prove $\emptyset = -1$ we use theorem (III.1). If G is cyclic of order q with generator α, we may write $\emptyset(\alpha) = m\alpha$ for some m relatively prime to q. If $< , >$ is the symmetric, bilinear form which exists on G by theorem (III.1), then :

$$< \alpha,\alpha > \quad < \emptyset(\alpha), \emptyset(\alpha) > =< m\alpha, m\alpha > =< \alpha, m^2\alpha >$$

Thus, by non-singularity of $<, >$, $\alpha = m^2\alpha$, i.e. $m = \pm 1 \bmod q$. If $m = + 1$, then G cannot be of type K; thus $m = -1$. Now $1 - \emptyset$ is multiplication by 2, so q must be odd if G is of type K. This completes the proof of theorem IV.

V. Lemma IV gives a criterion for a semi-direct product. $G \times_{\emptyset} Z$ to be a 3-knot group which reduces to the Λ-modules G and $\Lambda^2 G$ being of type K for G finitely-generated abelian. We give another formulation in terms of certain polynomial invariants of $(G, \emptyset)$.

If T is a finite Λ-module, we define, for every prime p, the <u>p-order</u> of T as follows.

$T_p = T/_{pT}$ can be regarded as a module over the principal ideal domain $\Lambda_p = \Lambda/p\Lambda$. Since T is finite, T_p is a finitely-generated Λ_p-torsion module. As such it has an order $\Delta(T_p) \in \Lambda_p$, well-defined up to multiplication by a unit, which can be defined, for example, as the product of the orders of the cyclic summands of T, or aas the characteristic polynomial of t regarded as a linear automorphism of T_p, a finite dimensional vector space over Z/p. If A is a Λ-module whose Z-torsion submodule T is finite, then $\Delta(T_p)$ is the p-order of A.

If A is a finitely-generated Λ-module, then $A_o = A \otimes_Z Q$ is a finitely-generated module over the principal ideal domain $\Lambda_o = \Lambda \otimes_Z Q$. A_o is a Λ_o-torsion module if and only if it is finite-dimensional as a vector space over Q. In this case we may define the <u>order</u> $\Delta(t)$ of A as the product of the orders of the cyclic summands over Λ_o, or as the characteristic polynomial of t as a linear endomorphism of A_o. We may choose the order to be a primitive polynomial in Λ, in which case it is well-defined up to a unit (in Λ) multiple. If A is finitely generated, as an abelian group, then $\Delta(t)$ is monic (i.e. its leading coefficient is ± 1) and, since t is an automorphism of A, its terminal coefficient is also ± 1.

The criterion for $\pi = G \times_\emptyset Z$ to be a 3-knot group will use these orders of the Λ-module G. The necessary conditions require a definition. A polynomial $\lambda(t) \in \Lambda_p$ will be called <u>p-asymmetric</u> if the greatest common divisor of $\lambda(t)$ and $\lambda(t^{-1})$ is either 1 or $(t + 1)$. If $p = 2$, we only allow 1 .

<u>Theorem V</u>. Suppose $\pi = G \times_\emptyset Z$, where G is finitely-generated abelian and $\emptyset$ an automorphism of G. Let $\Delta(t)$ be the order of the Λ-module G, and $\Delta_p(t)$ the p-order of G. Then π is a 3-knot group if and only if the product $\Delta(t)\, \Delta_p(t)$ is p-asymmetric, for every prime p.

Conversely, given such polynomials, where $\Delta(t)$ has leading and terminal coefficients ± 1, we can construct such a G by using companion matrices, for example. Kervaire-Haussmann [KH] have, moreover, determined which finitely-generated abelian groups G can be realized as the commutator subgroup of a 3-knot group - the only restrictions on G are on the 2 and 3-primary components and that rank $G \geq 3$. One gets, therefore, many 3-knot groups which are not among the 2-knot groups enumerated in theorem IV.

In § VI, we apply theorem V to give a sufficient (but not necessary) criterion for a Λ-module of type K to be a 3-knot module.

Proof of theorem V : Let T denote the Z-torsion submodule of G, and $F = {}^{G}/_{T}$. Then $\emptyset$ induces automorphisms $\emptyset_T$ and $\emptyset_F$ of T and F ; G is of type K if and only if T and F are of type K. Since $G = F \oplus T$, additively, $H_2(G) \simeq \Lambda^2 G$ splits, additively, as $\Lambda^2 F \oplus (F \otimes T) \oplus \Lambda^2 T$. As a Λ-module, $\Lambda^2 G$ has a filtration by submodules :

$$\Lambda^2 G \supseteq (F \otimes T) \oplus \Lambda^2 T \supseteq \Lambda^2 T .$$

The successive quotient modules are isomorphic to $\Lambda^2 F$, $F \otimes T$ and $\Lambda^2 T$ with the Λ-module structure defined by $t = \Lambda^2 \emptyset_F$, $\emptyset_F \otimes \emptyset_T$ and $\Lambda^2 \emptyset T$, respectively. Then $\Lambda^2 G$ is of type K if and only if $\Lambda^2 F$, $F \otimes T$ and $\Lambda^2 T$ are all of type K.

We first consider T and $\Lambda^2 T$.

Lemma (V.1) : Let T be a finite Λ-module, with p-order $\Delta_p(t)$. Then T and $\Lambda^2 T$ are of type K if and only if $\Delta_p(t)$ is p-asymmetric, for every prime p.

Proof : Since T splits, as a Λ-module, into the direct sum of its p-primary components, we may assume T is p-primary, for some prime p.

Case 1 : $pT = 0$.

We may regard T as a Λ_p-module and $\Delta_p(t)$ is its order. Clearly T is of type K if and only if $\Delta_p(1) \neq 0$. The analogous criterion for

$\Lambda^2 T$ to be of type K is $\Gamma(1) \neq 0$, where $\Gamma(t)$ is the order of $\Lambda^2 T$ as a Λ_p-module. We may interpret $\Delta_p(t)$ and $\Gamma(t)$ as characteristic polynomials of t and $\Lambda^2 t$. But it is a standard fact that the eigenvalues of $\Lambda^2 t$ are the products of the distinct (counting multiplicities) eigenvalues of t (see [B : chap. III). Thus if $\Delta_p(t) = \prod_i (t - \xi_i)$, then $\Gamma(t) = \prod_{i<j} (t - \xi_i \xi_j)$. The condition $\Gamma(1) \neq 0$ becomes $\xi_i \xi_j \neq 1$, for every $i \neq j$. Since the roots of $\Delta_p(t^{-1})$ are $\{ \xi_i^{-1} \}$, $\Gamma(1) \neq 0$ if and only if $\Delta_p(t)$ and $\Delta_p(t^{-1})$ have no common roots except +1 and -1 which can be simple roots. Adding the condition $\Delta_p(1) \neq 0$, we obtain precisely the p-asymmetry of $\Delta_p(t)$. This proves Lemma (V.1) for case 1.

We now obtain the general p-primary case by :

Lemma (V.2) : Let A be a finitely-generated Λ-module and m a positive integer. Then $\Lambda^2 A$ is of type K if and only if $\Lambda^2(mA)$ and $\Lambda^2(A/mA)$ are both of type K.

To prove Lemma (V.2) we require :

Lemma (V.3) : Let A be a finitely-generated Λ-module. Then $\Lambda^2 A$ is of type K if and only if A supports no non-trivial skew-symmetric bilinear forms $< , >$ satisfying $< t\alpha, t\beta > = < \alpha,\beta >$, for all $\alpha,\beta \in A$.

Proof of Lemma (V.3) : Such a form corresponds to an additive homomorphism $\emptyset : \Lambda^2 A \longrightarrow V$ (= group in which $< , >$ takes values) such that $\emptyset(t\alpha) = \emptyset(\alpha)$, for every $\alpha \in \Lambda^2 A$. Therefore $\emptyset$ induces a homomorphism $\Lambda^2 A/(t-1)\Lambda^2 A \longrightarrow V$. Conversely any homomorphism $\Lambda^2 A/(t-1)\Lambda^2 A \longrightarrow V$ defines a form of the required type. Thus no such form exists if and only if $\Lambda^2 A = (t-1)\Lambda^2 A$, which is equivalent to $\Lambda^2 A$ of type K when $\Lambda^2 A$ is finitely generated.

Proof of Lemma (V.2) : Suppose $\Lambda^2 A$ is of type K. Since mA and A/mA are both quotient modules of A, $\Lambda^2(mA)$ and $\Lambda^2(A/mA)$ are quotient modules of $\Lambda^2 A$. But a quotient module of a module of type K is again of type K.

Conversely, suppose $\Lambda^2(mA)$ and $\Lambda^2(A/mA)$ are of type K. If $\Lambda^2 A$ is not of type K, then, by Lemma (V.3), there exists a non-trivial skew symmetric bilinear form $< , >$ on A, with $< t\alpha, t\beta > = < \alpha,\beta >$ for all $\alpha,\beta \in A$. Since $\Lambda^2(mA)$ is of type K, the restriction of this form to mA is trivial. Therefore :

(*) $m^2 < \alpha,\beta > = < m\alpha, m\beta > = 0$, for all $\alpha,\beta \in A$.

Now define a form $[,]$ on A/mA by the formula $[\bar{\alpha}, \bar{\beta}] = m < \alpha,\beta >$, for $\bar{\alpha},\bar{\beta} \in A/mA$, where α,β are representatives of $\bar{\alpha},\bar{\beta}$, respectively. By (*) $[,]$ is well-defined and it is clearly bilinear, skew-symmetric, satisfying $[t\bar{\alpha}, t\bar{\beta}] = [\bar{\alpha}, \bar{\beta}]$ for all $\bar{\alpha},\bar{\beta} \in A/mA$. Since $\Lambda^2(A/mA)$ is of type K, $[,]$ must be trivial by Lemma (V.3). Thus $m < \alpha,\beta > = 0$, for all $\alpha,\beta \in A$. This implies that we can define a new form $\{,\}$ on A/mA by $\{ \bar{\alpha},\bar{\beta} \} = < \alpha,\beta >$ for $\bar{\alpha},\bar{\beta} \in A/mA$ and α,β representatives of $\bar{\alpha},\bar{\beta}$. Again we conclude $\{,\} = 0$, by Lemma (V.3), which means $< , > = 0$. Thus $\Lambda^2 A$ is of type K.

We now can conclude the proof of Lemma (V.1). Since T is a finite p-primary Λ-module, $p^n T = 0$, for some integer n. It follows from Lemma (V.2), by an inductive argument, that $\Lambda^2 T$ is of type K if and only if $\Lambda^2(p^{iT}/p^{i+1}T)$ is of type K, for every $i = 0,1,\dots n-1$. It is clear that, given any submodule T' of T, T is of type K if and only if T' and T/T' are both of type K. Thus T is of type K if and only if $p^i T/p^{i+1}T$ is of type K, for every $i = 0,1,\dots n-1$. By case 1, $p^{iT}/p^{i+1}T$ and $\Lambda^2/p^i T/p^{i+1}T)$ are both of type K if and only if its order $\Delta_i(t)$ is p-asymmetric.

Now $p^i T/p^{i+1}T$ is a quotient module of T/pT-multiplication by p^i defining the required epimorphism. Thus $\Delta_i(t)$ divides $\Delta_o(t)$ = the p-order of T. So p-asymmetry of $\Delta_o(t)$ is equivalent to p-asymmetry of every $\Delta_i(t)$, $i = 0, \ldots n-1$, which, as we have proved is equivalent to T and $\Lambda^2 T$ being of type K. This completes the proof of Lemma (V.1).

As a consequence of Lemma (V.1), we can prove that F and $\Lambda^2 F$ are of type K if and only if $\Delta(t)$ is p-asymmetric for every prime p. In fact F and $\Lambda^2 F$ are of type K if and only if F/pF and $\Lambda^2 F/p(\Lambda^2 F)$ are of type K, for every prime p-recall F is free abelian. Since $\Lambda^2 F/p(\Lambda^2 F) \approx \Lambda^2(F/pF)$, as Λ-modules, and $\Delta(t)$ is the p-order of F/pF, the desired conclusion follows.

It remains to consider $F \otimes T$. Since $(F \otimes T)/p(F \otimes T) \approx F/pF \otimes T/pT$, as Λ-modules, the p-order $\theta_p(t)$ of $F \otimes T$ is obtained from the order $\Delta(t)$ of F and the p-order $\Delta_p(t)$ of T as follows. Let $\alpha_1, \ldots, \alpha_k$ be the roots, with multiplicity, of $\Delta(t)$; since $\Delta(t)$ is monic and integral, the $\{a_i\}$ are algebraic integers. Let K be the splitting field of $\Delta(t)$ generated by $\{\alpha_i\}$, and P a prime ideal of K dividing p . If β_i is defined to be the reduction of α_i in the residue field F defined by P, then $\{\beta_i\}$ are the roots of $\Delta(t)$ mod p. Let $\gamma_1, \ldots, \gamma_l$ be the roots of $\Delta_p(t)$ in some extension of F. It is a standard fact (see [B]) that $\theta_p(t)$ has roots $\{\beta_i \gamma_j\}$.

It follows from the proof of Lemma (V.1) that $F \otimes T$ is of type K if and only if $\theta_p(1) \neq 0$, for every prime p. Thus $\beta_i \neq \gamma_j^{-1}$, for every i,j , or, equivalently, $\Delta(t)$ and $\Delta_p(t^{-1})$ are relatively prime over Z/p.

We have now shown that G and $\Lambda^2 G$ are of type K if and only if the following hold :

(i) $\Delta_p(t)$ is p-asymmetric, for every prime p,

(ii) $\Delta(t)$, reduced mod p, is p-asymmetric for every prime p,

(iii) $\Delta(t)$(mod p) is relatively prime to $\Delta_p(t^{-1})$, for every prime p .

Theorem V now follows by observing that these 3 conditions are equivalent to the requirement that $\Delta(t)\Delta_p(t)$ be p-asymmetric, for every prime p.

VI. Let A be a finitely-generated Λ-module of type K. We will prove :

Theorem VI. If the irreducible factors of the p-order of A are p-asymmetric, and -1 is either not a root or a simple root, for every prime p, then A is the module of some 3-knot group.

Proof : Let T be the Z-torsion submodule of A. We first show T is the module of a 3-knot group. Since T is the direct sum of its p-primary components, as Λ-module, we may, by using connected sum, assume T itself is p-primary. Let $\Delta(t) = \gamma_1(t)^{e_1} \ldots \gamma_k(t)^{e_k}$ be the factorization of the p-order $\Delta(t)$ of T, where $\gamma_i(t)$ is irreducible in Λ_p and, by assumption, p-asymmetric. It follows easily that $\gamma_i(t)^{e_i}$ is also p-asymmetric, since $\gamma_i(t) = t + 1$ implies $e_i = 1$. Let $\Gamma_i(t)$ be any integral polynomial such that $\Gamma_i(t) \equiv \gamma_i(t)$ mod p.

Let T_i be the $\Gamma_i(t)$-primary submodule of T, i.e. $T_i = \{\alpha \in T : \Gamma_i(t)^m\alpha = 0 \text{ for some } m > 0\}$.

Claim : $T = \bigoplus_i T_i$

Let $\tilde{\Delta}(t) = \prod_i \Gamma_i(t)^{e_i}$; then $\tilde{\Delta}(t)\, T \subseteq pT$, by definition of p-order If we choose m such that $p^m T = 0$, then $\tilde{\Delta}(t)^m T = 0$. If we write $\theta_i(t) = \tilde{\Delta}(t) / \Gamma_i(t)^{f_i}$ then $\{ \theta_i(t)^m \}$ are relatively prime mod p.

It follows that, for any integer n, there exist $\lambda_i(t)$ such that $\sum_i \lambda_i(t)\ \theta_i(t)^m \equiv 1 \bmod p^n$. If we take n = m, then $\sum_i \lambda_i(t)\theta_i(t)^m\alpha = \alpha$, for any $\alpha \in T$. Now $\theta_i(t)^m\alpha \in T_i$, since $\Gamma_i(t)^{me_i}\ \theta_i(t)^m = \tilde{\Delta}(t)^m$. Thus we see T is spanned by $\{T_i\}$.

Suppose $\alpha_i \in T_i$; then $\theta_i(t)^n\ \alpha_j = 0$ for $i \neq j$ and some integer n. If $\sum_i \alpha_i = 0$, then $\theta_i(t)^n\ \alpha_i = 0$ for suitably large n. As above, we may find $\lambda_i(t)$ so that $\sum_i \lambda_i(t)\ \theta_i(t)^n\ \alpha = \alpha$, for any $\alpha \in T$. Since $\theta_i(t)^n\ \alpha_j = 0$ for every i, j, we have :

This proves the Claim.

$$\alpha_j = \sum_i \lambda_i(t)\ \theta_i(t)^n\ \alpha_j = \sum_i \lambda_i(t) \cdot 0 = 0\ .$$

The p-order of T_i is $\gamma_i(t)^{e_i}$, since the p-order of T is the product of those of the T_i and the p-order of T_i is certainly some power of $\gamma_i(t)$.

Thus y the use of connected sum, again, we may assume $\Delta(t)$ is itself p-asymmetric. But now, by theorem V, T is the module of some 3-knot group.

A/T is a Z-torsion free Λ-module of type K, and, so, by [L 1 (3.5)], A/T has a presentation :

$\{ X_i : \sum_j \lambda_{ij}(t)X_j = 0 \}$ with equal number of generators and relations. The property that A/T of type K is equivalent to the matrix $(\lambda_{ij}(1))$ being unimodular.

Now A can be contructed from T by adjoining generators $\{X_i'\}$ and relations $\sum_i \lambda_{ij}(t)X_j' = \alpha_i$, for suitable elements $\alpha_i \in T$.

Let π_0 be a 3-knot group with $A(\pi_0) \approx T$. Define π from π_0 by adjoining generators $\{x_i\}$ and relations $r_i = \prod_{j,k} (\mu^k x_j^{a_{ijk}} \mu^{-k}) = \tau_i$, where $\lambda_{ij}(t) = \sum_k a_{ijk}t^k$, μ is a meridian element of π_0, and $\tau_i \in [\pi_0, \pi_0]$ is some representative of α_i (see [L 1 : § 11]). It is easy to see that μ is still a meridian of π if we choose $\lambda_{ij}(t)$ so that $\lambda_{ij}(1) = \delta_{ij}$ and order the letters in r_i appropriately. By the arguments of [L 1: § 11], adding the relation $\mu = 1$, will reduce $r_i = \tau_i$ to $x_i = \tau_i$. Since $\mu = 1$ kills π_0, it will kill π. Furthermore, $H_1(\pi)$ is infinite cyclic, since commutativity reduces $r_i = \tau_i$ to $x_i = \tau_i$ once again and so $H_1(\pi_0) \approx H_1(\pi)$.

It is clear that $A(\pi) \approx A$. It remains to prove $H_2(\pi) = 0$. Since $H_2(\pi_0) = 0$, there exists a finite cell-complex, such that $\pi_1(K_0,*) \approx \pi_0$ and $H_2(K_0) = 0$. Now adjoin to K_0 new 1-cells $\{\sigma_i\}$ to obtain K_1, with $\dot{\sigma}_i = *$, one for each new generator x_i of π. Adjoin to K_1 new 2-cells E_i where ∂E_i is a path representing $r_i\tau_i^{-1} = r_i(x_j,\mu)\tau_i^{-1}$ in $\pi_1(K_1,*)$. The resulting complex K has fundamental group $\pi_1(K,*) \approx \pi$. We will show $H_2(K) = 0$ which, by Hopf's theorem [H], implies $H_2(\pi) = 0$.

Let $Z = \sum_i m_i E_i + C$, be a 2-cycle of K, where C is a 2-chain of K_0. Now

$$\partial E_i = \sum_{j,k} a_{ijk}\sigma_j - \hat{\tau}_i \text{ , where } \hat{\tau}_i \text{ is some 1-chain in } K_0$$

corresponding to τ_i. Since $\sum_k a_{ijk} = \lambda_{ij}(1) = \delta_{ij}$, we have : $\partial E_i = \sigma_i - \hat{\tau}_i$. Thus $0 = \partial z = \Sigma m_i \partial E_i + C = \Sigma m_i \sigma_i - \Sigma m_i \hat{\tau}_i + \partial C$. Since $\partial C - \Sigma m_i \hat{\tau}_i$ is a chain in K_0, $\sum_i m_i \sigma_i = 0$ which implies every $m_i = 0$. Thus $z = C$ and, since every 2-cycle of K_0 is a boundary, z is a boundary.

This completes the proof of theorem VI. It is easy to see that theorem VI is far from a necessary property of 3-knot groups. For example, this property of A is not preserved under direct sums. Therefore, by taking connected sums we can obtain a 3-knot whose module has p-order with -1 as a root of any multiplicity.

More significantly we can produce, by twist-spinning, 2-knot modules with irreducible symmetric p-orders $\Delta_p(t)$ (i.e. $\Delta_p(t) = \Delta_p(t^{-1})$). For example, 3-twist-spinning the knots K_m and L_m of § IV produce 2-knots whose modules are cyclic $\approx \Lambda/I$, where

$$I = (at^2 + (1 - 2a)t + a,\ t^3 - 1),\quad a = \begin{cases} m & \text{for } K_m \\ -m & \text{for } L_m \end{cases}$$

We compute readily that $I = (t^2 + t + 1,\ 1 - 3a)$. By choosing a correctly, we thus produce knots with p-order $= t^2 + t + 1$ for any prescribed prime $p \neq 3$. If $p \equiv 2 \bmod 3$, $t^2 + t + 1$ is irreducible, mod p.

R E F E R E N C E S

[AC] J. ANDREWS, M. CURTIS : Free groups and handlebodies, Proc. A.M.S. 16 (1965), p. 192-5 .

[B] N. BOURBAKI : Eléments de mathématique, algèbre ; Hermann, Paris, 1970.

[C] R. CROWELL : Corresponding group and module sequences, Nagoya Math. S., 19(1961), p. 27-40.

[CS] S. CAPPELL, J. SHANESON : There exist inequivalent knots with the same complement, Annals of Math. 103 (1976), p.349-53.

[D] M.J. DUNWOODY : Relation modules, Bulletin London Math.Soc. 4 (1972), p. 151-55.

[F] M. FARBER : Linking coefficients and two-dimensional knots, Soviet Math. Dokl., 16 (1975), No 3, p. 647-50.

[G] M. GUTIERREZ : On Knot modules, Inventiones Math. 17(1972), p. 329-35.

[H] H. HOPF : Fundamentalgruppe und zweite Bettische gruppe, Commentarii Math. Helv. 14 (1942), p.257-309.

[HS] P. HILTON, U. STAMMBACH : A course in homological algebra, Springer-Verlag, New-York, 1971.

[K] M. KERVAIRE : Les noeuds de dimensions supérieures, Bull. Soc. Math. France, 93 (1965), p. 225-271.

[Ki] R. KIRBY : Problems in low-dimensional manifold theory, Proceeding Stanford Topology Conference, Summer 1976 to appear

[KH] M. KERVAIRE, J.-Cl. HAUSMANN : Sous-groupes dérivés des groupes de noeuds, L'Ens. Math. XXIV (1978), 111-123.

[KMS] A. KARRASS, W. MAGNUS, D. SOLITAR : Combinatorial group theory, Interscience Publishers, John Wiley and Sons, New-York, 1966.

[L 1] J. LEVINE : Knot Modules: I, Transactions A.M.S., 229 (1977), 1 - 50.

[L 2] J. LEVINE : Knot Modules : II (to appear)

[Sc] SCHUBERT : Knoten mit Zwei Brucken, Math. Zeitschrift 65 (1956), p. 133-70.

[S] S. SMALE : Structure of manifolds, Amer. J. Math., 84, (1962), p. 387-99.

[Z] E.C. ZEEMANN : Twisting spun knots, Trans. Amer. Math. Soc. 115 (1965), p. 471-95.

APPENDIX

to J.Levine's paper.

by

Claude WEBER

There is another way to look at J.Levine's results contained in his § V, which gives slightly different and a bit stronger results. I take this opportunity to thank J.Levine for kindly agreeing to the publication of this little appendix. We shall try and use his notations as much as possible.

Denote by $\mathbb{F}_p$ the field with p elements, and by $\otimes$ the tensor product over the integers $\mathbb{Z}$.

Let A be a finitely generated Λ-module. Then $A \otimes \mathbb{F}_p$ is a finitely generated Λ_p-module. We shall denote by $\Delta(A \otimes F_p) \in \Lambda_p$ the order of this module, i.e. a generator of its first elementary ideal.

<u>Theorem</u> : Let $\pi = G \times_\Phi Z$ be a finitely presented group, with G abelian. (G is then naturally a finitely generated Λ-module. We do not suppose that it is finitely generated as an abelian group). Then π is a 3-knot group if and only if $\Delta(G \otimes F_p)$ is different from zero and asymmetric (in the sense of Levine) for all primes p.

<u>Remark</u> : If G is finitely generated as an abelian group, then π is finitely presented and we recover Levine's hypothesis of theorem V. The comparison between the conclusions of the theorems is done at the end of this paper.

Proof : It results from J. Levine's paper or from M. Kervaire and J.-C. Hausmann : "Sous-groupes dérivés des groupes de noeuds" (L'Ens. Math. XXIV (1978), pp. 111-123) that, under the hypothesis of the theorem, π is a 3-knot group if and only if :

$$G \text{ and } \Lambda^2 G \text{ are both } \Lambda\text{-modules of type K.}$$

It is immediate that the theorem is then a consequence of the following three lemmas :

Lemma 1 : Let A be a finitely generated Λ-module. Then A is of type K if and only if $A \otimes \mathbb{F}_p$ is of type K for every prime p.

Lemma 2 : Let B be an abelian group. Then :

$$(\Lambda^2 B) \otimes \mathbb{F}_p \approx \Lambda^2 (B \otimes \mathbb{F}_p)$$

Lemma 3 : Let V be a finitely generated Λ_p-module. Then V and $\Lambda^2 V$ are simultaneously of type K if and only if $\Delta(V) \in \Lambda_p$ is non-zero and asymmetric.

Proof of lemma 1 : A being of type K is equivalent to the abelian group $P = A \otimes_\Lambda \mathbb{Z}$ being trivial ($\mathbb{Z}$ is given the trivial Λ-module structure).

Now A being finitely generated over Λ implies P being finitely generated over $\mathbb{Z}$. So $P = 0$ is equivalent to $P \otimes F_p = 0$ for every prime p.

But : $P \otimes \mathbb{F}_p = (A \otimes_\Lambda \mathbb{Z}) \otimes \mathbb{F}_p = (A \otimes \mathbb{F}_p) \otimes_\Lambda \mathbb{Z}$

So $P \otimes \mathbb{F}_p = 0$ is equivalent to $A \otimes \mathbb{F}_p$ being of type K.

Note : A challenging discussion with J.-C. Hausmann has been useful to me to pass from the case A $\mathbb{Z}$-finitely generated to the case A Λ-finitely generated.

Proof of lemma 2 : See Bourbaki Algèbre chap. III § 7 nb.5 .

Proof of lemma 3 : This is essentially case 1 in the proof of Levine's theorem V, with the added grain of salt about $\Delta(V)$ being non-zero. But there is nothing mysterious there, because V of type K implies $\Delta(V) \neq 0$ and, conversely, $\Delta(V) \neq 0$ implies V finite dimensional over $\mathbb{F}_p$ and so J.Levine's argument applies.

Q.E.D.

We conclude this appendix with a remark which should shed some light on the link between J.Levine's point of view and ours.

Remark : Using the notations and assuming the hypothesis of the theorem suppose moreover that G is of type K . Then what Levine calls

$$\Delta(t) \, . \, \Delta_p(t) \in \Lambda_p \quad \text{is equal to our} \quad \Delta(G \otimes \mathbb{F}_p) \, .$$

Proof : Let T be the $\mathbb{Z}$-torsion Λ-submodule of G, and let $F = {}^{G}/_{T}$. Then the sequence of Λ-modules :

$$0 \longrightarrow T \longrightarrow G \longrightarrow F \longrightarrow 0$$

splits as a sequence of abelian groups, because T is finite. Hence

$$0 \longrightarrow T \otimes \mathbb{F}_p \longrightarrow G \otimes \mathbb{F}_p \longrightarrow F \otimes \mathbb{F}_p \longrightarrow 0$$

is an exact sequence of Λ_p-modules. As Λ_p is principal, one has :

$$\Delta(T \otimes \mathbb{F}_p) \, . \, \Delta(F \otimes \mathbb{F}_p) = \Delta(G \otimes \mathbb{F}_p).$$

It is clear that $\Delta(T \otimes \mathbb{F}_p)$ is Levine's $\Delta_p(t)$. We claim that $\Delta(F \otimes \mathbb{F}_p)$ is Levine's $\Delta(t)$ (coefficients reduced mod.p). This follows from my paper : "Torsion dans les modules d'Alexander" published in this book. G being of type K is here essential.

ALGEBRAIC COMPUTATIONS OF THE INTEGRAL CONCORDANCE AND DOUBLE NULL CONCORDANCE GROUP OF KNOTS

Neal W. Stoltzfus
Dept. of Math, Louisiana State University
Baton Rouge, LA 70803/USA

I. INTRODUCTION

The first half of this note is a summary of the algebraic results concerned with the classification of isometric structures of the integers arising in knot theory under the concordance or metabolic equivalence relation. A detailed expostion can be found in the author's Memoir [St]. Briefly, the Seifert linking pairing, L, on the free submodule, M, of the middle dimensional homology of a Seifert manifold for an odd dimensional knot defines an endomorphism, t, of M by the equation

$$\text{i)} \qquad L(x,y) = b(t(x),y)$$

where b is the bilinear unimodular intersection pairing on M. From the symmetries satisfied by L we obtain the following relation

$$\text{ii)} \qquad b(t(x),y) = b(x,(Id - t)(y))$$

Such objects (M,b,t) are called isometric structures over the integers. It is called metabolic if there is a submodule N which is invariant under t and equal to its own annihilator under b, $N = N^{\perp} = \{m \text{ in } M | b(m,n) = 0 \text{ for each } n \text{ in } N\}$.

The algebraic technique of localization allows us to relate the integral case to the rational case which was computed by Levine [L]. Unlike the rational field case there are obstructions to the decomposition of an integral isometric structure accordin to the Z[X]-module structure induced by t which are measured by the coupling exact sequence. This reduces the explicit computation to modules over orders in some algeraic number field, where the final computations are made.

When (M,b,t) is metabolic on N, there is an exact sequence of Z[X]-modules

$$\text{iii)} \qquad 0 \longrightarrow N^{\perp} \longrightarrow M \longrightarrow Hom_Z(N,Z) \longrightarrow 0.$$

When this sequence splits, the isometric structure is called hyperbolic. This is

a necessary condition for the geometric condition of double null concordance of a knot studied by Dewitt Sumners [S] . Stabilization with this relation defines a new group of knots under the operation of connected sum which is much larger than the knot concordance group. In fact, a simple knot is trivial in this group only if it is (stably) isotopic to the connected sum of a knot with its inverse. Furthermore, the even dimensional group is non-trivial contrasting with the even-dimensional knot concordance group, which is zero [K]. The above techniques and ideas also apply to isometries of integral inner product spaces which arise geometrically in the bordism of diffeomorphsim question solved by Kreck [Kr]. The application may also be found in[St].
Grateful acknowledgement is made for the supportive assistance of Pierre Conner, Michel Kervaire, "Le Troisieme Cours" and the National Science Foundation.

II. THE METABOLIC CASE

Let R be a Dedekind domain, in particular the integers, Z, the rational field, Q, or a finite field with q elements, F_q. Let $\varepsilon = \pm 1$.

Definition 2.1 An ε-symmetric isometric structure over R is a triple (M,b,t) where M is a finitely generated R-module, b is an ε-symmetric bilinear form on M with values in R and t is an R-linear endomorphism of M satisfying:

i) (M,b) is an inner product space, that is the adjoint homomorphism, Ad b:M $\longrightarrow$ $Hom_R(M,R)$ given by Ad b(m) = b(m,-), is an isomorphism.

ii) b(t(x),y) = b(x,(Id - t)(y))

Let K denote the field of fractions of R. we will relate isometric structures over R and K by means of the following:

Definition 2.2 An ε-symmetric torsion isometric structure over R is a triple (T,b,t) where T is a finitely generated torsion R-module, b is an ε-symmetric bilinear form on T with values in the R-module K/R and t is an R-linear endomorphism of T with:

i) Ad b:T $\longrightarrow$ $Hom_R(T,K/R)$ is an isomorphism

ii) b(t(x),y) = b(x,(Id - t)(y))

An isomorphism of isometric structures must preserve the inner product and commute with the endomorphism. The isomorphism classes form a semigroup under the operation of orthogonal direct sum. We now define an equivalence relation so that the equivalence classes form a group.

<u>Definition 2.3</u> An isometric structure is <u>metabolic</u> if there is an R-submodule N

i) N is t invariant, that is $t(N)\subseteq N$, and

ii) $N = N^{\perp} = \{m \text{ in } M: b(m,N) = \{ b(x,n): n \text{ in } N \} = \{0\}\}$, the annihilator of N under the inner product b. We call N a metabolic submodule or simply, metabolizer.

<u>Examples</u> 2.4

i) The diagonal D in (M,b,t) + (M,-b,t) is a metabolic submodule

ii) Given an R-module N and an R-linear endomorphism s, where R is torsion free or completely torsion, the hyperbolic isometric structure $H(N,s) = (N + N^{*}, b, t)$ where $N^{*} = \mathrm{Hom}_R(N,R)$ in the torsion free case and $\mathrm{Hom}_R(N,K/R)$ is the torsion case with $b((x,f),(y,g)) = f(y) + \varepsilon g(x)$ and $t(x,f) = (s(x), f\circ(Id-s))$ has metabolic summands N and N^{*}.

iii) The torsion isometric structure $H(Z/(m^2), Id)$ has a metabolizer mT which is not a direct summand.

<u>Definition 2.5</u> Two isometric structures, M and N, are <u>Witt-equivalent</u> (or concordant) if there are metabolic isometric structures, H and K, such that M + H is isometric with N + K. The set of equivalence classes form a group, denoted $C^{\varepsilon}(R)$ ($C^{\varepsilon}(K/R)$ in the torsion case), under orthogonal direct sum. The inverse of (M,b,t) is (M,-b,t) as in example 2.4 i).

$C^{\varepsilon}(Z)$, which was first defined by Kervaire in [K], is well-known to be isomorphic to the geometric knot concordance group in dimensions above one and to have infinitely many elements of each possible order, two, four and infinite, [K,L] . We wish to further elucidate its structure. The first question we will solve is which rational isometric structures contain unimodular integral isometric structures?

Let (V,B,T) be an isometric structure on the field K, the fraction field of the Dedekind domain R. An R-lattice in V is a finitely generated R-submodule of V. An obvious necessary condition for an R-lattice L to be invariant under T, is that T satisfy a monic polynomial with coefficients in R. (If R = Z, then L is a free Z-module and this is the theorem of Cayley and Hamilton, in a general Dedekind domain, this applies to each localization.) Let $C_0^{\varepsilon}(K)$ be the Witt group of isometric structures over K satisfying $f(T)=0$ for some monic polynomial with coefficients in R. If $\{x_i\}$ is a basis for V, then the R-module generated by $T^j x_i$, $j < \deg f$, is an R-lattice invariant under T. Since K is the fraction field of R, we may scale the lattice L by the product of the denominators of the inner product B on the finite set $T^j x_i$ to obtain a new lattice dL on which the inner product B is R-valued. Define the lattice $L^{\#}$ = {v in V: B(v,l) is in R, for every l in L} .This lattice is also invariant under T, so that T induces an endomorphism of the quotient $L^{\#}/L$, say t. Finally we define a K/R valued inner product on $L^{\#}/L$ by $b(\overline{u},\overline{v}) = B(u,v) \bmod R$. This is a torsion isometric structure and its equivalence class in $C^{\varepsilon}(K/R)$ is independent of the choice of the lattice L. We have defined a homomorphism $\partial : C^{\varepsilon}(K) \to C^{\varepsilon}(K/R)$, called the boundary.

LOCALIZATION THEOREM 2.6 the following sequence is exact:

$$0 \longrightarrow C^{\varepsilon}(R) \xrightarrow{\ i\ } C_0^{\varepsilon}(K) \xrightarrow{\ \partial\ } C^{\varepsilon}(K/R)$$

In the case R = Z, the boundary is onto, making the sequence short exact. The map i is given by localization at zero (or tensor product with the field of fractions.)

Example: Let R = Z, and consider the rational isometric structure, given with respect to a basis, e_1 and e_2, by the matrices (V= Q + Q, $B = \begin{pmatrix} 0 & 5 \\ -5 & 0 \end{pmatrix}$, $t = \begin{pmatrix} 0 & -1 \\ 1 & 1 \end{pmatrix}$

Then $L^{\#} = 1/5\, e_1 + 1/5\, e_2$ and $L^{\#}/L = Z/(5) + Z/(5)$ with the form $\begin{pmatrix} 0 & 1/5 \\ -1/5 & 0 \end{pmatrix}$. The form has a one-dimensional self-annihilating subspace generated by $e_1 - e_2$, but no one dimensional line in $L^{\#}/L$ over the field F_5 can be invariant since the minimal polynomial is X^2-X+1, which is irreducible mod 5.

We now introduce a very powerful notion which simplifies the computation of the last two terms in the sequence.

Definition 2.7 An isometric structure is anisotropic, if for any t invariant R-submodule N, $N \cap N^{\perp} = 0$

Proposition 2.8 Every Witt equivalence class has an anisotropic representative.

Proof: Let L be an invariant pure R-submodule with $L \subset L^{\perp}$ .Then $L^{\perp}/L$ inherits a quotient isometric structure and N = { (x,x+L): x is in L} is a metabolizer.

We now note that the endomorphism t must have kernel 0 in any anistropic representative, for $b(x,(1-t)y) = b(x,y) = b(tx,y) = b(0,y) = 0$ if x and y are in the kernel. Hence the kernel is invariant and self-annihilating. If R is a field or if (M,b,t) is a torsion structure on a finite module M, then t is invertible and $s = Id - t^{-1}$ is an isometry of b, by an easy verification.

Let Λ denote the polynomial ring $R[X]$. It has an involution induced by $X^* = 1-X$. An isometric structure over R can be viewed as a Λ-module in the traditional way, by allowing the indeterminate X to act as the endomorphism t. We will denote restrictions on the module structure by subscripts, for example, if I is an ideal in R, $C_{R/I}(K/R)$ will denote the Witt group of torsion isometric structures on R/I-modules, that is torsion R modules annihilated by I. As a second application of the existence of anisotropic representatives, we have the following:

Theorem 2.9 The following inclusions are isomorphisms:

$$\bigoplus_{I,\text{ maximal ideal in } R} C^{\varepsilon}_{R/I}(K/R) \subset C^{\varepsilon}(K/R)$$

$$\bigoplus_{I=I^*,\text{ invariant maximal ideal}} C^{\varepsilon}_{\Lambda/I}(F) \subset C^{\varepsilon}(F)$$

where R is a Dedekind ring of finite rank over Z and F is a field.

Proof: If N is a t invariant subspace in M, an anisotropic structure, then $N \cap N^{\perp} = 0$. If the module M is a vector space over a field or if M is a torsion module with finite cardinality (this is assured by the condition on R in the torsion case) then the exact sequence: $0 \longrightarrow N^{\perp} \longrightarrow M \longrightarrow \mathrm{Hom}_R(N,R) \longrightarrow 0$, where the second map is the adjoint homomorphism restricted to N, implies that $M = N + N^{\perp}$ by a dimension or cardinality argument. Hence the annihilator of an irreducible anisotropic must be a maximal ideal in the appropriate ring, R or Λ. Furthermore in the field case, the annihilator must be invariant under the involution for otherwise, if a is in the annihilator I, then $0 = b(ax,y) = b(x,a^*y)$ and if $a^*y \neq 0$ for some y then the form is singular. Injectivity is verified in [St].

In the field case the invariant maximal ideals in F[X]are generated by self-dual monic polynomials $p = p^*$, where $p^*(X) = (-1)^{\deg p}p(1-X)$. In the torsion case there is a further reduction. If Iis a maximal ideal, R/I is a field and if the torsion R-module is annihilated by I it must take its values in the submodule $I^{-1}/R = \{k : kI \subset R\}/R$ in K/R which is isomorphic to R/I by a choice of uniformizer. (If R = Z we make this choice canonical by choosing the positive prime element.) We then have $C^{\varepsilon}_{R/I}(K/R) = C^{\varepsilon}(R/I)$, and we may apply the field case of (2.9) to obtain a further decomposition according to the R/I[X]structure.

Corollary 2.10: The following inclusion is an isomorphism:

$$\bigoplus_{I=I^*,\ R[X]/I\ \text{finite}} C^{\varepsilon}_{R[X]/I}(R/(I\cap R)) \longrightarrow C^{\varepsilon}(K/R)$$

in the integral case, R = Z, the involution on Z[X]/I is trivial only if $X = X^*$ = 1-X mod I, that is 2X = 1 mod I, so 2 is in I and $Z[X]/I = F_p$ for some odd prime p.

$$C^{\varepsilon}_{R[X]/I}(R/I)=\begin{cases} Z/(2)+Z/(2) & p = 1 \bmod 4 \\ Z/(4) & p = 3 \bmod 4 \end{cases}\ \varepsilon=+1;\quad 0 \quad \varepsilon=-1 \quad \text{Trivial involution};\quad Z/(2) \text{ if the involution is non-trivial}$$

The first three computations are the well-known computations of the Witt groups of finite fields, the last is a computation of the witt group of Hermitian forms over a finite field which will follow from the theory of the succeeding paragraphs.

We must now make use of the R[X]-module structure to make further computations. The setting we desire is specified as follows: Let S be an R-algebra, finitely generated as an R-module, Δ an S-module and L an R-module. Suppose also there is an R-linear map $s:\Delta \to L$ such that $s_*: S \times \Delta \to L$ given by $s_*(a,k) = s(ak)$ is non-singular (that is, both adjoints are R-module isomorphisms). Then we have the following trace lemma of Milnor [M]and Knebusch and Scharlau [KS]:

<u>TRACE LEMMA 2.11</u>: s_* induces an equivalence of the category of non-singular L-valued bilinear forms, (,), on R-modules with an S-module structure and the category of Δ-valued forms $\langle\ ,\ \rangle$ on S-modules. The equivalence is given by $s_*\langle\ ,\ \rangle = (\ ,\ )$. Furthermore, if there is an involution * on S, trivial on R such that $s(a^*) = s(a)$, then if (,) is ε-symmetric and satisfies $(ax,y) = (x,a^*y)$ then (,) corresponds to a ε-Hermitian form $\langle\ ,\ \rangle$.

<u>Examples 2.12 i)</u> If R is a field and S is a finite degree separable field extension, then $\Delta = S$, L = R and $s = \text{trace}_{S/R}$ satisfies the conditions of the trace lemma (whence its name). This induces an isomorphism $C^{\varepsilon}_{R[X]/I}(R) = H^{\varepsilon}$ (S = R[X]/I), the Witt equivalence classes of ε-Hermitian forms over S. This completes the computation of $C^{\varepsilon}(Q)$ using Landherr's Theorem 2.14 (following).

ii) Let P be a prime ideal in R[X]and let S = R[X]/P. This is an R-order in a finite degree extension E of the fraction field K, provided that $P\cap R = 0$. Making this assumption, let $\Delta = \Delta^{-1}(S/R) = \{e \text{ in } E: \text{trace}_{E/K}(eS) \subset R\}$, the inverse different of S. By its very definition, the trace induces a nonsingular pairing $S \times \Delta \to R$. This example is very important in the further computation of $C^{\varepsilon}(Z)$.

iii) Let $P = (X^2-X+1)$ and $S = Z[X]/P$ with an integral basis 1 and $a = (1+\sqrt{-3})/2$. Then Δ is $(1/p'(a))S = (1/\sqrt{-3})S$ by a classical theorem of Euler. Consider the Δ-valued Hermitian form on the rank one free module S, given by $[x,y] = xy^*/\sqrt{-3}$. This is a non-singular form and corresponds to the symmetric isometric structure (S, b = trace $[\ ,\ ]$, t = multiplication by a) $= \left(Z + Z, \begin{pmatrix} 0 & 1 \\ -1 & 0 \end{pmatrix}, \begin{pmatrix} 0 & -1 \\ 1 & 1 \end{pmatrix}\right)$ in matrix notation.

The beneficial decompostion of M as $N + N^{\perp}$ for any invariant submodule N of an anisotropic structure in the field and torsion case fails in the integral case. In $R[X]$the prime ideals result from an interaction between the primes in R and those in $K[X]$. This is featured in a second exact sequence which measures the failure of the above-mentioned decomposition.

<u>COUPLING EXACT SEQUENCE 2.13:</u> There is an exact sequence

$$0 \longrightarrow \bigoplus_{\substack{P=P^*,\ \text{prime in } R[X] \\ R\cap P = 0}} C^{\varepsilon}_{R[X]/P} \longrightarrow C^{\varepsilon}(R) \xrightarrow{\ C\ } \bigoplus_{p} C^{\varepsilon}_{R[X]/P}(K/R)$$

The coupling map is defined as follows: Given (M,b,t) let $M_p = \{m : P^i m = 0 \text{ for some } i\}$. The form b restricted to M is non-degenerate since if x is in M_p, there is a y in M so that $b(x,y) \neq 0$. But $0 = b(ax,y) = b(x,a^*y)$ hence y must be in M_p. Therefore M_p is a non-degenerate sub-lattice of M and we may define $c_p(M) = (M_p^{\#}/M_p, \bar{b}, \bar{t})$ as in the definition of boundary. The coupling map C is the direct sum of the c_p. In general the cokernel of C is known but it is not well-understood.

Now, by the trace lemma and the localization sequence, we have a commutative diagram:

$$\begin{array}{ccccccc}
0 & \longrightarrow & C^{\varepsilon}_{R[X]/P} = S^{(R)} & \longrightarrow & C^{\varepsilon}_{R[X]/P}(K) & \longrightarrow & C^{\varepsilon}_{R[X]/P}(K/R) \\
 & & \uparrow & & \uparrow & & \uparrow \\
0 & \longrightarrow & H^{\varepsilon}(\Delta^{-1}(S/R) = \Delta) & \longrightarrow & H^{\varepsilon}(E = K[X]/P) & \longrightarrow & H^{\varepsilon}(E/\Delta)
\end{array}$$

where the vertical arrows are isomorphisms. We will make a further reduction and compute the boundary homomorphism only for the maximal order in E, the ring of algebraic integers D, and obtain the computation for S from the appropriate commutative diagram of forgetful maps.

We now recall Landherr's theorem on the computation of the Witt group of Hermitian forms over an algebraic number field E with involution $*$. Denote by $Q(E,*)$ the semi-direct product $Z/(2) \times F^{\cdot}/NE^{\cdot}$ where $NE^{\cdot} = \{ee^*\}$is the image of the norm map $N_{E/F}: E^{\cdot} \to F^{\cdot}$

and the multiplication is given by $(e_1,d_1)(e_2,d_2) = (e_1 + e_2,(-1)^{e_1+e_2} d_1d_2)$. To each conjugate pair of equivariant embeddings $(E,*) \to (C, -)$ where C is the complex numbers with complex conjugation -, there is an associated signature homomorphism: σ_i: $H(E) \to H(C) = Z$.

Landherr's Theorem 2.14: There is an exact sequence:

$$0 \longrightarrow (4Z)^s \longrightarrow H(E,*) \longrightarrow Q(E,*) \xrightarrow{w} 0$$

where w is (rank mod 2, disc = $(-1)^{n(n-1)/2}$det) and n is the dimension of the form. The kernel of w is detected by the s signature homomorphisms which are $\equiv 0$ mod 4.

The group $F^{\cdot}/NE^{\cdot}$ is computed by the Hilbert symbols $(d,z^2)_{P_0} = \pm 1$ at all the prime ideals P_0 in the fixed field F, where $E = F(\sqrt{z^2})$, by the Hasse Cyclic Norm Theorem and can be realized arbitratily subject to Hilbert Reciprocity that only an even finite number can be - 1. For the next result we will suppose that no dyadic prime ramifies in the extension E/F. This is true in all the number fields arising in the knot concordance group. After performing the technical task of relating the boundary homomorphism to the Hilbert symbols we obtain:

Theorem 2.16: Let D be the ring of integers in an algebraic number field with involution. Then $H^{+1}(\Delta^{-1}(D/Z))$ is computed as follows:

i) there is a rank one form if and only if no prime ramifies in the extension E/F.

ii) Let JH be the subgroup of elements of even rank and H, the Hilbert reciprocity homomorphism. Then the following sequence is exact:

$$0 \longrightarrow JH^{+1}(\Delta^{-1}(D/Z)) \xrightarrow{\sigma} (2Z)^s \xrightarrow{H} (Z^{\cdot} = \pm 1) \longrightarrow 1$$

In particular there are no elements of order four and this is true in general for $H(\Delta^{-1}(S/Z)) \approx C_S^{+1}(Z)$.

Corollary 2.17: An element of order four in $C^{+1}(Z)$ must have a non-trivial coupling invariant and an Alexander polynomial with distinct factors.

We now give a complete description of two important subgroups of the knot concordance group: those whose minimal polynomials are a product of quadratic polynomials (this is related to low genus knots) and those whose Alexander polynomials are a product of cyclotomic polynomials (this is case for Milnor-Brieskorn knots).

Let λ_n be a primitive n^{th} root of unity and, denote by $p_n(X)$ the minimal polynomial of $(1 - \lambda_n)^{-1}$. This is a monic polynomial with integer coefficients if and only if n is composite. Let C be the semigroup of polynomials generated by the p_n under multiplication. Applying the coupling exact sequence 2.13) to this case, we have:

Cyclotomic Coupling Theorem 2.18: $C^{\varepsilon}_{\Lambda/(p_n)}(Z)$ is torsion free and the following is exact: $0 \to + C^{\varepsilon}_{\Lambda/p_n}(Z) \longrightarrow C^{\varepsilon}_{C}(Z) \longrightarrow + C^{\varepsilon}_{\Lambda/p_n}(Q/Z) \to C^{\varepsilon}(Q/Z) \to 0$

Any monic self-dual quadratic polynomial must be of the form: $X^2 - X + b$. Let d be the largest square free divisor of the discriminant and let a be the number of prime divisors of 1-4b not dividing d and b be the number of prime divisors of d. Denoting by T the semigroup generated by the quadratic self-dual polynomials, we have:

Quadratic Coupling Theorem 2.19: The following sequence is exact:

$$0 \longrightarrow \bigoplus_{p=p^*,\ \deg = 2} C^{\varepsilon}_{\Lambda/p}(Z) \longrightarrow C^{\varepsilon}_{T}(Z) \longrightarrow \oplus\ C^{\varepsilon}_{\Lambda/p}(Q/Z) \longrightarrow C^{\varepsilon}_{T}(Q/Z) \oplus \oplus \text{ Coker } c_p \to 0$$

ε	Sign(1-4b)	$C^{\varepsilon}_{\Lambda/p}(Z)$	Coker c_p
+1	-	2Z	$(Z/(2))^{a+b-1}$
+1	+	0	$(Z/(2))^{a+b-1}$
-1	-	$Z + (Z/(2))^{a+b-1}$	0
-1	+	order = 2^{a+b} two torsion except single element of order 4 iff prime ≡ 3(4) divides d	0

$$C^{\varepsilon}_{\Lambda/p}(F_q) = \begin{cases} Z/(2) & (1-4b\,|\,q) = -1 \\ 0 & (1-4b\,|\,q) = 1 \\ (Z/(2))^2 & p \equiv 1(4),\ 1-4b \equiv 0(q),\ \varepsilon = +1 \\ Z/(4) & p \equiv 3(4),\ 1-4b \equiv 0(q),\ \varepsilon = +1 \\ 0 & 1-4b \equiv 0(q),\ \varepsilon = -1 \end{cases}$$

where (|) is the Legendre symbol.

This has been a brief outline of the structure of $C^{\varepsilon}(Z)$ elucidated in complete detail in [St], together with many explicit examples of a more arithmetic nature and a geometric analog of the localization sequence.

III. THE HYPERBOLIC CASE

We now introduce another relation on the semigroup of isometric structures under the operation of orthogonal direct sum. Although this relation was first conceived algebraically, we will give a geometric interpretation in Theorem 3.13.

Definition 3.1: The hyperbolic ε-symmetric isometric structure on (N,s) where N is either a torsion or a torsion free R-module and s is an R-module endomorphism is the structure (M,b,t) where:

i) $M = N + \mathrm{Hom}_R(N,R)$ $(\mathrm{Hom}_R(N,K/R)$ in the torsion case$) = N + N^*$

ii) $b((x,f),(y,g)) = f(y) + \varepsilon g(x)$

iii) $t(x,f) = (s(x), f \circ (\mathrm{Id} - s))$

As we have seen, the $R[X]$- module structure plays a strong role in the analysis of isometric structures. The following proposition relates the above definition to this structure.

Proposition 3.2: An ε-symmetric isometric structure over R is hyperbolic if and only if there is a metabolic submodule N such that the short exact sequence

$$0 \longrightarrow N^{\perp} \longrightarrow M \xrightarrow{\mathrm{Ad}|_N = p} \mathrm{Hom}_R(N, R \text{ or } K/R) \longrightarrow 0$$

splits as a sequence of $R[X]$- modules.

Proof: The isomorphism will be constructed in two stages. First, define the split isometric structure $H(M,q)$ on $N + N^*$ depending on $q:N \to N^*$ satisfying $q = q^*$ $:N^* \to N^{**} = N$ by $b((x,f),(y,g)) = f(y) + \varepsilon g(x) + q(x)(y)$. If $a:N^* \to M$ splits the sequence there is an isometry $H(N,q) \to M$ where $q(x)(y) = b(a(x),a(y))$ given by $(x,f) \to x + a(f)$ because $b(x+a(f),y+a(g)) = b(x,y) + b(a(f),y) + b(x,a(g))$ $+ b(a(f),a(g)) = 0 + f(y) + \varepsilon g(x) + q(f)(g) = b((x,f),(y,g))$ since N is self-annihilating and a splits the adjoint map.

Next, we use the evenness property built into the fundamental relation of isometric structures: $b(x,y) = b(tx,y) + b(x,ty) = b(tx,y) + \varepsilon b(ty,x)$ to demonstrate the isomorphism: $H(N,q) = H(N,0) = H(N)$, given by $(x,f) \to (x, b(t(a(x)),a(-)) + g)$. For $b((x,b(t(a(x)),a(-)) + f), (y,b(t(a(y)),a(-)) + g)) = b(t(a(x),a(y)) + f(y) + \varepsilon g(x) +$ $b(t(a(y)),a(x)) = f(y) + \varepsilon g(x) + q(x)(y)$.

We now define an equivalence relation on the semigroup and demonstrate that this give a new group of isometric structures.

Definition 3.3: Two isometric structures M and N are (stably) hyperbolic equivalent if there are hyperbolic isometric structures H and K so that M + H is isometric to N + K.

As in the metabolic case, this is an equivalence on the isometry classes of isometric structures. Note that the trivial calss is the set of stably hyperbolic structures. Now the diagonal in M + -M is a metabolizer with an invariant complement (M, for instance). Hence by the proposition, the form on M + -M is hyperbolic, so inverses exists and the equivalences classes form a group under the orthogonal direct sum, which will be denoted $CH^{\varepsilon}(R)$. While the localization machinery of the metabolic case will give some results on this new group, many of the results fail to generalize.

When R is a field, we can make a complete analysis by using ancient and often reproven results concerning the classification of isometries of inner product spaces. To reduce to the case of isometries, we need the following:

Proposition 3.4: If F is a field, then every equivalence class in $CH^{\varepsilon}(F)$ has a representative (V,B,T) with T injective (and therefore an isomorphism.) This is also true for $CH^{\varepsilon}(K/R)$ when R is finitely generated over Z.

Proof: Let $H = \{v: T^N(v) = 0$ for some integer $N\}$ and let $K = \{v:(Id-T)^N(v) = 0\}$ Now, Id - T is invertible on H and T on K. Furthermore, $B(T^N x,y) = B(x,(Id-T)^N y)$. Hence, both H and K are self-annihilating. Furthermore, the above relation shows that H and K are dually paired, so that the form on H + K is hyperbolic. The orthogona complement of H + K is the desired isometric structure.

By the standard verification, the monodromy, $s = Id - t^{-1}$ exists and is an isometry: $B(sx,sy) = B(x,y)$. We now consider a new involution on the polynomial ring: $P^*(X) = a_0 X^{\deg p} p(x^{-1})$, where a_0 is the zeroth coefficient. An isometry of an inner product space satisfies a self-dual polynomial, $p = p^*$. Fixing some self-dual polynomial p, define $V_k = \{v: p^k(T)(v) = 0\}$. and consider the form $b_k(v,w) = b(p^{k-1}v,$

Claim: $V_k^{\perp} = V_{k-1} + pV_{k+1}$

Proof: If v + pw is in the right hand side then $b(p^{k-1}u,v+pw) = b(u,p^{k-1}v + p^k w)$ $= 0$ so that the inclusion of the RHS in the left is obvious. Using the cyclic decomposition of F[X]- modules and computing dimensions, equality must hold. ($\dim V_k + \dim V_k^{\perp} = \dim V$.).

From the claim, it follows that the structure ($W_k = V_k/V_{k-1} + pV_{k+1}, B_k, T$) is an inner product space under the induced form B_k and isomorphism T. Furthermore, if the original structure was hyperbolic, so is W_k. Observing that W_k is annihilated by p, we have defined a homomorphism: $\Phi: CH^{\varepsilon}(F) \longrightarrow +_{p = p^*} (\Sigma_{n=1} CH_p^{\varepsilon}(F))$

By the trace lemma, $CH_p^{\varepsilon}(F) = H^{\varepsilon}(F[X]/P(X))$. The following theorem computes $CH^{\varepsilon}(F)$.

Theorem 3.5 [Milnor[M]]Φ is an isomorphism.

One may also consider the nth component of the right hand side as the hyperbolic equivalence classes of structures over F which are projective as $F[X]/p^n(X)$- modules.

This computation shows that the hyperbolic relation is infinitely finer than the metabolic relation. Furthermore, it is possible to show that stably hyperbolic implies hyperbolic in the field case. Hence, if a structure is in the kernel of Φ, it is isometric to a hyperbolic structure. This is an unknown and interesting question in the integral case.

In the integral case, the localization sequence degenerates to:

Theorem 3.6: $CH^{\varepsilon}(R) \longrightarrow CH^{\varepsilon}(K) \longrightarrow C^{\varepsilon}(K/R)$ is exact.

Proof: Proceed as in the localization theorem 2.6. The boundary is well-defined since metabolic is weaker than hyperbolic and its vanishing is the necessary condition (and sufficient also) for a rational structure to contain a unimodular lattice.

The previous injection of the localization homomorphism from the Dedekind Ring R to its field of fractions K is false. However, when the order $S = R[X]/P$ is maximal that is, S is the Dedekind ring of integers in its field of fractions, we have:

Theorem 3.7: $CH^{\varepsilon}_{R[X]/P}(R) = C^{\varepsilon}_{R[X]/P}(R) = H^{\varepsilon}(\Delta^{-1}(D/R))$

Proof: If M is metabolic then any metabolizer is torsion free over R and hence over D, the maximal order. Therefore the metabolizer is projective and the sequence of 3.2 splits.

Therefore no counterexample to the injection of (3.6) exists when the order S is maximal. The following example shows that the localization homomorphism fails to be injective in a slightly more restricted situation. We restrict ourselves to structures on projective S-modules and consider the case when S is local. Note then, the sequence of (3.2) splits if and only if the metabolizer is projective as an S-module.

Lemma 3.8: Let S be a local ring and E its field of fraction with a non-trivial involution *. Consider the homomorphism $Sym(S^{\cdot})/Norm(S^{\cdot}) = \{s = s^*,\ s \text{ a unit}\}/\{uu^*\}$ $\to F^{\cdot}/NE^{\cdot}$ induced by the inclusion i. If $i(a) = 0$ then the form $<1> + <-a>$ is metabolic but not hyperbolic, even after the addition of a projective hyperbolic S-module.

Proof: The hermitian space is metabolic because it has rank two and its discriminant $= -(1)(-a) = a = ee^*$, a norm in $F^{\cdot}$ by assumption. An isotropic vector is given by (e,a). However the space annot be stably hyperbolic since the discriminant of a hyperbolic space is a norm in $S^{\cdot}$. Note that the discrinimant is defined because projective modules over a local ring are free.

The above space defines a self-dual S-lattice in a metabolic space V over E. Hence the boundary of this lattice is zero and we can find a non-singular lattice which localizes to the above lattice.

Example 3.9: Let $P(X) = X^2 - X - 11$ and consider the localization of the order S at the ideal $(3, X + 1)$. Then $i(-5) = 0$ because $-5 = (\sqrt{5})(-\sqrt{5})$, but $\sqrt{5}$, while integral, is not in S which is not a maximal order. By a straight-forward computation -5 is non-trivial in $Sym(S^{\cdot})/Norm(S^{\cdot})$. Therefore the form $<1> + <-5>$ is metabolic but not hyperbolic at this localiztion and, by the above remarks, there is a unimodular lattice over the integers realizing this lattice.

In the hyperbolic case, the coupling exact sequence disintegrates. However, it still yields necessary conditions for a structure to be hyperbolic which are independent of the condition that the rational form be hyperbolic. Let $p(X)$ be a self-dual polynomial. Let $L_p = \{1 : p^N(t)(1) = 0\}$ and define $c_p(L) = (L_p^{\#}/L_p, b, t)$ be the usual alttice construction. If L had a hyperbolic splitting $L \simeq H + H^*$ and π_p be the projection of the vector space V containing L onto its P-primary component, then $L_p^{\#} = \pi_p(L)$, hence π_p induces a hyperbolic splitting of $c_p(L)$. This gives a well-defined homomorphism $c_p : CH^{\varepsilon}(R) \to CH^{\varepsilon}(K/R)$. As $CH^{\varepsilon}(F_q)$ is contained in $CH^{\varepsilon}(Q/Z)$ this is a much sharper invariant. Using these invariants it is possible to construct elements in the kernel of $CH^{\varepsilon}(Z) \to CH^{\varepsilon}(Q)$. We mention several other results that can be proven using these techniques. Let p and q be self-dual polynomials such that the associated orders are maximal.

Theorem 3.10: $0 \to CH_p^{\varepsilon}(Z) + CH_q^{\varepsilon}(Z) \to CH_{pq}^{\varepsilon}(Z) \xrightarrow{c_p + c_q} CH_p^{\varepsilon}(Q/Z) + CH_q^{\varepsilon}(Q/Z)$ is exact.

This is the one non-trivial case (in addition to the Dedekind case (3.7)) in which the integral hyperbolic case is computable.

A further complication occurs when the minimal polynomial of t is a power of an irreducible polynomial. In this case we have obtained the following partial results, without any Dedekind assumptions on the orders. Let $CH^{\varepsilon}_{p^n,0}(Z)$ be the subgroup of the group of isometric structures satisfying the condition that the condition that the rational structure is projective (and hence free) over $Q[X]/(P^n(X))$.

<u>Theorem 3.11</u>: There are homomorphisms $CH^{-\varepsilon}_{p}(Z) \to CH^{\varepsilon}_{p^{2n},0}(Z)$ and $CH^{\varepsilon}_{p}(Z) \to CH^{\varepsilon}_{p^{2n+1},0}(Z)$. The second homomorphism is split by the map $s(L) = (L_{p^n}/L_{p^n}, b, t)$ where L_{p^n} is the subgroup annihilated by p^n.

Hence the second map is an injection (the first is if the order is Dedekind). The surjectivity of either map is unknown for n greater than zero.

In answering a question posed by R.H. Fox at the Georgia Topology Conference in 1961, DeWitt Sumners introduced the notion of double null concordant knots [S]. Although the hyperbolic relation in the context of isometric structures first occurred in the algebraic context, the connection with the geometric notion of double null concordance was soon realized.

<u>Definition 3.12</u>: A knot (S^{n+2}, Σ^n) is doubly null concordant if it is the cross section of the trivial knot (S^{n+3}, S^{n+1}). (Equivalently, there is a smooth function $f: S^{n+3} \to R$ such that $f^{-1}(t)$ is the knot (S, Σ) for some regular value t.)

<u>Theorem 3.13</u> (Sumners [S], Kearton [Ke]) A simple knot $(S^{2n+1}, \Sigma^{2n-1})$ is doubly null concordant if and only if the Seifert isometric structure $S(S, \Sigma)$ is hyperbolic, provided $n \neq 1$.

In the geometric setting, we can form a new group of knots from the semigroup of knots under connected sum by introducing the equivalence relation:

<u>Definition 3.14</u>: Two knots (S^{n+2}, Σ_0) and (S^{n+2}, Σ_1) are equivalent if there are doubly null concordant knots H and K such that the connected sums $\Sigma_0 \# H$ and $\Sigma_1 \# K$ are isotopic.

This is an equivalence relation and Corollary 2.9 of Sumners [S] verifies that -K is the inverse to K in the new group, denoted $\mathcal{CH}_n(Z)$. Let $\mathcal{CH}_n^{(q-1)}(Z)$ denote the group of knots with (q-1)-connected Seifert manifolds ((q-1)-simple) under the above

The above theorem and the realizeability of isometric structures gives:

Theorem 3.15: $\mathcal{CH}^{(n-1)}_{2n-1}(Z) \rightarrow CH^{\varepsilon}(Z)$ is an isomorphism $(\varepsilon = (-1)^n)$.

In the even dimensional case we have the surprising result that $\mathcal{CH}_{2n}(Z)$ is non-trivial, contrasting the triviality of the even dimensional knot concordance group due to M. Kervaire [K]. This result was prompted by a remark made to me by J. Levine at Les Plans. Dewitt Sumners has also communicated to me that he had an example (the double spun trefoil) of an even dimensional knot that could not be double null concordant.

According to Levine [L1], if (S^{n+2}, Σ^n) is an n-knot with complement $X = S^{n+2}\setminus\Sigma$ and universal abelian cover $\tilde{X}$, the torsion subgroup T_q of the knot module $A_q = H_q(\tilde{X};Z)$ has a nonsingular pairing $T_q \times T_{n-q} \rightarrow Q/Z$ satisfying $b(tx,y) = b(x,t^{-1}y)$ where t is the automorphism of T_q induced by the oriented generator of the group of covering translations. Furthermore, since X has the homology of a circle by Alexander duality, $Id - t$ is invertible (called type K by Levine). Then the isomorphism $T = (Id - t)^{-1}$ satisfies: $b(Tx,y) = b((1-t)^{-1}x,y) = b(x,(I-t^{-1})^{-1}y) = b(x,1-(1-t)^{-1}(y)) = b(x,(1-T)y)$ If $n = 2q$ then (T_q,b,T) is an $\varepsilon = (-1)^{q+1}$-symmetric torsion isometric structure, and taking its equivalence class in $CH^{\varepsilon}(Q/Z)$ we have a map S.

Theorem 3.16: There is a well-defined morphism $S: \mathcal{CH}_{2q}(Z) \rightarrow CH^{\varepsilon}(Q/Z)$ which is an epimorphism for q greater than one.

Proof; By Theorem 13.1 of [L1] the map S is an epimorphism provided q is greater than one. To prove well-definedness it suffices to show that S vanishes on double null concordant knots. If (S^{2q+2}, Σ) is a cross-section of the trivial knot, the complement X of Σ splits the complement of the trivial knot, $S^1 x D^{2q+2}$ into two components, V and W. By the Mayer-Vietoris sequence: $H_{n+1}(S^1 x D^{2q+2}) \rightarrow H_n(X) \rightarrow H_n(V) + H_n(W) \rightarrow H_n(S^1 x D)$ we compute that V and W are also homology circles. Furthermore the Mayer-Vietoris sequence of the infinite cyclic covers:

$$0 = H_{n+1}(R x D^{2q+2}) \rightarrow H_n(\tilde{X}) \rightarrow H_n(\tilde{V}) + H_n(\tilde{W}) \rightarrow H_n(R \times D^{2q+2}) = 0$$

degenerates to an isomorphism at the middle morphism. Hence the $(i_*,-j_*)$ induced by the respective inclusions of X into V and W is an isomorphism. Let $H = \text{Kernel } i_* \cap T_q$ and $K = \text{kernel } j_* \cap T_q$. By the exactness of the homology exact sequence of the respective pairs, H and K are the torsion subgroup of the images of the respective boundaries. Note that H and K are invariant under t (and hence T) because the infinite cyclic cover of X is induced from that of V or W. Now H and K are disjoint and $T_q = H + K$ since $(i_*,-j_*)$ is an isomorphism. To complete the demonstration that our invariant is hyperbolic on double null concordant knots, it suffices to show that H and K are self-annihilating under the inner product b.

Let $\Theta = Z[X,X^{-1}]/(X^k-1)$ where $t^k = \mathrm{id}$ on T_q and let $I(\Theta)$ be the injective hull of the ring Θ. In Section Six of [L1] (particularly(6.4f) the pairing b is related to a pairing $\{\ ,\ \}: T_q \to \mathrm{Hom}T_q, I(\Theta)/\Theta)$ with the following definition.

Consider the following sequence (numbered (6.4) in [L1]).

$$H_e^{q+2}(\tilde{X}) \xrightarrow{\delta_4} H_e^{q+1}(\tilde{X},\Lambda_m) \xrightarrow{\delta_1} H_e^{q}(\tilde{X},\Theta/m\Theta) \xrightarrow{e'} \mathrm{Hom}\,(A_q,\Theta/m\Theta)$$

If α is in T_{n-q} and β is in T_q then $\{\alpha,\beta\} = ire'(\alpha')(\beta)$ where α' in $H_e^q(\tilde{X},\Theta/m\Theta)$ satisfies $-\delta_4\delta_1(\alpha') = \bar{\alpha}$, the dual of α. (Here, δ_1 and δ_4 are coboundaries induced by the appropriate coefficient sequences.) The map r is given by restriction to T_q and i is induced by the coeffieient map $\Theta/m\Theta \xrightarrow{1/m} Q(\Theta)/\Theta \subset I(\Theta)/\Theta$.

By the commutativity (up to sign) of the following diagram of duality maps

$$\begin{array}{ccc} H_e^{q+2}(\tilde{W}) & \xrightarrow{i^*} & H_e^{q+2}(\tilde{X}) \\ \uparrow & & \uparrow \\ H_{q+1}(\tilde{W},\tilde{X}) & \xrightarrow{\partial} & H_q(\tilde{X}) \end{array}$$

if α,β are in the image of ∂ then their duals are in the image of i^*.

Now consider the following commutative diagram induced by inclusion and which exists since W is also a homology circle (so that δ_1 and δ_4 are defined).

$$\begin{array}{ccccc} & & & & H_e^{q+1}(\tilde{W},\tilde{X};\Theta/m\Theta) \\ & & & & \uparrow \delta \\ H_e^{q+2}(\tilde{X}) & \xrightarrow{\delta} & H_e^{q+1}(\tilde{X},\Lambda_m) & \xrightarrow{\delta} & H_e^{q}(\tilde{X},\Theta/m\Theta) \\ \uparrow i^* & & \uparrow i^* & & \uparrow i^* \\ H_e^{q+2}(\tilde{W}) & \xrightarrow{\delta} & H_e^{q+1}(\tilde{W},\Lambda_m) & \xrightarrow{\delta} & H_e^{q}(\tilde{W},\Theta/m\Theta) \end{array}$$

Therefore if α and β are in the image of the $\partial = \ker i_*$, then $\bar{\alpha}$, the dual of α is in the image of i^*, $\alpha = i^*(\gamma)$. Let γ' satisfy $-\delta_4\delta_1(\gamma') = \gamma$ so that $i^*(\gamma') = \alpha'$. Also let $\beta = \partial n$. Then we have $\{\alpha,\beta\} = ire'(\alpha')(\beta) = ire'(i^*\gamma')(\beta) = ire'(i^*\gamma')(\partial n)$
$= ire'(\delta i^*\gamma')(n)$ by the duality of δ and ∂ under the Kronecker pairing,
$= 0$ by the exactness of the cohomology sequence of $(\tilde{W},\tilde{X})$. This completes the proof of the well-definedness of S.

This theorem should be compared with the stronger classification of Kearton of (q-1)-simple 2q-knots when T_q is of odd order in [Ke2]. There is also a geometric long exact sequence for these new groups that can be developed according to that in Section Six of [St]. Finally, we raise the question of the relation between stably doubly null concordant and doubly null concordance or (equivalently, in the simple case), the relation between stably hyperbolic and hyperbolic isometric structures.

BIBLIOGRAPHY

[K] Kervaire, M. Les noeuds de dimension superieure. Bull. Soc. Math. de France 93(1965), 225.

[Ke] Kearton, C. Simple knots which are doubly null-concordant. Proc. Amer. Math. Soc. 52(1975), 471.

[Ke2] Kearton, C. An algebraic classification of some even-dimensional knots. Topology 15(1976), 363.

[Kr] Kreck, M. Bordism of Diffeomorphisms. Bull. A.M.S. 82(1976), 759.

[KS] Knebusch, M. and Scharlau, W. Quadratische Formen und quadratische Reziprozitätsgesetze über algebraischen Zahlkörpen. Math Zeit. 121(1971), 346.

[L] Levine, J. Invariants of knot cobordism. Inventiones Math. 8(1969), 98

[L1] Levine, J. Knot Modules, I. (to appear)

[M] Milnor, J. On isometries of inner product spaces. Invent. Math. 8(1969).

[S] Sumners, D. Invertible Knot Cobordisms. Comm. Math. Helv. 46(1971), 240.

[St] Stoltzfus, N. Unraveling the integral knot concordance group. Memoirs A.M.S. 192(1977)

KNOT MODULES AND SEIFERT MATRICES

H.F. Trotter, Princeton University
Princeton, New Jersey 08540, U.S.A.

Introduction

In theorem 12.1 of [L], Levine has algebraically characterized the middle-dimensional Alexander module with Blanchfield pairing of an odd-dimensional knot. It is a corollary of his geometric realizability theorem that any module with pairing satisfying his conditions can be presented by a Seifert matrix. The main purpose of this paper is give a direct algebraic proof of this fact. Much of the argument amounts to reformulating Levine's conditions in a different framework. I have included some discussion not strictly required for the main result that helps to explain the trace function used in [T1]. To use trace functions derived from field extensions, one must decompose the Alexander module according to irreducible factors of the Alexander polynomial, but it turns out that one can handle the sort of generalities discussed here without worrying about reducibility.

A precise statement of the main result requires several definitions. Let $\Lambda_0 = Z[t,t^{-1}]$, viewed as a subring of $Q(t)$, the field of rational functions over Q. We use an overbar to denote the involution on $Q(t)$ and Λ_0 characterized by $\bar{t} = t^{-1}$. Fix $e = 1$ or -1. An e-hermitian pairing on a Λ_0-module A is a pairing $\langle\ ,\ \rangle : A \times A \longrightarrow Q(t)/\Lambda_0$ which is linear in the first argument and satisfies $\langle a,b\rangle = e\overline{\langle b,a\rangle}$, (and hence is conjugate-linear in the second argument).

The <u>Levine conditions</u> on a Λ_0-module A with pairing $\langle\ ,\ \rangle$ are:

(L1) A is finitely generated over Λ_0.

(L2) A is torsion-free as a Z-module. (See footnote p. 299)

(L3) Multiplication by $1-t$ is an automorphism on A.

(L4) $\langle\ ,\ \rangle$ is an e-hermitian pairing.

(L5) $\langle\ ,\ \rangle$ is non-singular, i.e., the adjoint map from A to $\mathrm{Hom}_{\Lambda_0}(A, Q(t)/\Lambda_0)$ which sends a to the homomorphism h_a defined by $h_a(x) = \langle x,a\rangle$, is a conjugate-linear isomorphism.

A <u>Seifert matrix</u> is a square matrix V of integers such that $\det(V - eV') = \pm 1$, where V' is the transpose of V. Given a Seifert matrix

V , define A_V to be the Λ_o-module with relation matrix $R_V = tV - eV'$, i.e., if V is an n-by-n matrix then A_V is the quotient of the module of column vectors (n-by-1 matrices) over Λ_o by the submodule generated by the columns of R_V . A pairing with values in $Q(t)//\Lambda_o$ is defined by setting $\langle a,b\rangle_V = a_o'(t-1)R_V^{-1}\bar{b}_o$, where a_o and b_o are column vectors over Λ_o representing a and b , and the right side is evaluated in $Q(t)$ and then mapped into $Q(t)//\Lambda_o$. We omit the fairly straightforward verification that the pairing is well-defined, and that A_V with the given pairing satisfies the conditions (L1 - L5). (See [T1] for more details. The definitions in this paragraph are equivalent to those in [T1], except that we have reversed the sign convention to be compatible with [L] and e here corresponds to $-\varepsilon$ in [T1].)

THEOREM. Any Λ_o-module with pairing that satisfies the Levine conditions is isometric to $(A_V, \langle\ ,\ \rangle_V)$ for some Seifert matrix V .

A universal trace function

Milnor [M1] showed that the trace map associated with an algebraic field extension could be used to give a correspondence between hermitian forms and isometries of spaces with an inner product, and a generalization of this idea plays an important role in [S]. An ad hoc analogue of the trace map is used similarly in [T1], and the object of this section is to explain how this analogue is more closely related to the ordinary trace than may at first appear. We are ultimately concerned with Λ_o-modules, but we begin with the simpler case of modules over a polynomial ring.

Let $F = Q(z)$ be the field of rational functions over Q , and $P = Q[z]$ the subring of polynomials. Let V be a vector space of finite dimension n over Q , and T an endomorphism of V . Make V a P-module in the usual way by defining $zv = T(v)$. Let c in P be the characteristic polynomial of T , and define R as the quotient ring P/cP . Because V is annihilated by c , it can be viewed as an R-module.

Consider any h in $\mathrm{Hom}_P(V,F/P)$. For any v , $ch(v) = h(cv) = 0$, so if w in F represents $h(v)$ in F/P , then cw is in P . There is clearly a unique choice for w of the form $w(z) = (u_{n-1}z^{n-1} + \dots + u_o)/c$ representing a given $h(v)$. Thus h in fact takes values in $(c^{-1}P)/P$, which as a P-module is isomorphic to $R = P/cP$ via multiplication by c . We thus have maps of P-modules $R \approx c^{-1}P/P \longrightarrow F/P$ with the last map an injection, and the composition induces an isomorphism of $\mathrm{Hom}_P(V,R)$ with $\mathrm{Hom}_P(V,F/P)$.

Now suppose that the endomorphism T is represented by a matrix of integers relative to some basis. Let K be the lattice on V (i.e., finitely generated Z-submodule that generates V over Q) generated by such a basis.

The integrality of the matrix of T is equivalent to the condition that T maps K into K, so K becomes a P_0-module, where $P_0 = Z[z]$ is the ring of polynomials with integer coefficients. The characteristic polynomial c lies in P_0, and K can equally well be viewed as a module over $R_0 = P_0/cP_0$. By the same argument as above, any h_0 in $\mathrm{Hom}_{P_0}(K,F/P_0)$ actually takes its values in $c^{-1}P_0/P_0 \simeq R_0$ and we get an isomorphism between $\mathrm{Hom}_{P_0}(K,R_0)$ and $\mathrm{Hom}_{P_0}(K,F/P_0)$. (One has to use the fact that if cw is in P_0 then, because c is monic and so has content 1, w is also in P_0.)

We define a trace function $s:F \longrightarrow Q$ by setting $s(f)$ equal to the coefficient of z^{-1} in the Laurent expansion of f at infinity. Since s vanishes on P, it is well-defined on F/P. If f is a proper fraction, so $f(z) = (u_{m-1}z^{m-1} + \ldots + u_0)/(z^m + v_{m-1}z^{m-1} + \ldots + v_0)$, then $s(f) = u_{m-1}$. As a Q-module, F is the direct sum of P and the proper fractions, so this formula gives another characterization of s.

If we define s on R by identifying R with the submodule $c^{-1}P/P$ of F/P, then for $w = u_{n-1}z^{n-1} + \ldots + u_0$ in R, we have $s(w) = u_{n-1}$. The Q-valued bilinear form defined on R by $\{x,y\} = s(xy)$ is obviously non-singular, since $s(xy) = 0$ for all y only if $x = 0$. (If $x = u_k z^k + \ldots + u_0$, with $u_k \neq 0$, $0 \leq k < n$, then $s(xz^{n-k-1}) \neq 0$.) If c has integer coefficients, we can consider $R_0 = P_0/cP_0$ as a lattice embedded in R. The form $\{\ ,\ \}$ is integer-valued on R_0, and in fact R_0 is easily seen to be self-dual with respect to it. (If $\{\ ,\ \}$ is a Q-valued bilinear form on a vector space V over Q, the <u>dual</u> of a lattice K on V consists of all v such that $\{k,v\}$ is in Z for all k in K.) This in turn is equivalent to the statement that $\{\ ,\ \}$ is non-singular as a Z-valued form on R_0.

Multiplication by an element w of R gives an endomorphism M_w of the Q-module structure of R, and defining $tr_{R/Q}(w)$ to be the trace of M_w gives the usual trace function $R \longrightarrow Q$. By a classical formula, $tr_{R/Q}(w) = s(wc'(r))$, where c' is the derivative of c and r is the image in $R = P/cP$ of z in P. (If c is a separable polynomial, $c'(r)$ is invertible in R and the formula may be written $s(w) = tr(wc'(r)^{-1})$. For a proof in this case see, for instance, chapter 7 of [A].) The formula shows that when c has integer coefficients and is irreducible, the pairing $\{\ ,\ \}$ of R_0 with itself induced by s is equivalent to the pairing of R_0 with its inverse different induced by the classical trace, as considered by Stoltzfus [S].

We now introduce the involution on F, denoted by an overbar, which takes z to $1-z$, and assume that V is furnished with a non-singular e-hermitian form $<\ ,\ >:V \times V \longrightarrow F/P$. (P and F/P inherit the involution from F.) The trace map s is related to the involution by the formula $s(\bar{f}) = -s(f)$, and if we define $[u,v] = s(\langle u,v\rangle)$, we get a $-e$-symmetric form which satisfies

$$[pu,v] = s(\langle pu,v\rangle) = s(p\langle u,v\rangle) = s(\langle u,\bar{p}v\rangle) = [u,\bar{p}v]$$

for all p in P. If $\langle u,v\rangle = 0$ for all v, then obviously $[u,v] = 0$ for all v. Conversely, if $[u,v] = 0$ for all v, then for any v, $[u,pv] = s(\bar{p}\langle u,v\rangle) = 0$ for all p in P, so $\langle u,v\rangle = 0$. Thus, $[\ ,\ ]$ is non-singular if and only if $\langle\ ,\ \rangle$ is.

Since $[\ ,\ ]$ is non-singular it determines a dual basis for any given basis for V. The relation $[zu,v] = [u,(1-z)v]$, i.e., $[T(u),v] = [u,(1-T)v]$, implies that if M is the matrix of T in some basis, then $(I-M)'$ is the matrix of T in the dual basis. Then $c(z) = \det(zI-M) = \det(zI-(I-M')) = (-1)^n \det((1-z)I-M') = (-1)^n c(1-z)$, so $c = (-1)^n \bar{c}$. Thus the ideal cP is symmetric and $R = P/cP$ inherits an involution from P. If n is odd, $c(\frac{1}{2}) = \bar{c}(1-\frac{1}{2}) = -c(\frac{1}{2})$, so $c(\frac{1}{2}) = 0$. Since c is monic, this is impossible if c has integer coefficients, which is always the case in our application. To avoid some minor complications *we shall assume from here on that n is even and hence* $c = \bar{c}$. Under this assumption, multiplication by c^{-1} respects the involution, and so does our earlier identification of R with the (symmetric) submodule $c^{-1}P/P$ of F/P. This induces a bijection between R-valued and F/P-valued non-singular e-hermitian forms on V, just as we earlier obtained an isomorphism between $\mathrm{Hom}_P(V,R)$ and $\mathrm{Hom}_P(V,F/P)$. In the integral case we get a bijection between R_0-valued and F/P_0-valued non-singular forms on a lattice K in V.

The important point, however, is that the trace map s gives a bijection between F/P-valued non-singular e-hermitian forms $\langle\ ,\ \rangle$ and Q-valued $-e$-symmetric forms $[\ ,\ ]$ satisfying $[pu,v] = [u,\bar{p}v]$ on a vector space V, and between F/P_0-valued non-singular e-hermitian forms $\langle\ ,\ \rangle$ and Z-valued $-e$-symmetric forms satisfying the same condition on a lattice K. These facts follow from our earlier remarks on the non-singularity of $\{x,y\} = s(xy)$ as a form on R and R_0. The proof is an application of Stoltzfus's formulation of the "trace lemma" (lemma 2.6 in [S1]), to which we refer the reader for details. (The lemma in [S1] applies to arbitrary bilinear forms, but the additional conditions relating to the hermitian symmetry are trivial to check.)

Application to knot modules

In this section we show how the material of the previous section can be adapted to apply to modules satisfying the Levine conditions. It will be convenient to set up notation for several subrings of the field of rational functions.

Let $F = Q(t) = Q(z)$ where $z = (1-t)^{-1}$. As before, we use a bar to denote the involution such that $\bar{t} = t^{-1}$ and $\bar{z} = 1-z$. We shall be concerned with the following subrings of F.

$$\begin{aligned} P &= Q[z] & P_0 &= Z[z] \\ \Lambda &= Q[t,t^{-1}] & \Lambda_0 &= Z[t,t^{-1}] \\ L &= Q[t,t^{-1},z] & L_0 &= Z[t,t^{-1},z] \\ &= \Lambda[z] & &= \Lambda_0[z] \\ &= P[z^{-1},\bar{z}^{-1}] & &= P_0[z^{-1},\bar{z}^{-1}] \end{aligned}$$

(The equivalence of the various descriptions of L and L_0 follows from the relation $t = -\bar{z}z^{-1}$. Note that all the subrings are self-conjugate under the involution.)

Let A be a Λ_0-module satisfying L1,L2,L3, and let $V = Q \otimes A$. V is a vector space over Q and has a Λ-module structure. Because A is Z-torsion free it embeds naturally in V as a Λ_0-submodule.

It is shown in [T2] that A has a presentation matrix over Λ_0 of the form $tM + (I-M)$, with M a square matrix of integers such that both M and $(I-M)$ have non-zero determinant. (The theorem in [T2] refers to Λ_0-modules which are isomorphic to the commutator subgroup of some metabelian group with infinite cyclic commutator quotient group. The proof proceeds, however, by showing that such modules satisfy L1, L2, and L3, and derives the stated result from these conditions. It is a corollary that the first three Levine conditions characterize such group modules.) The same matrix presents V over Λ, and it is then quite easy to see that the dimension of V over Q is equal to the number of rows of M, and multiplication by t is given by the matrix $-(I-M)M^{-1} = I-M^{-1}$ with respect to a suitable basis B. Furthermore, B generates A as a Λ_0-submodule of V. (See proposition 2.5 in [T1] for a more detailed discussion of essentially the same situation.)

Because $1-t$ gives an automorphism on A, the action of $(1-t)^{-1}$ is defined, and A is an L_0-module and V an L-module. The action of $1-t$ on V is given by $I-(I-M^{-1}) = M^{-1}$, so the action of $z = (1-t)^{-1}$ is given by M, which is known to be an integer matrix. Let K be the lattice generated over Z by the elements of the basis B mentioned above. Then K is a P_0-module and generates A as a Λ_0-module.

Enlarging the ring of coefficients from Λ_0 to L_0 obviously has no real effect on A as a module. Suppose now that A is furnished with a pairing $\langle\ ,\ \rangle$ satisfying L4,L5. Composing with the map $F/\Lambda_0 \longrightarrow F/L_0$ (induced by the inclusion of Λ_0 in L_0 as a Λ_0-module) gives a F/L_0-valued form which is obviously e-hermitian. Non-singularity (L5) is not quite so obvious, but follows at once from the following lemma, which establishes a bijection between $\mathrm{Hom}_{L_0}(A,F/L_0)$ and $\mathrm{Hom}_{\Lambda_0}(A,F/\Lambda_0)$. The lemma also shows that the image of the pairing in $F/\!/\Lambda_0$ is mapped one-to-one onto the image in F/L_0 . We can thus use $\langle\ ,\ \rangle$ for the new F/L_0-valued form without confusion.

Lemma. The map $r:\mathrm{Hom}_{\Lambda_0}(A,F/\!/\Lambda_0) \longrightarrow \mathrm{Hom}_{\Lambda_0}(A,F/L_0)$ induced by the natural map $F/\!/\Lambda_0 \longrightarrow F/L_0$ is a bijection. The map $\mathrm{Hom}_{L_0}(A,F/L_0) \longrightarrow \mathrm{Hom}_{\Lambda_0}(A,F/L_0)$ obtained by forgetting the L_0-module structure is also a bijection.

Proof. The last assertion is obvious. To prove the first, suppose $h:A \longrightarrow F/\!/\Lambda_0$ is in the kernel of r . For any fixed x in A , let y in F represent $h(x)$ mod Λ_0 . Since $r(h) = 0$, y lies in L_0 , and hence for some n (depending on x), $(1-t)^n y$ is in Λ_0 and $h((1-t)^n x) = 0$. Let N be the maximum value of n as x ranges over a finite set of generators of A . Then $h((1-t)^N x) = 0$ for all x , and (by L3) $h = 0$. Hence r is injective.

Now let $p:X \longrightarrow A$ be a map of a finitely-generated free Λ_0-module onto A , and for a given $h:A \longrightarrow F/L_0$, let $\bar{h}:X \longrightarrow F$ be a Λ_0-homomorphism such that $\bar{h}$ followed by reduction mod L_0 is equal to hp . Let C be the kernel of p , so that $\bar{h}(C)$ is contained in L_0 . Since Λ_0 is Noetherian, C is finitely generated, and arguing as in the preceding paragraph we see that for some N , $\bar{h}((1-t)^N C)$ is contained in Λ_0 . Define $\bar{h}'(x) = \bar{h}((1-t)^N x)$ for x in X , so $\bar{h}'(C)$ is contained in Λ_0 and $\bar{h}'$ induces h' in $\mathrm{Hom}_{\Lambda_0}(A,F/\!/\Lambda_0)$. Now define $h''(a)$ as $h'((1-t)^{-N}a)$. Then $r(h'') = h$, and r is surjective, as was to be proved.

(Note that the conclusion of the lemma is false for $A = F$, so the hypothesis that A is finitely generated is necessary.)

We have seen that the Levine conditions are equivalent to the same conditions with L_0 replacing Λ_0 throughout. (Condition L3 becomes superfluous but remains true with the new coefficients.) From now on we treat A as an L_0-module and V as an L-module, with forms valued in F/L_0 and F/L .

The trace function s of the preceding section does not vanish on L , and must be slightly modified for application here. By the theory of partial fractions, F is the direct sum over Q of L (polynomials in z , z^{-1} , and $\bar{z}^{-1} = (1-z)^{-1}$) and the proper fractions with denominators prime to z and $\bar{z}$. We can define a modified trace s' by setting it equal to s on this restricted class of proper fractions and equal to zero on L . Just as in the previous section, the values of bilinear forms on V and A will lie in $c^{-1}L/L$ and $c^{-1}L_0/L_0$, where c is the characteristic polynomial of the matrix M representing the action of z . Proceeding as in the previous section, we define $S = L/cL$ and identify $c^{-1}L/L$ with it, and identify $c^{-1}L_0/L_0$ with $S_0 = L_0/cL_0$, and consider forms as taking values in these rings. Because both M and $I-M$ have non-zero determinants, c does not vanish at either 0 or 1 and so is prime to z and $\bar{z}$. Hence when we use canonical representatives of values of $\langle\ ,\ \rangle$, both s and s' give the same result. (Note: s' is the same as the map χ in [T1], defined in terms of z instead of t .)

As before, by defining $[x,y]$ as $s'(\langle x,y\rangle)$, we derive a $-e$-symmetric form from an e-hermitian one. It is easy to see that $\langle\ ,\ \rangle$ is non-singular on V if and only if $[\ ,\ ]$ is. Indeed, because c is prime to z and $\bar{z}$, the inclusion of P in L induces an isomorphism of $R = P/cP$ with $S = L/cL$, and the situation is the same as in the previous section.

With integer coefficients there is in general a difference. Since $L_0 = P_0[z^{-1},\bar{z}^{-1}]$, L_0 consists of those w in L for which there exist integers M, N such that $z^M\bar{z}^Nw$ is in P_0 . Every w in S_0 has a standard representative mod c of the form $w_{n-1}z^{n-1} + \ldots + w_0$, but the coefficients may not be integers. (Specifically, if $c = z^{n-1} + c_{n-1}z^{n-1} + \ldots + c_0$, and w is as above, then $v = z^{-1}w$ has coefficients v_i given by $v_{n-1} = -u_0c_0^{-1}$ and $v_j = u_{j+1} - u_0c_0^{-1}v_j$ for $j = 0,1,\ldots,n-2$. To get a formula for $\bar{z}^{-1}w$, note that it is equal to $(z^{-1}\bar{w})^-$.) Unless $c_0 = \pm 1$, there will be elements of S_0 whose standard representatives do not have integer coefficients. Thus although the form $\{x,y\} = s'(xy)$ is non-singular as a Q-valued form on S , it is in general not even Z-valued, let alone non-singular, as a form on S_0 . Instead, we have the following statement.

Proposition. An element x of L is in L_0 if and only if for every y in L_0 there exist M_0 and N_0 such that $\{x,z^M\bar{z}^Ny\}$ is in Z for all $M \geq M_0$, $N \geq N_0$.

Let K be as defined above, a Z-lattice on V which is a P_0-module (i.e. closed under multiplication by z) and generates A as a L_0-module. If $\langle\ ,\ \rangle$ is non-singular as an S_0-pairing on A , its extension to an S-pairing on V is also non-singular and hence so is the associated Q-pairing $[\ ,\ ]$ on V . Assume only that the pairings on V are non-singular. Then any Z-lattice J has a dual $J^\#$, defined to consist of those x in V such that $[x,y]$ is in Z for all y in J . The following lemma is a replacement for the trace lemma, in that it allows us to characterize non-singularity of $\langle\ ,\ \rangle$ as an S_0-module on A in terms of $[\ ,\ ]$.

Lemma. Let J be a lattice on V which is a P_0-module, suppose V carries a non-singular S-valued e-hermitian form $\langle\ ,\ \rangle$, and let $[\ ,\ ] = s'(\langle\ ,\ \rangle)$. Let $C = L_0J$ be the L_0-module generated by J . Then the adjoint map $V \longrightarrow \mathrm{Hom}_L(V,S)$, which is a bijection by assumption, restricts to a bijection $L_0J^\# \longrightarrow \mathrm{Hom}_{L_0}(C,S_0)$, where $J^\#$ is the dual of J with respect to $[\ ,\ ]$.

Proof. We first remark that tensoring with Q gives an injection of $\mathrm{Hom}_{L_0}(C,S_0)$ into $\mathrm{Hom}_L(V,S)$ because S_0 is Z-torsion free. Let C' denote the inverse image of $\text{Hom}_{L_o}(C,S_o)$ under the adjoint map. We want to show that $C' = L_oJ^{\#}$. For any x in C, $z^M\bar{z}^N x$ is in J for all sufficiently large M and N, so $[z^M\bar{z}^N x,y]$ is in Z for any y in $J^{\#}$. Hence $s'(z^M\bar{z}^N\langle x,y\rangle)$ is in Z for all sufficiently large M and N, so $\langle x,y\rangle$ is in S_o. Hence $J^{\#}$ is contained in C' and therefore so is $L_oJ^{\#}$. Conversely, if y is in C' then for any x in J, $\langle x,y\rangle$ is in S_o and there exist M_x, N_x such that $s'(z^M\bar{z}^N\langle x,y\rangle)$ is in Z for all $M \geq M_x$, $N \geq N_x$. Taking the maximum of M_x, N_x as x runs over a Z-basis for J shows that for some M, N we have $z^M\bar{z}^N y$ in $J^{\#}$, so y is in $L_oJ^{\#}$, and the proof is complete.

Corollary. The form $\langle\ ,\ \rangle$ is non-singular on $A = L_oK$ if and only if $L_oK = L_oK^{\#}$.

Construction of a Seifert matrix

We have shown that if A and $\langle\ ,\ \rangle$ satisfy the Levine conditions then, after changing to L_o as coefficient ring, A embeds in an L-module V containing a lattice K that is a P_o-module such that $A = L_oK = L_oK^{\#}$, where $K^{\#}$ is the dual of K with respect to $[\ ,\ ] = s'(\langle\ ,\ \rangle)$. We want to show that A and $\langle\ ,\ \rangle$ can be presented by a Seifert matrix. By lemma 2.14 of [T1], if $K = K^{\#}$ then the desired matrix can be constructed using any Z-basis for K. It therefore suffices for us to show that given $A = L_oK = L_oK^{\#}$ as above, we can find J with $A = L_oJ$ and $J = J^{\#}$.

To do so we look at the associated p-adic lattices K_p as in [T1], to which the reader is referred for background information. For almost all p, $K_p = K_p^{\#}$, and we construct J as a lattice which has $J_p = K_p$ for these p. If we can specify J_p for each remaining p so that $J_p = J_p^{\#}$, we are done.

Fix such a p. To simplify notation we drop the p subscript from here on, and write V, K, etc., to denote the p-adifications V_p, K_p, etc.. By theorem 3.4 of [T1], V splits as an L-module into subspaces V_o, V_+, V_- such that $z^m v \longrightarrow 0$ as $m \longrightarrow \infty$ for all v in V_+, $\bar{z}^m v \longrightarrow 0$ as $m \longrightarrow \infty$ for all v in V_-, K is the direct sum (as a P_o-module) of its intersections K_o, K_+, K_- with the corresponding subspaces of V, and $L_oK = K_o + V_+ + V_-$ as a direct sum of L_o-modules. (Note: The theorem is stated for "admissible" lattices, which are defined to be P_o-modules and self-dual. The self-duality, however, is not used in the proof of theorem 3.4 .)

Take a Z-basis B for K that is a union of bases B_o, B_+, B_- for K_o, K_+, K_- respectively, and let $B^{\#}$ be the dual basis for $K^{\#}$. From the relation $[zx,y] = [x,\bar{z}y]$ it follows that V_o is orthogonal to V_+ and V_- while V_+ and V_- are isotropic and dual to each other with respect to $[\ ,\ ]$ (see lemma 3.5 of [T1]). It follows that $B_o^{\#}$ (consisting of the elements of $B^{\#}$ dual to those of B_o) is a Q-basis for V_o, $B_+^{\#}$ for V_-, and $B_-^{\#}$ for

V_+ , so that they are Z-bases for the intersections $K_o^\#$, $K_-^\#$, $K_+^\#$ of $K^\#$ with V_o, V_-, V_+ respectively. From $L_oK = L_oK^\#$ we have $K_o = K_o^\#$ (corollary 3.4a in [T1]). Define J as the direct sum of K_o, K_+, and $K_-^\#$. It is a P_o-module, and is self-dual because $K_o = K_o^\#$, and K_+ and $K_-^\#$ are isotropic, orthogonal to K_o , and in duality with each other because they have dual bases B_+ and $B_+^\#$ by construction. Also, $L_oJ = K_o + V_+ + V_- = L_oK = A$, so J satisfies the required conditions and the proof is complete.

References

[A] E. Artin, Theory of Algebraic Numbers, Göttingen 1957.

[L] J. Levine, Knot modules I . Trans. AMS 229 (1977)p. 1-50.

[S] N. Stoltzfus, Unraveling the integral knot concordance group.

[T1] H. Trotter, On S-equivalence of Seifert matrices, Inv. Math. 20 (1973), 173-207.

[T2] ------ Torsion-free metabelian groups with infinite cyclic quotient groups, Proc. Second Internat. Conf. Theory of Groups, Canberra 1973, 655-666.

Footnote : I am indebted to the referee for the observation that the condition (L2) follows from the others. For a proof, see C. Kearton, Blanchfield duality and simple knots, Trans. AMS 202 (1975), 141-160 (on p. 155).

TORSION DANS LES MODULES D'ALEXANDER

par

Claude WEBER (Genève)

§ 1. Introduction

Soit $\Lambda = Z\,[t, t^{-1}]$ l'anneau des polynômes de Laurent à une variable et à coefficients entiers. Un Λ-module A sera appelé un module d'Alexander s'il est de type fini et si la multiplication par (1-t) est un isomorphisme. En ce qui concerne les modules d'Alexander les faits suivants sont désormais classiques :

1. Si X est le complémentaire dans S^{n+2} d'une sphère de dimension n, localement plate et orientée, et si $\tilde{X} \longrightarrow X$ désigne le revêtement cyclique infini, alors $H_i(\tilde{X}, Z)$ est un module d'Alexander pour tout $i > 0$.

2. Un module d'Alexander est de Λ-torsion. C'est pourquoi, quand on parlera de la torsion d'un tel module, il s'agira toujours de sa Z -torsion. Suivant la tradition, nous désignerons par t(A) le sous-module de Z -torsion de A et par f(A) le module quotient $A/_{t(A)}$. t(A) et f(A) sont aussi d'Alexander, si A l'est.

3. Pour un module d'Alexander A, t(A) est fini.

4. Un module d'Alexander est sans torsion si et seulement si il admet une présentation carrée.

Ces résultats ont été démontrés par Kervaire dans [4]. Pour une démonstration purement algébrique (et bien plus courte) de 4, voir [6].

Soient maintenant R un anneau commutatif avec 1, M un R-module de présentation finie et soit :

$$R^r \xrightarrow{\varphi} R^s \longrightarrow M \longrightarrow 0$$

une présentation finie de M . Désignons par E(M) l'idéal de R engendré par tous les $s \times s$ mineurs de la matrice de φ . Il est bien connu que E(M) ne dépend pas de la présentation (finie) choisie. E(M) est appelé traditionnellement le premier idéal élémentaire de M. Si R est factoriel, on désignera par $\Delta(M)$ un pgcd des éléments de E(M). Si, de plus, $R = S[t,t^{-1}]$, $\Delta(M)$ n'est rien d'autre que le "polynôme d'Alexander" de M. Comme il n'est défini qu'à une unité près, il est utile de donner les précisions suivantes :

Soit x un élément non nul de $S[t,t^{-1}]$. On a :

$$x = \sum_{j=u}^{v} a_j t^j \quad , \quad u \text{ et } v \in Z \, , \quad u \leqslant v \, , \quad a_u \neq 0 \neq a_v \ .$$

L'entier (v-u) est le degré de l'élément x. Il ne dépend pas du choix de x dans sa classe d'associés. Les éléments a_u et a_v s'appelleront les coefficients extrémaux de x et a_u le premier coefficient extrémal.

Supposons maintenant que S = Z et que A soit un module d'Alexander. Alors $\Delta(A)$ est différent de 0. En fait, sa valeur pour t = 1 est égale à ± 1 . Voir [4] ou [6]. Comme dans $Z[t,t^{-1}]$ les unités sont les éléments $\pm t^{1}$, les coefficients extrémaux de $\Delta(A)$ sont bien définis au signe près. Enfin, d'après 4, si A est sans torsion, l'idéal E(A) est principal. Le but de cette note est essentiellement de démontrer la réciproque de cette dernière affirmation.

§ 2 . Un théorème de fibration.

Dans ce paragraphe, nous indiquons rapidement comment les faits énoncés au paragraphe précédent permettent de donner une démonstration concise d'un théorème de fibration dû essentiellement à Sumners [7].

Théorème : Soit Y une variété compacte de dimension $n+2$, $n \geq 4$, ayant l'homologie entière d'un cercle et telle que $\pi_1(Y) = Z$. Soit $g : Y \longrightarrow S^1$ une équivalence d'homologie entière telle que $g|\partial Y$ soit une fibration localement triviale et soit $\tilde{Y} \longrightarrow Y$ le revêtement universel de Y. Alors g est homotope à une fibration localement triviale si et seulement si $\Delta(H_i(\tilde{Y}\, ;Z))$ a des coefficients extrémaux égaux à $\pm$ 1, pour $i = 2,\ldots, n+1$. ($H_i(\tilde{Y}, Z)$ est un Λ-module une fois qu'on a choisi une orientation de S^1).

Applications : Prenons pour Y le complémentaire X d'un noeud différentiable dans S^{n+2}, $n \geq 4$, et supposons que $\pi_1(X) = Z$. On obtient alors que le noeud est "fibrant" si et seulement si les polynômes d'Alexander du noeud ont des coefficients extrémaux égaux à $\pm$ 1 . D'après les résultats de Levine dans [5], cette dernière condition est équivalente à demander que le premier coefficient extrémal de $\Delta(H_i(\tilde{Y}\, ,Z)$ soit égal à $\pm$ 1 pour $i = 2,\ldots,n-1$.

Preuve du théorème : En vertu des hypothèses, $\tilde{Y}$ est également : i) le revêtement cyclique infini de Y; ii) la "fibre homotopique" de l'application g.

Comme cette fibre est simplement connexe, on peut appliquer le théorème de fibration dans la version Browder-Levine [1] : g est homotope à une fibration si et seulement si $\pi_i(\tilde{Y})$ est de type fini, comme groupe abélien, pour tout i.

En vertu de la théorie de Serre, comme $\tilde{Y}$ est simplement connexe, $\pi_i(\tilde{Y})$ est de type fini comme groupe abélien pour tout i si et seulement si $H_i(\tilde{Y}\, ; Z)$ l'est, pour tout i.

Maintenant, comme Y a l'homologie entière d'un cercle et est compacte, les Λ-modules $H_i(\tilde{Y}\, ; Z)$ sont d'Alexander; Cf [4] ou [6]. Comme $t(H_i(\tilde{Y}\, ; Z)$ est un groupe fini, il suffit de voir à quelles conditions

$f(H_1(\tilde{Y} ; Z))$ est un groupe abélien de type fini.

Lemme : Si A est un module d'Alexander, $\Delta(A)$ et $\Delta(f(A))$ sont associés dans Λ .

Preuve du lemme :

1) Comme $f(A)$ est un quotient de A, $\Delta(f(A))$ divise $\Delta(A)$.

2) Comme $A \otimes_Z Q$ et $f(A) \otimes_Z Q$ ont même rang sur Q, le degré de $\Delta(f(A))$ est égal au degré de $\Delta(A)$.

3) Comme A est d'Alexander, la valeur en $t = 1$ de $\Delta(A)$ est ± 1, ce qui montre que le pgcd des coefficients de $\Delta(A)$ est 1.

1), 2) et 3) impliquent immédiatement que $\Delta(A)/\Delta(f(A))$ est une unité de Λ . Ceci achève la preuve du lemme.

D'après 4 de l'introduction, le Λ-module $f(H_1(\tilde{Y} ;Z))$ possède une présentation carrée; par définition, le déterminant de cette présentation est un associé de $\Delta(f(H_1(\tilde{Y} ;Z)) = \Delta(H_1(\tilde{Y} ;Z))$. Dans ces circonstances, un théorème de Crowell [2] nous apprend que le groupe abélien est de type fini si et seulement si les coefficients extrémaux de $\Delta(f(H_1(\tilde{Y} ; Z)))$ sont égaux à ± 1.

§ 3. A la recherche de la p-torsion dans un module d'Alexander.

Si A est un module d'Alexander et si p est un nombre premier, nous désignerons par $r_p(A)$ la dimension sur F_p de $t(A) \otimes_Z F_p$, où F_p est le corps Z/pZ . $r_p(A)$ est le rang de la p-torsion de A. Nous noterons Λ_p l'anneau $F_p[t,t^{-1}]$ et ρ_p l'homomorphisme naturel (réduction modp. des coefficients) : $\Lambda \longrightarrow \Lambda_p$. Dans ce qui suit les produits tensoriels sont sur Z .

<u>Proposition</u> : Soit A un module d'Alexander. Alors
$r_p(A) = \text{degré}\ \Delta(A \otimes F_p) - \text{degré}\ \rho_p(\Delta(A))$.

<u>Preuve de la proposition</u> : On a la suite exacte de Λ-modules :

$$0 \longrightarrow t(A) \longrightarrow A \longrightarrow f(A) \longrightarrow 0 .$$

Comme f(A) est sans torsion, $\mathrm{Tor}_Z(f(A) ; F_p) = 0$ et on a donc une suite exacte de Λ-modules :

$$(*) \qquad 0 \longrightarrow t(A) \otimes F_p \longrightarrow A \otimes F_p \longrightarrow f(A) \otimes F_p \longrightarrow 0$$

<u>Affirmation 1</u> : La dimension sur F_p de $A \otimes F_p$ est égale au degré du polynôme $\Delta(A \otimes F_p) \in \Lambda_p$.

En effet, Λ_p est un anneau principal et $A \otimes F_p$ est de type fini sur Λ_p. Le théorème de classification donne alors immédiatement le résultat.

<u>Affirmation 2</u> : La dimension sur F_p de $f(A) \otimes F_p$ est égale au degré du polynôme $\rho_p(\Delta(A)) \in \Lambda_p$.

En effet, nous avons vu au paragraphe précédent que f(A) possède une présentation carrée :

$$\Lambda^S \xrightarrow{\ \varphi\ } \Lambda^S \longrightarrow f(A) \longrightarrow 0$$

et que le déterminant de φ est associé à $\Delta(A)$. Par exactitude à droite du produit tensoriel, on obtient une suite exacte de Λ_p-modules

$$\Lambda_p^S \xrightarrow{\ \overline{\varphi}\ } \Lambda_p^S \longrightarrow f(A) \otimes F_p \longrightarrow 0 .$$

La matrice de $\overline{\varphi}$ n'est rien d'autre que la matrice de φ dont on a réduit les coefficients modp. Donc le déterminant de $\overline{\varphi}$ est associé à $\rho_p(\Delta(A))$. Appliquant encore une fois le théorème de classification, l'affirmation 2 est démontrée.

Les deux affirmations et l'exactitude de la suite (*) entraînent la validité de la proposition.

Essayons maintenant de rejouer avec A le jeu que nous venons de jouer avec f(A), dans l'affirmation 2. Soit :

$$\Lambda^m \xrightarrow{\Phi} \Lambda^n \longrightarrow A \longrightarrow 0$$

une présentation finie de A. Par tensorisation, on obtient la suite exacte de Λ_p-modules :

$$\Lambda_p^m \xrightarrow{\overline{\Phi}} \Lambda_p^n \longrightarrow A \otimes F_p \longrightarrow 0 .$$

Par définition, E(A) est l'idéal de Λ engendré par tous les $n \times n$ mineurs $u_1,\dots,u_k$ de Φ . De même, $E(A \otimes F_p)$ est l'idéal de Λ_p engendré par tous les $n \times n$ mineurs $v_1,\dots,v_n$ de $\overline{\Phi}$. Comme $\overline{\Phi}$ est la réduction modp de Φ : $v_i = \rho_p(u_i)$ $i = 1,\dots,k$.

Par définition, $\Delta(A \otimes F_p)$ est le pgcd de $\{v_1,\dots,v_k\}$ et $\Delta(A)$ est le pgcd de $\{u_1,\dots,u_k\}$.

La remarque suivante est vraie chaque fois que l'on a affaire à des anneaux factoriels. Nous l'énonçons dans le cadre plus restreint qui nous intéresse :

<u>Remarque</u> : Soit α un idéal de Λ et soit (a) le plus petit idéal principal qui contient α . Alors ρ_p(a) est un idéal principal contenant $\rho_p(\alpha)$, mais ce n'est pas nécessairement le plus petit.

Si l'on applique la remarque à $\alpha = E(A)$, on a $(a) = (\Delta(A))$ et la proposition précédente nous apprend que t(A) a de la p-torsion chaque fois que l'idéal $\rho_p(\Delta(A))$ est "trop grand" pour être le pgcd de $E(A\otimes F_p)$.

<u>Théorème</u> : Soit A un module d'Alexander. Alors A est sans torsion si et seulement si E(A) est principal.

<u>Preuve</u> : La nécessité est bien connue : en effet, si A est sans

torsion, d'après 4 de l'introduction il possède une présentation carrée et donc $E(A)$ est engendré par le déterminant de cette présentation.

Réciproquement, si $E(A)$ est principal, il résulte de la discussion ci-dessus que $r_p(A) = 0$ pour tout p puisque $\alpha = (a)$ et qu'ainsi le phénomène décrit dans la remarque ne peut se produire.

§ 4. Exemples.

Dans ce paragraphe, nous illustrons la proposition du § 3 pour des situations simples.

Il est bien connu que le premier module d'Alexander du noeud 6_1 est cyclique d'ordre $\Delta = 2t^2 - 5t + 2$ tandis que celui du noeud 9_{46} est somme de deux modules cycliques d'ordre $\lambda = t-2$ et $\lambda^* = 2t-1$, $\lambda\lambda^* = \Delta$. Si l'on fait subir un g-twist spin à chacun de ces noeuds le premier module d'Alexander A_g du premier noeud a pour matrice de présentation :

$$\begin{pmatrix} 2t^2 - 5t + 2 & \dfrac{t^g - 1}{t - 1} \end{pmatrix}$$

tandis que le premier module d'Alexander B_g du second noeud a pour matrice de présentation :

$$\begin{pmatrix} t-2 & 0 & \dfrac{t^g - 1}{t - 1} & 0 \\ 0 & 2t-1 & 0 & \dfrac{t^g - 1}{t - 1} \end{pmatrix}$$

Pour $g \geqslant 1$, les polynômes d'Alexander de A_g et de B_g sont égaux à 1. Il est connu, d'après les calculs de Fox[3] qu'un nombre premier intéressant pour ces deux noeuds est $p = 3$. Dans Λ_3, les matrices de présentation deviennent :

$$\begin{pmatrix} (t+1)^2 & t^{g-1}+ \dots + 1 \end{pmatrix} \quad \text{et}$$

$$\begin{pmatrix} t+1 & 0 & t^{g-1}+\dots+1 & 0 \\ 0 & t+1 & 0 & t^{g-1}+\dots+1 \end{pmatrix}$$

Pour $g=2$, $\Delta(A_2 \otimes F_3)$ est égal à $(t+1)$, tandis que $\Delta(B_2 \otimes F_3)$ est égal à $(t+1)^2$. Donc :

$$r_3(A_2) = 1 \quad \text{et} \quad r_3(B_2) = 2 \, .$$

Un peu plus de calcul montre que pour $g = 2h$:

$r_3(A_g) = 1$ lorsque $h \not\equiv 0 \mod 3$

$r_3(A_g) = 2$ lorsque $h \equiv 0 \mod 3$

tandis que $r_3(B_g) = 2$ pour tout g pair.

Bien sûr, beaucoup de calculs du même genre peuvent être faits. On y découvre, par exemple, une version affaiblie du théorème de Plans.

BIBLIOGRAPHIE

[1] W. Browder et J. Levine : "Fibering manifolds over a circle". Comment. Math. Helv. 40(1966) p. 153-160.

[2] R. Crowell : "The group G'/G'' of a knot group G". Duke Math. Journal 30(1963) p. 349-354.

[3] R. Fox : "The homology characters of the cyclic coverings of the knots of genus one". Annals of Math. 71(1960) p. 187-196.

[4] M.Kervaire : "Les noeuds de dimension supérieure" . Bull. Soc. Math. France 93(1965) p. 225-271.

[5] J. Levine : "Polynomial invariants of knots of codimension two". Annals of Math. 84(1966) p. 537-554.

[6] J. Levine : "Knot Modules I". Trans. AMS 229 (1977) p.1-50 .

[7] D. Sumners : "Polynomial invariants and the integral homology of coverings of knots and links". Inventiones Math. 15(1972) p. 78-90.

Institut de mathématiques

Université de Genève

2-4, rue du Lièvre

1211- Genève 24

PROBLEMS

The following problems were raised at a problem session which was held during the conference.

Problem 1. The crossing number $c(K)$ of K is the minimum number of crossings in any projection of K.

Is $c(K)$ additive with respect to connected sum?

Problem 2. The unknotting number $u(K)$ of K is the minimum number (over all projections) of overcrossing-undercrossing changes required to unknot K.

Is $u(K)$ additive with respect to connected sum?

Problem 3. Let $p(n)$ be the number of prime knots with $c(K) \leq n$, and $t(n)$ the number of all knots with $c(K) \leq n$.

Can one say anything about the asymptotic behavior of $p(n)$, $t(n)$, or $p(n)/t(n)$?

Problem 4. Calculate $u(K)$ for the knots which arise as links of complex algebraic plane curve singularities. In particular, for the torus knots (p,q), where the expected answer is $\frac{(p-1)(q-1)}{2}$. (This is known in the case $p=2$; see K. Murasugi, On a certain numerical invariant of link types, Trans. Amer. Math. Soc. 117(1965), 387-422.)

For background, see J. Milnor, Singular Points of Complex Hypersurfaces, Annals of Math. Studies 61, 1968.

Problem 5. Are there ways of deciding exactly how many minimal genus surfaces a knot has, (up to either isotopy leaving the knot fixed, or homeomorphism), that will include cases where this number is finite but >1?

A particular example is the following. The pretzel knot $(3,1,5)$ is the same as the pretzel knot $(3,5,1)$, and has corresponding 'obvious' minimal Seifert surfaces with Seifert matrices $\begin{pmatrix} 2 & 1 \\ 0 & 3 \end{pmatrix}$ and $\begin{pmatrix} 3 & 1 \\ 0 & 2 \end{pmatrix}$. Are these the only minimal Seifert surfaces? In particular, does it have one with Seifert matrix $\begin{pmatrix} 1 & 1 \\ 0 & 6 \end{pmatrix}$ (which is S-equivalent to the others)? (See H. F. Trotter, Some knots spanned by more than one unknotted surface of minimal genus, Annals of Math. Studies 84, 51-62.)

Problem 6. Classify knot modules.

When $\Delta(t)$ is irreducible and $\mathbb{Z}[t,t^{-1}]/(\Delta(t))$ is Dedekind, the classification is reduced to classical numbers theory (Levine, Stoltzfus). What happens if one or both of these fail?

Problem 7. What $\mathbb{Z}[t,t^{-1}]$-modules can occur as G'/G'' for a higher-dimensional knot group G?

This is the one problem remaining in the characterization of the higher-dimensional knot modules $H_q(\tilde{X};\mathbb{Z})$. Furthermore, the only difficulty here is with the $\mathbb{Z}$-torsion submodule. (See J. Levine, Knot modules, Annals of Math. Studies 84, 25-34, and Knot modules I, to appear.)

Problem 8. When are the homotopy groups of a higher-dimensional knot finitely-generated over $\mathbb{Z}\pi_1$? In particular, if $\pi_1 \cong \mathbb{Z}$?

Problem 9. What manifolds can occur as fibres of higher-dimensional fibred knots?

Problem 10. Do there exist, for all $n \geq 3$, matrices $A \in SL(n,\mathbb{Z})$ such that $\det(\Lambda^r A - I) = \pm 1$ for $1 \leq r \leq [\frac{n-1}{2}]$, and A has no negative eigenvalues?

Such matrices provide examples of inequivalent higher-dimensional knots with homeomorphic complements. (See S. E. Cappell and J. L. Shaneson, There exist inequivalent knots with the same complement, Annals of Math. 103 (1976), 349-353.)

Problem 11. A weak form of the Smith conjecture. Can a knot with Arf invariant $\neq 0$ be the fixed-point set of a $\mathbb{Z}_p$-action on S^3?

(See S. E. Cappell and J. L. Shaneson, Branched cyclic coverings, Annals of Math. Studies 84.)

Problem 12. Relate the Arf invariant of a link (when defined) to other link invariants. (For the case of a knot, see, for example, K. Murasugi, The Arf invariant for knot types, Proc. Amer. Math. Soc. 21(1969), 69-72.) In particular, find practical ways of computing the Arf invariant of a link.

Problem 13. Does every knot concordance class contain a prime knot?

Problem 14. Let C_n be the concordance group of knots of S^n in S^{n+2}. Find non-trivial homomorphisms from $\ker(C_1 \to C_{4n+1}, n > 0)$ to some abelian group of algebraic objects.

(Casson-Gordon have shown that this kernel is non-zero.)

Problem 15. Do there exist elements in C_1 of finite order $\neq 2$?

Problem 16. Does every element of order 2 in C_1 have an amphicheiral representative?

Problem 17. Is every knot in S^3 (resp. a homology 3-sphere with zero Rohlin invariant) with $\Delta(t) = 1$ slice (resp. slice in a homology 4-ball)?

Problem 18. Let g denote the genus of a knot, and h the 4-genus (the minimal genus of an orientable surface in the 4-ball with the knot as boundary).

Is $h(K) = g(K')$ for some K' concordant to K?

Problem 19. Is $h(K) = g(K)$ $(= \frac{(p-1)(q-1)}{2})$ for K a torus knot of type (p,q)?

This is known in the case $p = 2$ (see K. Murasugi, On a certain numerical of link types, Trans. Amer. Math. Soc. 117(1965), 387-422). Signature arguments also give some other cases.

Problem 20. Find invariants of concordance of links which are not just invariants of homology-cobordism, (especially for 1-(dimensional) links).

Problem 21. Is every n-link concordant to a boundary link- for $n \geq 2$?

Problem 22. Which 1-links are concordant to boundary links? (Not all are.)

Problem 23. A knot is doubly null-concordant if it is a cross-section of the trivial knot.

Does stably doubly null-concordant (i.e. there exists D such that $K\#D$ and

D are doubly null-concordant) imply doubly null-concordant?

(In higher dimensions, this is equivalent to a purely algebraic question.)

Problem 24. Let D_n be the abelian group of equivalence classes of n-knots unde: the equivalence relation: $K_1 \sim K_2$ if and only if $K_1 \# -K_2$ is stably doubly null-concordant.

Compute D_n.

(Stoltzfus has some results, both for n odd and n even. In particular, in contrast to the case of ordinary concordance, $D_{2n} \neq 0$.)

Problem 25. Do there exist links in S^3 with the same complement which are distinguished by the first Betti number of their 2-fold branched covers?